Renewable Energy Power Generation and Power Demand Side Management

Guest Editors

Bingtuan Gao
Xiaofeng Liu
Lixia Sun

MDPI

Basel • Beijing • Wuhan • Barcelona • Belgrade • Novi Sad • Cluj • Manchester

Guest Editors

Bingtuan Gao
School of Electrical Engineering
Southeast University
Nanjing
China

Xiaofeng Liu
Department of Electrical Engineering
Nanjing Normal University
Nanjing
China

Lixia Sun
School of Electrical and Power Engineering
Hohai University
Nanjing
China

Editorial Office
MDPI AG
Grosspeteranlage 5
4052 Basel, Switzerland

This is a reprint of the Special Issue, published open access by the journal *Energies* (ISSN 1996-1073), freely accessible at: https://www.mdpi.com/journal/energies/special_issues/P2K4WHA0IJ.

For citation purposes, cite each article independently as indicated on the article page online and as indicated below:

Lastname, A.A.; Lastname, B.B. Article Title. *Journal Name* **Year**, *Volume Number*, Page Range.

ISBN 978-3-7258-4427-2 (Hbk)
ISBN 978-3-7258-4428-9 (PDF)
https://doi.org/10.3390/books978-3-7258-4428-9

Contents

energies

MDPI

Review

Demand-Side Flexibility in Power Systems, Structure, Opportunities, and Objectives: A Review for Residential Sector

Hessam Golmohamadi *, Saeed Golestan, Rakesh Sinha and Birgitte Bak-Jensen *

Department of Energy, Aalborg University, 9220 Aalborg, Denmark; sgd@energy.aau.dk (S.G.); rsi@energy.aau.dk (R.S.)

* Correspondence: hessamgolmoh@energy.aau.dk (H.G.); bbj@energy.aau.dk (B.B.-J.)

Abstract: The integration of renewable energy sources (RESs) is rapidly increasing within energy systems worldwide. However, this shift introduces intermittency and uncertainty on the supply side. To hedge against RES intermittency, demand-side flexibility introduces a practical solution. Therefore, further studies are required to unleash demand-side flexibility in power systems. This flexibility is relevant across various sectors of power systems, including residential, industrial, commercial, and agricultural sectors. This paper reviews the key aspects of demand-side flexibility within the residential sector. To achieve this objective, a general introduction to demand flexibility across the four sectors is provided. As a contribution of this paper, and in comparison with previous studies, household appliances are classified based on their flexibility and controllability. The flexibility potential of key residential demands, including heat pumps, district heating, electric vehicles, and battery systems, is then reviewed. Another contribution of this paper is the exploration of demand-side flexibility scheduling under uncertainty, examining three approaches: stochastic programming, robust optimization, and information-gap decision theory. Additionally, the integration of demand flexibility into short-term electricity markets with high-RES penetration is discussed. Finally, the key objective functions and simulation software used in the study of demand-side flexibility are reviewed.

Keywords: battery storage; demand flexibility; electric vehicle; heat pump; residential; uncertainty

1. Introduction

1.1. Why Demand Flexibility?

In the last decade, the penetration of renewable power, including wind and solar, has been increasing in power systems worldwide. Some countries have scheduled plans to increase the penetration of renewable energy sources (RESs) up to 100%. For example, Denmark has committed to a 100% renewable power supply by 2050 [1]. In this way, many fossil fuel power plants are retired and replaced with renewable ones. Increasing RES penetration, the intermittency and volatility of the supply side increase noticeably. In addition to the phase-out of fossil fuel power plants, due to growing concerns about environmental problems, the use of fossil fuel vehicles is decreasing in the world. Instead, the penetration of electric vehicles (EVs) is increasing gradually. EVs are highly mobile electrical demands with variable connection points to power grids [2]. Therefore, the imperfect data of the connection points introduces significant uncertainty in power systems. As a result, future power systems will encounter considerable uncertain data not only on the supply side but also on the demand side. To hedge against intermittent generation and consumption, future power systems need demand-side flexibility in power system management.

Based on the above-mentioned facts, there are two main reasons to emphasize demand-side flexibility in power systems. The reasons can be summarized as follows:

1. The phase-out of fossil fuel power plants and vehicles.
2. The phase-in of RESs and EVs with intermittent power generation and consumption.

1.2. Flexibility in Demand Sectors

Demand-side flexibility is the portion of demand that can be changed, i.e., reduced, increased, shifted, or curtailed, in response to external stimuli. The stimuli can be price-based or incentive-based demand response programs (DRPs) [3] to encourage consumers to change their electricity consumption. In power systems, there is a wide variety of electricity consumption on the demand side. To unlock power flexibility, the demand side should first be split into different sectors, including residential, commercial, industrial, and agricultural. This classification makes it possible to schedule DRPs for each sector based on its electricity consumption characteristics. Expert knowledge is therefore required to understand the key characteristics of electricity consumption. Flexibility studies should be conducted for each sector individually because demand profiles, load patterns, and flexibility potentials exhibit significant variability across the different sectors. Moreover, each sector has different operational constraints, energy consumption behaviors, and regulatory frameworks, which impact their response to demand-side management strategies. It is worth mentioning that multi-carrier energy systems, e.g., integrated gas and electricity [4] and hydrogen systems [5], exhibit significant flexibility potentials.

In the residential sector, there is significant flexibility potential in the electricity consumption of Heat, Ventilation, and Air Conditioning (HVAC). Additionally, in recent years, electrically operated heat pumps (HPs) have been replacing conventional fossil fuel heating systems. HPs are an economical alternative, not only reducing energy consumption costs but also facilitating the integration of renewables into power systems [6]. Therefore, much more flexibility potential can be extracted from residential buildings. In addition, by increasing the use of smart household appliances, the electricity consumption of shiftable devices, such as washing machines, can be scheduled to align with the flexibility requirements of the supply side [7]. Moreover, heating and cooling systems [8], water heaters [9], EVs, household appliances, and home batteries [10] are introduced as the main sources of demand flexibility in the residential sector. Furthermore, small renewable self-generation facilities, such as photovoltaic panels [11] and micro wind turbines can facilitate demand flexibility in the residential sector.

In the industrial sector, many processes in both light and heavy industries can offer considerable demand flexibility to the grid. Generally, industrial plants consist of energy-intensive processes. By unlocking power flexibility in these processes, significant demand response (DR) can be injected into the supply side. Some heavy industries, such as cement manufacturing plants, include interruptible processes [12]. Therefore, these processes can be interrupted in response to the flexibility requirements of the power system. In this way, uninterruptible processes can be supplied by feedstock from storage. Storage facilities can increase the flexibility of such industries. In the case of light industries, there is significant flexibility potential in food industries, such as dairy processing plants, which consume both power and heat in their internal processes [13]. Recently, cement manufacturing plants [14], metal and aluminum smelting [15], pulp and paper milling [16], and oil refineries [17] have been extensively discussed in the literature for their potential demand flexibility. In addition, many industries are equipped with Combined Heat and Power (CHP) units that can be scheduled to provide heat-to-power flexibility for both industrial operations and the upstream power grid [18].

In the agricultural sector, there are many flexibility opportunities in open-air farms, such as water irrigation systems and poultry farms. On farms and in greenhouses, water irrigation pumps can be operated based on the flexibility requirements of the supply side. Energy-intensive groundwater pumps can be turned off or down during renewable power shortages. Water booster pumps with Variable Frequency Drives (VFDs) can adjust water pressure to provide up/down-regulation for the power system during surpluses or shortages of renewable power [19]. In poultry farms, electrically operated HPs can be installed to adjust electricity consumption in response to renewable power availability on the supply side. In these farms, heat controllers play a key role in flexible power consumption. To integrate the flexibility potential of the agricultural sector into power systems, a major restructuring is required in this sector [20].

Related to the commercial sector, recently, the use of EVs has increased considerably. EVs are mobile electrical storages that can act as prosumers in power systems. Since EVs are parked for long hours during the day, the electrical storage capacity of their batteries can be used to provide power flexibility. The charging and discharging capacity of EVs can be used for the up- and down-regulation of the power system during their dwell time [21]. Smart charging stations, including those in private and public parking lots, can optimize the charging and discharging strategies of EVs, not only to provide flexibility to the power system but also to generate profit for EV owners [22]. Many parking lots in commercial shopping centers and airports now accommodate EVs for extended periods. When equipped with smart charging stations, these parked EVs can serve as a reliable flexibility resource for power systems with high-RES penetration [23]. In a recent study, public parking lots were considered virtual power plants supplying the commercial sector and shopping malls [24]. Furthermore, hypermarket refrigerators [25] can provide demand flexibility to local grids in the commercial sector.

Based on the abovementioned facts, power flexibility in demand sectors requires expert knowledge. Therefore, flexibility opportunities should be aggregated separately in each demand sector. Segregated flexibilities may not meet power system requirements. To address this, a Demand Response Aggregator (DRA) is introduced to coordinate demand-side flexibilities based on the technical requirements of the supply side. The DRA is defined as a marketer, broker, public agency, city, county, or special district that combines loads of multiple end-use customers to facilitate the sale and purchase of electrical energy, transmission, and other services on their behalf [26].

The DRA provides contracted consumers with opportunities to understand their demand flexibility on one side and integrate the aggregated flexibility potentials into power systems on the other side. Based on expert knowledge, residential, commercial, industrial, and agricultural DRAs, i.e., RDRA, CDRA, IDRA, and ADRA, are identified. Additionally, Parking Lot Aggregators (PLAs) are introduced to coordinate the charging and discharging strategies of EV fleets in response to power system requirements. Figure 1 depicts a schematic diagram showing the role of DRAs as intermediary agents in power systems. This figure illustrates how flexibility potentials are aggregated on the demand side and integrated into the supply side.

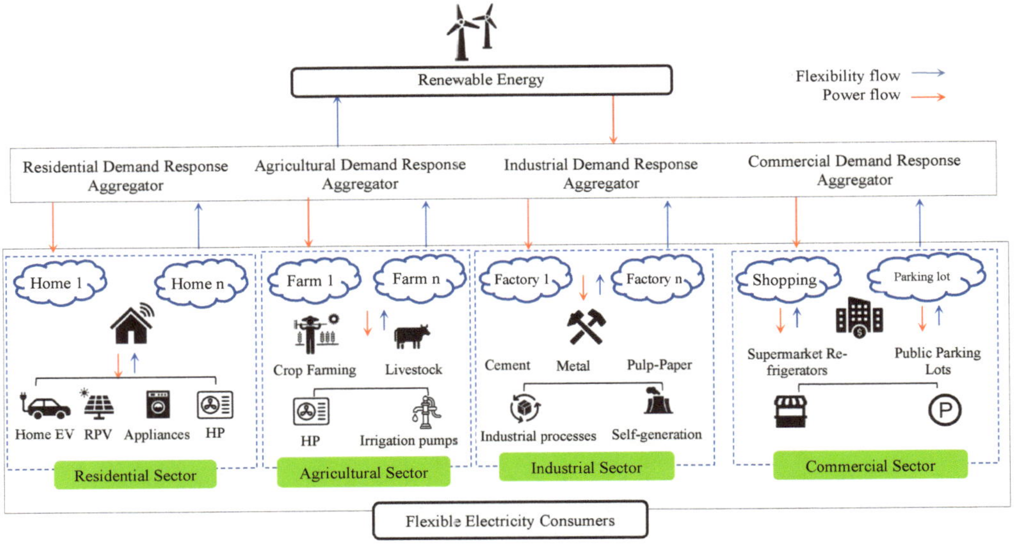

Figure 1. General overview of flexibility in four demand sectors [27].

1.3. Paper Structure and Contributions

This paper reviews the key concepts of demand-side flexibility with an emphasis on the residential sector. The main contributions of the paper are as follows:

(1) Classification of household appliances: a classification of household appliances from the perspectives of flexibility and controllability is provided, along with a comprehensive review of the most important residential demands, including HPs, district heating, EVs, and PV-battery systems.
(2) Survey of demand flexibility under uncertainty: a survey of demand flexibility under uncertainty is conducted, focusing on three recent models, including stochastic programming, robust optimization, and information-gap decision theory.
(3) Review of objectives and simulation software: a comprehensive review of the main objectives of demand flexibility and the software tools used for scheduling demand flexibility is carried out.

The rest of the paper is organized as follows: Section 2 reviews demand flexibility in the residential sector and home energy management systems. Section 3 addresses demand flexibility under uncertainty. Section 4 discusses the integration of demand flexibility into short-term electricity markets. Section 5 reviews the key objective functions, communication and data exchange, and software simulations. The challenges and limitations of current technologies for demand-side flexibility, as well as future works, are discussed in Section 6. Finally, Section 7 concludes this study. Figure 2 shows the main contents of the paper.

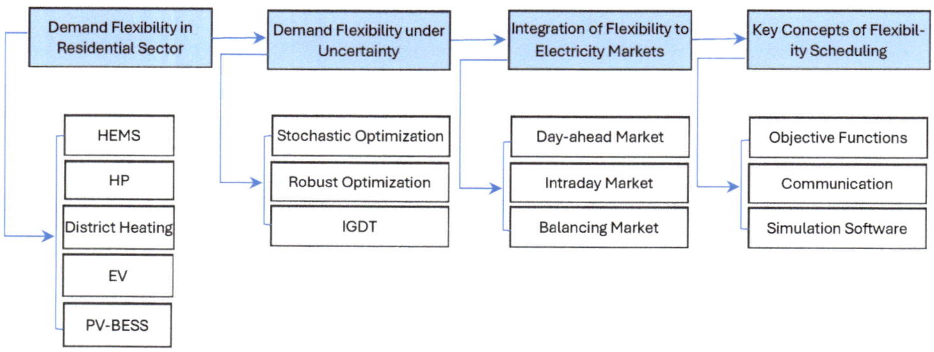

Figure 2. The main contents of the paper.

2. Demand Flexibility in Residential Sector

From 1974 until now, electricity consumption has been increasing across demand sectors worldwide. In 2018, the residential sector consumed 6008 TWh of electrical energy [28]. In comparison with other sectors, this represents 27% of the total electricity consumption. The high share of power consumption in the residential sector is a key factor contributing to power system flexibility.

To fully harness the flexibility potential of residential demand, it is first necessary to identify and classify flexible demand based on its controllability. In the following subsections, after providing a general classification, the flexibility potential of the most important residential demands is reviewed.

2.1. Classification of Household Appliances

In residential buildings, electricity consumption varies widely from energy-intensive appliances, such as HVAC systems, to low-power devices, such as LED lamps. Some household appliances, such as HVAC systems, can store energy during daily operations. This energy storage capability is reflected in the building's thermal dynamics [29]. The operation of certain appliances can be shifted to specific times without causing inconvenience. For example, washing machines and tumble dryers are shiftable appliances [30]. To investigate the flexibility potential of residential buildings, household appliances can be classified into three distinct categories, as follows:

(1) Thermostatically controllable appliances (TCAs)
(2) Controllable non-thermal appliances (CNTAs)
(3) Uncontrollable appliances (UAs)

TCAs are appliances whose energy consumption follows thermal dynamics. Examples include HVAC systems, electric water heaters (EWHs), HPs, and refrigerators. The flexibility potential of TCAs is reflected in their thermal dynamics, which are described by Differential Equations (DEs) [31]. Therefore, to unlock the power flexibility of TCAs, sophisticated controllers are needed to optimize these DEs [32]. The power consumption of HPs can be optimized to unlock the flexibility of both space heating and domestic hot water consumption [33]. In addition, the power consumption of household refrigerators can be optimized by addressing door-opening patterns to provide peak shaving for power grids [34].

CNTAs are electrical appliances whose operational flexibility is independent of thermal dynamics. Examples include lighting systems, dishwashers, and washing machines. This group can be divided into two subgroups: (1) deferrable CNTAs and (2) curtailable CNTAs. Deferrable CNTAs, such as washing machines and dishwashers, are devices whose operations can be shifted to times outside of DR events, such as during low-price hours. In this case, the CNTA operation is deferred to off-peak hours without changing the total electricity consumption [35]. In contrast, curtailable CNTAs include appliances whose power consumption can be adjusted (increased or decreased) in response to DR requests. In this situation, the power consumption changes, but there is no expectation to shift the remaining consumption to other times. For example, when the power consumption of a lighting system is adjusted in response to DR requests, there is no expectation to shift the power consumption to other hours to compensate for reduced lighting [36].

In contrast to the aforementioned appliances, UAs are inflexible devices, such as TVs, ovens, and computers. This group of appliances barely responds to DR plans because their daily operation is closely related to residents' comfort. Therefore, unlocking the flexibility of UAs without compromising occupants' comfort is challenging. To summarize, Figure 3 illustrates household appliances from the perspective of power flexibility.

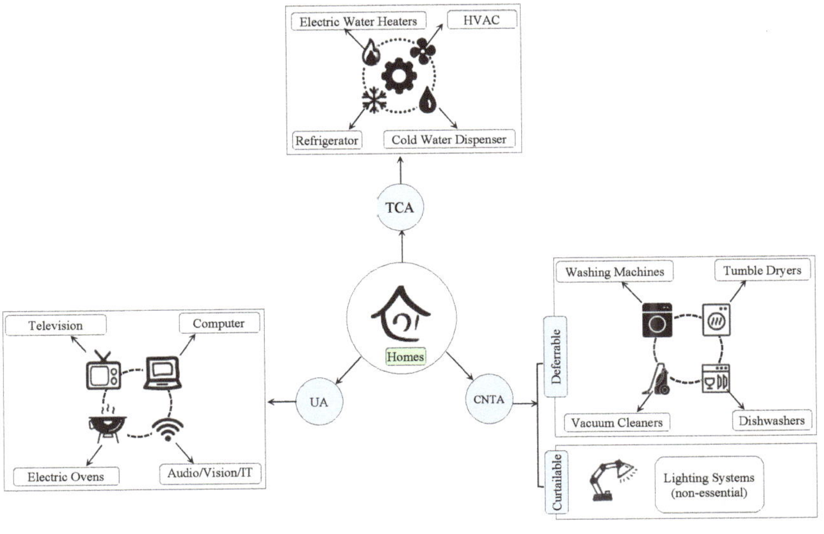

Figure 3. Classification of household appliances from the flexibility point of view.

2.2. Home Energy Management System

The home energy management system (HEMS) is an advanced framework designed to monitor, control, and optimize the operation of household appliances, including TCAs and CNTAs. The system aims to balance two competing objectives: (1) meeting flexibility requirements on the supply side and (2) ensuring residents' comfort on the demand side [37]. The HEMS features a communication system with the grid system operator, such as the Distribution System Operator (DSO). However, this setup varies by country, as different entities may be responsible for data exchange in some regions. The DSO informs the HEMS of DR events, which may include peak hours, off-peak hours, price-based, and incentive-based DR plans. Conversely, the HEMS receives comfort parameters from home residents, such as reference indoor temperature, hot water temperature, and laundry time. Thus, the main duty of the HEMS is to provide power flexibility for the supply side while maintaining the comfort conditions for the occupants.

Regarding the communication system, the HEMS typically uses a Power Line Carrier (PLC) or standard communication networks [38]. The PLC allows communication with the DSO in remote areas, where conventional networks are unavailable.

Smart controllers are the core components of the HEMS. These controllers, equipped with computer processors, optimize the operation of household appliances. Their primary function is to balance the flexibility requirements of the supply side with the comfort conditions of the demand side. Residents set their comfort preferences through a user-friendly interface. In this panel, occupants can specify their preferences for appliance operation, such as indoor temperature. Flexibility requirements, including price signals and critical hours, are received from the DSO. In [39], a smart HEMS is suggested to unlock the flexibility potentials of BESS, photovoltaic systems, and EVs. To provide a general overview of the HEMS, Figure 4 depicts a schematic diagram of a module-based HEMS.

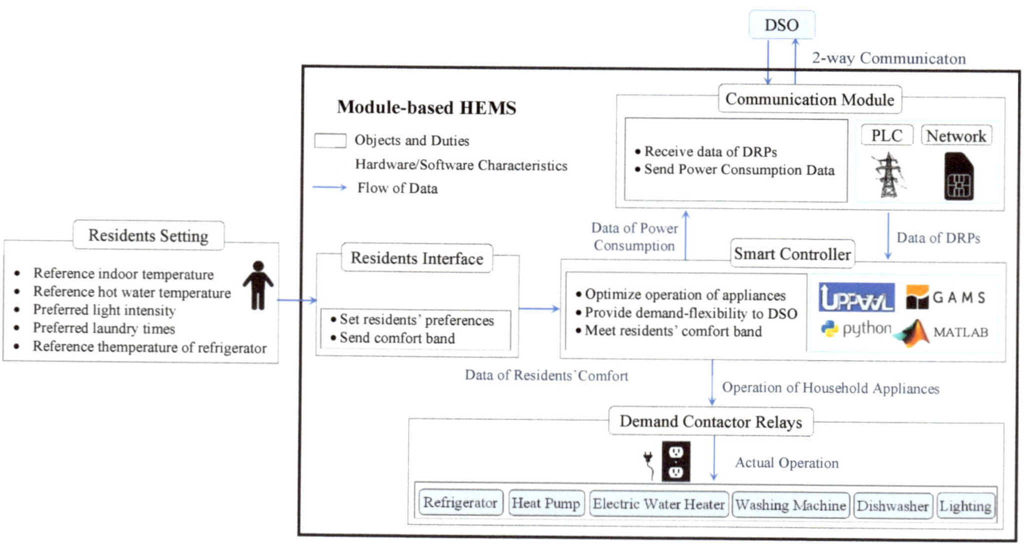

Figure 4. The module-based structure of the HEMS [40].

2.3. Heat Pumps

In recent years, HPs' popularity has increased due to their energy efficiency. They can typically produce 3–5 times more heat energy than the electricity they consume. Additionally, they can recover heat waste from the ground, outdoor environment, and sewage water [41]. The flexibility of HPs is addressed for both cooling and heating purposes [42].

The flexibility potential of HPs depends on the thermodynamic characteristics of buildings. Therefore, in many studies, the thermal dynamics of buildings are calculated to design an HP controller. Among them, the Continuous-Time Stochastic Model (CTSM) was developed by the DTU Compute [43] and was used in many studies to design HP controllers [44]. The HP controller aims to unlock HPs' flexibility in response to external signals, such as electricity prices, while maintaining the indoor air within the comfort range, e.g., between 20–25 °C. The controller receives the indoor air setpoint from the resident and communicates with the local DSO to obtain flexibility requirement data, electricity prices, and meteorological data such as the outdoor temperature and solar irradiation.

In many studies, model predictive control (MPC) is designed for HVAC, taking weather data forecasts into account to optimize future control actions [45]. Some studies suggest economic MPC (EMPC) for HPs to minimize household energy consumption costs [46].

HP compressors can be turned up or down in response to flexibility signals. Although the compressors can quickly respond to electricity price changes, the indoor temperature remains stable due to the longer time constant of thermal dynamics, compared to electrical dynamics. Many studies suggest using heat storage to increase the flexibility potential of HPs. Heat storage can be in the form of water tanks [47] or phase-change materials (PCMs) [48]. Regardless of the type, heat storage allows the HP to operate when electricity prices are low, i.e., down-regulation is required for power system imbalance, and then supply the heating system from the heat storage during high electricity prices when up-regulation is required for power system imbalance.

2.4. District Heating

District Heating (DH) consists of a distributed heat network where heat is generated at a central location and distributed to residential buildings through insulated pipes. The heat is used for space heating and hot water. Generally, DH heat sources can be classified into electricity, fossil fuels, solar-thermal, biomass, geothermal, and waste heat [49].

For electrically operated DH systems, HPs and electric boilers offer significant flexibility potential [50]. Fuel-based DH systems are divided into heat-only units and CHP units. Heat-only units, such as gas boilers, have no demand flexibility for power systems. In contrast, CHP units can provide power flexibility for the electricity grids [51]. Solar–thermal, biomass, and geothermal DH systems are non-electrical and thus offer limited power flexibility. In some cases, DH systems utilize waste heat from electrical consumers, such as industries and data centers [52]. Although this type of DH has no direct power flexibility, it may indirectly unlock power flexibility from the main heat source.

In DH systems with electricity-based heat sources, there are three types of heat flexibility:

(1) The flexibility of thermal inertia of buildings: This is reflected in the thermal dynamics of buildings. Buildings contain thermal mass, such as walls and windows, which can store heat energy [53]. Consequently, indoor temperatures can be adjusted in response to electricity price variations and/or renewable power availability on the supply side.
(2) The flexibility of thermal storage devices: These are specifically designed to store heat energy. Common storage devices in DH systems include water tanks, boreholes, chemical storage, and aquifers [54]. Heat energy can be stored during low-price hours (excess power) and used during high-price hours (power shortages).
(3) The flexibility of the heat network: This is reflected in the temperature of the heat carrier. Adjusting the heat carrier temperature can provide power flexibility. However, temperature variation can accelerate pipe aging and material fatigue, particularly at weak joints, which is a limiting factor [55].

To provide a general insight into the flexibility potential of DH systems, Figure 5 presents a schematic diagram from the flexibility perspective.

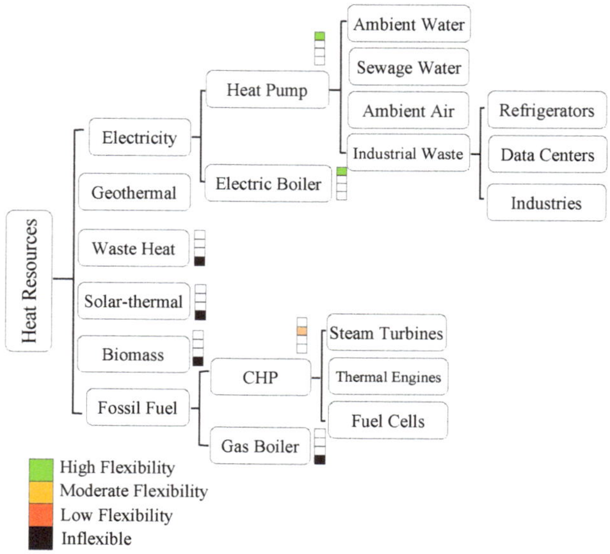

Figure 5. Qualification of flexibility potentials in a DH concerning the heat resources.

2.5. Electric Vehicles

EVs are mobile electric consumers and storage units which, unlike fixed demand, can be connected to different points of distribution grids. As the penetration of EVs increases and fossil fuel cars are retired, the distribution grid comes under pressure, causing transformer overload and congestion in power lines [56]. This can threaten the reliable operation of distribution grids. However, if the charging and discharging of EVs are coordinated, they will not compromise the reliability of the power system. Instead, they can enhance the system's reliability by providing a workable electrical storage backup for the power system.

To unlock EV flexibility, smart charging stations are required. These charging stations can work in both vehicle-to-grid (V2G) and grid-to-vehicle (G2V) modes. Common charging stations are G2V, while V2G is less common due to technical barriers. Smart G2V stations encourage charging during times when electricity prices are low, corresponding to an excess of renewable power availability. Therefore, charging tariffs are important signals to unleash EV flexibility. While G2V stations can provide down-regulation to the power network [57], V2G stations can provide up-regulation when the grid encounters a deficit of renewable energy or system reliability is jeopardized due to failure or unscheduled maintenance [58].

Nowadays, local DSOs/aggregators/retailers use mobile apps to inform EV owners of the charging and discharging prices for the next 24 h. Charging and discharging tariffs can be designed adaptively based on the flexibility requirements of the local grid. For example, during peak (off-peak) hours, charging tariffs are set high (low) to discourage (encourage) EV owners from charging, aiming to provide peak shaving (valley filling). Here, the IoT and communication play a key role in unlocking EV flexibility.

EV flexibility can be discussed for private [59] and public parking lots [60]. Private parking refers to residential parking where EVs are parked typically after working hours. Therefore, the dwell time is usually limited to nighttime hours, with capacities of 2.3, 7.4, 11, and even 22 kW. On the other hand, public parking lots serve a wide variety of EVs with different time characteristics, i.e., arrival time, dwell time, and departure time [61]. Among them, office parking lots typically see EVs arriving around 8 am, departing at 4 pm, and having an 8 h dwell time. Food court parking experiences two rush times, during lunch and dinner, with a one-hour dwell time. Entertainment complexes have arrival and departure times after working hours, for example, from 6 pm to 11 pm. Therefore, if the charging and discharging operations of public parking lots are coordinated, they can act as a virtual storage plant 24 h a day.

2.6. Photovoltaic and Battery Storage

Photovoltaic (PV) systems play a key role in enhancing power system flexibility. In residential areas, rooftop PV systems are practical solutions for decentralized power generation. They help reduce stress on the power system during peak demands and can supply local demands [62], preventing power line congestion in weak areas [63]. As a result, they can postpone power system reinforcement to increase the capacity of power transformers and line flows [64].

PV systems can operate alone or be integrated with battery energy storage systems (BESSs). In the former, PV power generation is injected into the grid or consumed by local demands directly. In the latter, the PV system is linked with a BESS and a smart energy management system (SEMS) [65]. This way, the PV-BESS can respond to real-time demand and supply conditions to either store solar power in the battery or inject it into the local grid. Therefore, the PV-BESS can provide peak shaving and valley filling [66]. Some studies suggest using the PV-BESS to provide frequency regulation [67] and voltage support [68].

3. Demand Flexibility under Uncertainty

The long-term aim of demand-side flexibility is to facilitate the integration of renewables into power systems. The intermittency and volatility of renewable power introduce significant uncertainty into power systems. In this context, deterministic approaches often fail to fully leverage demand-side flexibility. Therefore, to manage the uncertainties associated with renewable power, non-deterministic methodologies should be adopted. Stochastic programming, robust optimization, and information-gap decision theory (IGDT) are commonly used in such studies. These non-deterministic methodologies allow for the incorporation of imperfect data regarding uncertain variables into the flexibility problem.

To select an appropriate non-deterministic approach, two key questions should be addressed:

(1) How complete is our information about the uncertain variables?
(2) How precise do the strategies need to be for the final plan?

The first question pertains to the information available about uncertain variables, such as the availability of historical data, Probability Distribution Functions (PDFs), and possible scenarios. For instance, in power system studies, wind velocity regimes are typically described using the Weibull PDF [69]. Therefore, when definite PDFs are available, non-deterministic approaches that utilize these distributions can be applied. Conversely, in some studies, there is no perfect information about the probability distribution of certain uncertain variables, such as electricity prices. In these cases, non-probabilistic quantification of the uncertain variables is required. Thus, non-deterministic approaches that accommodate non-probabilistic models of uncertainties are adopted.

The second question addresses the main objective of the operator in studying the problem under uncertainty. In some studies, the operator aims to determine the optimal responses of the system to all possible realizations of uncertain variables. For these scenarios, non-deterministic approaches that optimize the problem across various possible trajectories of the uncertain variables are used. On the other hand, if the operator is not concerned with all realizations of the uncertain variables, the focus shifts to optimizing strategies under best-case and/or worst-case scenarios. In such cases, non-deterministic approaches that address the gaps between favorable and unfavorable aspects of uncertain variables, rather than considering the entire PDF, are employed.

Based on the above facts, three non-deterministic approaches—stochastic programming, robust optimization, and IGDT—are commonly used in power system studies under uncertainty. These approaches are elaborated on in the following subsections.

3.1. Stochastic Programming

Stochastic programming is a probabilistic approach that provides the system operator with detailed information about different realizations of uncertain variables [70]. In this approach, different sets of scenarios are generated to cover the uncertain range of the stochastic variable. These scenarios are typically based on historical data and selected using the best-fitted PDFs. Each scenario is characterized by a magnitude and a probability value. When the problem involves multiple uncertain variables, different permutations of scenarios create possible trajectories. The problem is then optimized across all these trajectories, providing decision-makers with optimized strategies for all potential scenarios. Although this approach offers detailed information about optimized strategies, it requires accurate data on the distribution of uncertain variables. Additionally, due to the large number of scenarios, stochastic programming can impose a heavy computational burden, making it less suitable for near real-time analysis. In recent studies, stochastic programming has been extensively utilized in two-stage [71], three-stage [72], and multi-stage approaches [73] to provide demand flexibility in power systems. In [74], multilayer iterative stochastic dynamic programming is introduced to optimize energy management in smart residential areas considering EV systems.

3.2. Robust Optimization

In robust optimization, the uncertain range of the uncertain variable is defined by lower and upper thresholds. The problem is then optimized for the worst-case realization within these bounds [75]. In this approach, no PDF or probability is defined for the uncertain variables. Due to the absence of scenarios, this approach is very efficient in terms of execution time. However, it fails to determine optimal strategies for different realizations of the uncertain variables. Unlike stochastic programming, which addresses various uncertain variables, robust optimization focuses on determining robust strategies against the worst-case realization of the key uncertain variable. Therefore, the most critical uncertain variable is used to develop these robust strategies. In [76], robust optimization is employed to aggregate substantial power flexibility from a large number of distributed energy resources. This approach is further applied to optimize the flexible operation of active distribution grids with bidirectional EV fleets [77]. Robust optimization is also used to model electricity price uncertainty in flexibility-constrained smart HEMSs with roof-top PV systems and BESSs [78]. In [79], robust optimization is applied to building heating systems and HPs for demand response control considering the uncertainty of building demand. In [80], robust optimization is discussed to optimize the HEMS's operation with EVs, addressing the uncertainties associated with real-time electricity prices.

3.3. Information-Gap Decision Theory

IGDT is a non-probabilistic approach that addresses both the adverse and favorable aspects of severe uncertain variables. To establish two threshold strategies, IGDT uses two immunity functions: the robustness function and the opportuneness function [81]. Similar to robust optimization, IGDT considers an uncertain range, known as the envelope bound, for the uncertain variable. IGDT aims to find robust strategies that can handle both catastrophic failures and windfall successes. The working points of the problem are located within the gap between the adverse and favorable states. This approach has recently been used in several studies to facilitate the integration of flexibility in power systems. In [82], IGDT is employed to manage the risk associated with uncertain wind power generation in flexible power system operation. This approach is further addressed to model strategic decisions of a price-maker virtual power plant considering demand flexibility [83]. In [84], the IGDT is applied to optimize the charging and discharging of EVs to cope with the uncertainty of distributed generation in active distribution grids.

Although this paper discussed only the three abovementioned non-deterministic models, but they are not limited to these. For example, interval optimization is addressed in power system studies to model uncertainties associated with renewable energy output [85] and load demands [86].

Figure 6 depicts a general overview of scenario characterization for the three non-deterministic approaches. In subfigure (a), it is shown how different trajectories are formed by different scenarios through a scenario graph in stochastic programming. In subfigure (b), the lower and upper bounds of the robust optimization are shown to determine the worst-case realization of the uncertain variable. In subfigure (c), the pernicious and propitious facets of the uncertain envelope are depicted in the IGDT.

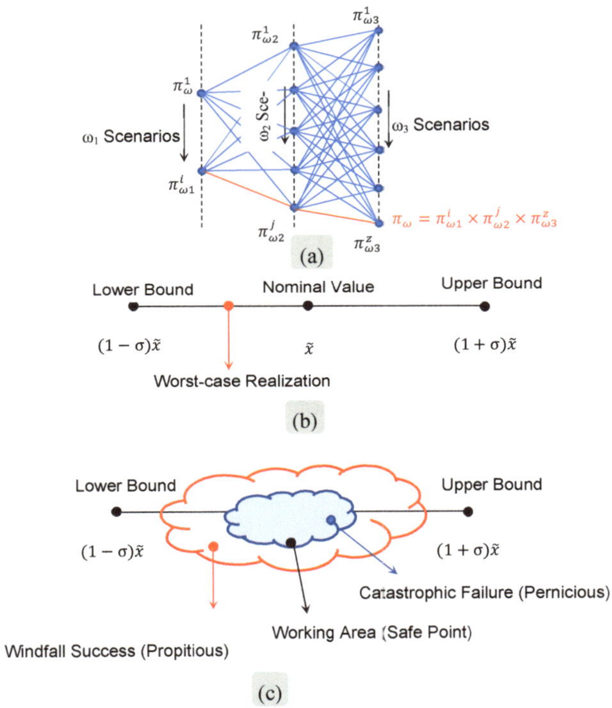

Figure 6. Uncertainty characterization in (**a**) stochastic programming (**b**) robust optimization (**c**) IGDT.

4. Integration of Demand Flexibility to Electricity Market

In power systems with high-RES penetration, the uncertainty of power generation decreases as the power delivery time approaches. As the delivery time nears, the supply-side uncertainty gradually diminishes. To address the intermittent nature of RES, power balancing between generation and consumption is managed hierarchically, from one day ahead to real-time. This approach allows for the unlocking of flexibility potentials in different time slots based on renewable power availability.

Typically, in short-term electricity markets, three trading periods are scheduled to manage RES intermittency over 24 h:

(1) Day-ahead market: conducted 24 h before the power delivery time, this market determines electricity prices based on the intersection of supply and demand curves [87].
(2) Intraday market: Held 60 to 10 min before the power delivery time, this market, also known as the adjustment market, allows participants to adjust their power procurement strategies based on updated flexibility requirements. Participants can buy or sell parts of their power portfolio, originally procured from the day-ahead market, in response to RES availability [88].

(3) Balancing market (real-time market): conducted a few seconds before the energy delivery time, this market provides final up-/down-regulation to the power system [89].

Demand-side flexibilities are integrated into these three trading periods. The challenge is to align the flexibility potentials of demand sectors with the timing of these market periods. The time response of different demand flexibilities varies: some can be unlocked on short notice and integrated into near real-time markets, while others may require longer notice and thus are suited for day-ahead markets. For example, in the metal smelting industry, the operation of smelting pots cannot be interrupted once the process has begun, making them unsuitable for short-notice flexibility; therefore, they are inappropriate for the intraday and balancing markets [90]. In contrast, crushers in cement manufacturing plants can be interrupted on short notice, allowing them to provide near real-time flexibility in balancing markets [91].

Based on these considerations, demand-side flexibilities should be categorized into different classes based on their response times to long, mid, and short advance notices. This categorization is fundamental to analyzing flexibility in demand sectors and determining which flexibility opportunities can be integrated into the hierarchical electricity market floors. Figure 7 illustrates the hierarchical integration of demand flexibilities into the three market floors according to long, mid, and short notice periods.

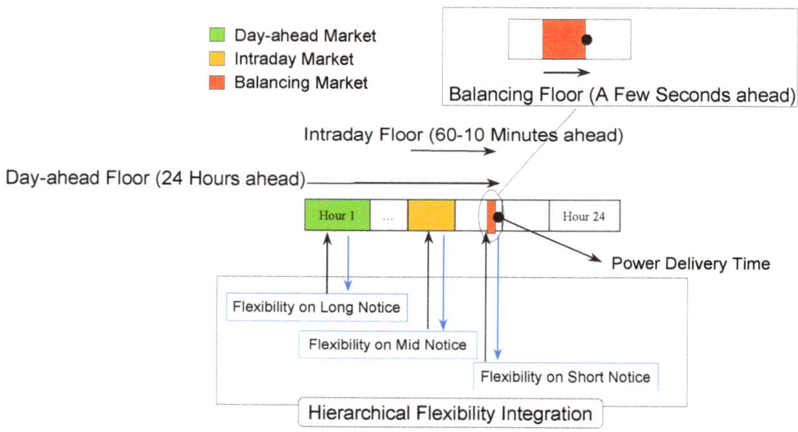

Figure 7. Hierarchical flexibility integration into three trading floors of short-term electricity markets [92].

5. Key Concepts of Flexibility Scheduling

In this section, some key concepts of demand flexibility scheduling are reviewed. They include objective functions of demand flexibility, communication and data exchange, and simulation software.

5.1. Objective Functions

The DR programs are designed to unleash the heat and power demand flexibility with the aim of achieving specific objectives. From a mathematical point of view, the objective function can be constructed in either a linear or non-linear function depending on the demand model and corresponding constraints. The linear functions can be optimized by mathematical approaches called linear programming (LP). The non-linear objective functions are optimized by non-linear programming (NLP) and metaheuristic algorithms, e.g., Genetic Algorithm.

In the demand response literature, more specific objective functions are discussed. Among them, the most important objectives include energy consumption cost minimization [93] and self-consumption maximization [94], maximization of renewable power penetration [95], minimization of greenhouse gas emissions [96] and carbon emission control [97], maximization of power system reliability [98], voltage regulation [99], frequency control [100], increasing power quality [101], congestion management [102], peak shaving and valley filling [103], and increasing energy efficiency [104].

In addition to the aforementioned objectives, some studies highlight grid balancing as the aim of demand flexibility. Demand flexibility primarily contributes to supply–demand balancing in power systems during periods of surplus and deficit of renewable energy [105]. This role becomes especially critical in power systems with high levels of RES penetration, where the supply side is subject to the intermittency of wind and solar power. Thus, demand flexibility is specifically coordinated to provide up- and down-regulation to counteract positive and negative imbalances in the power system.

In some studies, demand flexibility is designed to respond to operational needs within local distribution grids such as failures in power supply and equipment [106]. When anomalies such as equipment failures or operational abnormalities are detected in the local grid, the local DSO collaborates with the DRAs in different demand sectors to address the technical problems of the local grid. This involves responding to fault detection mechanisms with appropriate and timely actions aimed at restoring the distribution grid or preventing the spread of faults to other areas. These actions may include demand curtailment or adjustment, as well as maneuvers to reconfigure power flows within the distribution grid, thereby mitigating the risk of cascading failures. Thus, the MPC is extensively used to design a fault-tolerant energy management system (EMS) in power systems with high-RES penetration [107]. To sum up, Table 1 summarizes the main objective functions of demand-side flexibility in some recent studies.

Table 1. Some key objective functions of demand-side flexibility in recent studies.

References	The Key Objective Function
[93]	Energy consumption cost minimization
[94]	Self-consumption maximization
[95]	Maximization of renewable power penetration
[96,97]	Minimization of greenhouse gas emissions and carbon emission control
[98]	Maximization of power system reliability
[99]	Voltage regulation
[100]	Frequency control
[101]	Increasing power quality
[102]	Congestion management
[103]	Peak shaving and valley filling
[104]	Increasing energy efficiency

5.2. Communication and Data Exchange

To optimize the operation of local energy communities, the objective function may require real-time data from various sources. These include electricity market data such as electricity prices [108], meteorological data [109] such as forecasts of wind speed and outdoor temperature, and grid operator data such as voltage magnitude and line power flows. Additionally, data from measurement sensors, such as domestic hot water and indoor air temperatures in residential buildings, are essential for a HEMS [110]. Therefore, to unlock demand flexibility, either one-way or two-way communication units are required [111].

To provide a general overview of the communication and data exchange, Figure 8 depicts a schematic diagram of a smart irrigation system designed for agricultural farms. The predictive controller within this system gathers data and information from multiple sources, including the local grid operator, meteorological office, as well as local measurement sensors and setpoints provided by farmers. The objective function of the controller is to run the water pumps when the electricity price is low and the surplus of renewable power is available and to turn down the water pumps during the high electricity prices corresponding to the deficit of renewable power availability.

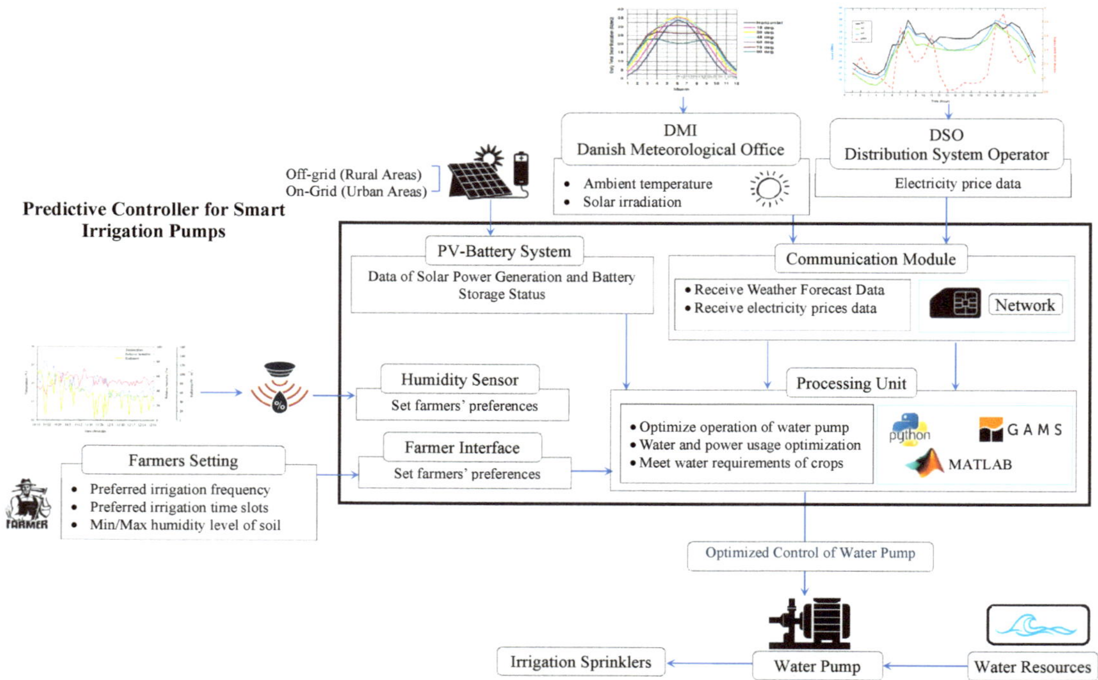

Figure 8. Schematic diagram of a controller for the smart irrigation of farms, with communication and data exchange units.

5.3. Simulation Software

In order to model demand flexibility, simulation software is used. On these software platforms, various energy sources and demands can be modeled. For example, to unlock demand flexibility in a smart building, a smart HEMS can be modeled in a simulation. This model can simulate the electric demand of HPs, EVs, and household appliances, including controllable ones, such as washing machines and tumble dryers, and non-controllable ones, such as audio systems, TVs, and lighting systems. In [112], to unlock the heat-to-power flexibility of a residential HP, sensor measurements are first exported to R software to simulate the building's thermal dynamics. These thermal dynamics are then exported to a model-checking software, UPPAAL-STRATEGO [112] and MATLAB [113], to simulate the HP controller and unleash the heat demand flexibility. "UPPAAL is an integrated tool environment for modeling, validation, and verification of real-time systems modeled as networks of timed automata and was developed in collaboration between the Uppsala University, Sweden, and the Aalborg University, Denmark" [114]. Recently, it was used at Aalborg University to design HP controllers.

In addition to demand, the model can include self-generation facilities, such as PV systems and BESSs. The simulation software can also simulate the comfort needs of residents. For example, for the HP controller, the simulation software can translate the preferred indoor air temperature into a range with lower and upper thresholds. For EVs, it can obtain the preferred departure time and minimum departure state of charge (SoC) from the PDF of in-driving and out-driving EV fleets. The software can then optimize the charging and discharging of the EVs during dwell time to minimize power costs and provide flexibility to power grids.

Various software tools are suggested in studies for simulating demand flexibility. Among them are MATLAB, Simulink [115], MATPOWER [116], GAMS [117], DIgSI-

LENT [118], GridLAB-D [119], OpenDSS [120], EnergyPlus [121], and EnergyPro [122]. Among them, MATPOWER and DIgSILENT can be used to model active distribution grids with the aim of congestion management and voltage regulation. GAMS is a general-purpose software for mathematical modeling and optimization, enabling the implementation of stochastic programming and robust optimization.

After simulation, the demand flexibility is tested in real-time simulators for real-world validation. This process allows for the flexibility potentials to be evaluated in a controlled and realistic environment to ensure their effectiveness. Subsequently, this demand flexibility undergoes limited real-world field tests to further confirm its performance and reliability in practical applications. In the literature, different real-time simulators are discussed to test the demand flexibility, including OPAL-RT [123], RTDS [124], Typhoon HIL [125], PLECS RT Box [126], dSPACE [127], and xPC [128].

To sum up, Table 2 compares the main features of some scoping studies in recent years in comparison to the current study.

Table 2. The main features of some recent scoping studies for demand-side flexibility.

Reference	Addressed Demand Flexibility and Related Concepts
[129]	- Building flexibility - HVAC - EWH - Refrigerators - Wet appliances - Lighting
[130]	Flexibility potentials for industrial, residential, agricultural and commercial sectors
[131]	- Demand flexibility in northern Europe, Sweden, Denmark, Norway, Finland, Estonia, Lativia, Lithuania - Flexibility of industrial, residential, commercial sectors - Flexibility of heating system, shopping centers, office buildings
[132]	- District heating - Heat resources - Control methods of flexible heat demands - Integration of heat flexibility into electricity markets
[133]	- Energy efficiency - Price-based and incentive-based demand response programs - Hardware and communication technology for demand flexibility - Soft computing such as neural network and fuzzy logic - Optimization approaches for scheduling demand flexibility

6. Limitations and Future Works

Although demand-side flexibility is discussed for the residential sector, there are some limitations and challenges for the actual implementation associated with current technologies. Among them, the most important limitations can be stated as follows:

(1) Consumer awareness and engagement in demand response programs;
(2) Lack of standardizations for smart HEMSs;
(3) Insufficient grid infrastructures for communication and data exchange;
(4) Lack of market and regulatory mechanisms;
(5) Vulnerability to cybersecurity attacks.

Although a comprehensive review is conducted for the residential demand flexibility, future directions can involve enhancing digitalization and smart technologies to overcome the current technological barrier. Among them, artificial intelligence and machine learning, as well as automated demand response, can be integrated to the future demand response technologies.

7. Conclusions

This paper reviewed demand-side flexibility in power systems, with a particular emphasis on the residential sector. To highlight the importance of demand flexibility in power systems, flexibility potentials were examined across four sectors: residential, industrial, commercial, and agricultural. It was concluded that harnessing sector-specific flexibility potential requires expert knowledge. Moreover, for demand flexibility to be effective in bulk power systems, sectoral demand flexibility should be aggregated through demand response aggregators.

The focus then shifted to demand flexibility within the residential sector. Household appliances were categorized into three groups based on flexibility and controllability: thermostatically controllable appliances, controllable non-thermal appliances, and uncontrollable appliances. The role of home energy management systems in enhancing building demand flexibility was also discussed. Heat pumps were identified as significant sources of flexibility in residential buildings, with their potential being controllable through thermal dynamics and model predictive controls. The district heating system was highlighted as a key factor in increasing renewable energy integration into the residential sector, particularly through the application of central and booster heat pumps. Additionally, the flexibility potential of EVs in residential and parking lots was examined, noting how EVs can address local distribution grid issues through smart charging and discharging schedules. The roles of PV systems and battery energy management systems were also reviewed, demonstrating how these technologies can facilitate demand flexibility and postpone the need for local distribution grid reinforcement.

The discussion then turned to demand flexibility scheduling under uncertainty, covering three common non-deterministic approaches: stochastic programming, robust optimization, and information-gap decision theory. The integration of demand flexibility into short-term electricity markets was also reviewed, with flexibility potentials classified into short-, mid-, and long-term advance notices for participation in day-ahead, adjustment, and balancing markets.

Finally, three key concepts of flexibility scheduling were examined: objective functions, communication and data exchange, and simulation software. It was concluded that communication systems are critical for unlocking demand flexibility, as they enable the necessary real-time or short-notice data exchange before integrating flexibility into upstream power grids.

Author Contributions: Conceptualization, H.G. and S.G.; methodology, H.G.; validation, H.G.; formal analysis, H.G.; investigation, H.G., S.G.; resources, H.G., S.G., R.S. and B.B.-J.; data curation, S.G., H.G., R.S. and B.B.-J.; writing—original draft preparation, S.G. and H.G.; writing—review and editing, H.G., S.G., R.S. and B.B.-J.; visualization, H.G. and S.G.; supervision, B.B.-J.; project administration, B.B.-J.; funding acquisition, B.B.-J. All authors have read and agreed to the published version of the manuscript.

Funding: This project received funding from the European Union's Horizon 2020 Research and Innovation Programme under grant agreement No. 957682 for SERENE and No.101022587 for the SUSTENANCE projects.

Data Availability Statement: The original contributions presented in the study are included in the article, further inquiries can be directed to the corresponding author/s.

Conflicts of Interest: The authors declare no conflict of interest.

Abbreviations

Battery energy storage systems	BESS
Combined Heat and Power	CHP
Continuous-Time Stochastic Model	CTSM
Controllable non-thermal appliances	CNTA
Demand response	DR
Demand Response Aggregator	DRA
Demand response program	DRP
Differential Equations	DE
Distribution System Operator	DSO
District heating	DH
Electric vehicle	EV
Electric water heaters	EWH
Energy management system	EMS
Grid-to-vehicle	G2V
Heat pump	HP
Heat, ventilation, and air conditioning	HVAC
Home energy management system	HEMS
Information-gap decision theory	IGDT
Model predictive control	MPC
Phase-change Materials	PCM
Photovoltaic	PV
Power Line Carrier	PLC
Probability Distribution Function	PDF
Renewable Energy Source	RES
Thermostatically controllable appliances	TCA
Uncontrollable appliances	UA
Variable Frequency Drives	VFD
Vehicle-to-grid	V2G

References

1. Lund, H.; Mathiesen, B. Energy system analysis of 100% renewable energy systems—The case of Denmark in years 2030 and 2050. *Energy* **2009**, *34*, 524–531. [CrossRef]
2. Dumlao, S.M.G.; Ishihara, K.N. Impact assessment of electric vehicles as curtailment mitigating mobile storage in high PV penetration grid. *Energy Rep.* **2022**, *8*, 736–744. [CrossRef]
3. Pourramezan, A.; Samadi, M. A novel approach for incorporating incentive-based and price-based demand response programs in long-term generation investment planning. *Int. J. Electr. Power Energy Syst.* **2022**, *142*, 108315. [CrossRef]
4. Zhang, S.; Chen, S.; Gu, W.; Lu, S.; Chung, C.Y. Dynamic optimal energy flow of integrated electricity and gas systems in continuous space. *Appl. Energy* **2024**, *375*, 124052. [CrossRef]
5. Saedi, I.; Mhanna, S.; Mancarella, P. Integrated electricity and gas system modelling with hydrogen injections and gas composition tracking. *Appl. Energy* **2021**, *303*, 117598. [CrossRef]
6. Klyapovskiy, S.; You, S.; Cai, H.; Bindner, H.W. Integrated Planning of a Large-Scale Heat Pump in View of Heat and Power Networks. *IEEE Trans. Ind. Appl.* **2019**, *55*, 5–15. [CrossRef]
7. Lezama, F.; Soares, J.; Canizes, B.; Vale, Z. Flexibility management model of home appliances to support DSO requests in smart grids. *Sustain. Cities Soc.* **2020**, *55*, 102048. [CrossRef]
8. Finck, C.; Li, R.; Kramer, R.; Zeiler, W. Quantifying demand flexibility of power-to-heat and thermal energy storage in the control of building heating systems. *Appl. Energy* **2018**, *209*, 409–425. [CrossRef]
9. Lakshmanan, V.; Sæle, H.; Degefa, M.Z. Electric water heater flexibility potential and activation impact in system operator perspective–Norwegian scenario case study. *Energy* **2021**, *236*, 121490. [CrossRef]
10. Luo, Z.; Peng, J.; Tan, Y.; Yin, R.; Zou, B.; Hu, M.; Yan, J. A novel forecast-based operation strategy for residential PV-battery-flexible loads systems considering the flexibility of battery and loads. *Energy Convers. Manag.* **2023**, *278*, 116705. [CrossRef]
11. Yang, S.; Gao, H.O.; You, F. Demand flexibility and cost-saving potentials via smart building energy management: Opportunities in residential space heating across the US. *Adv. Appl. Energy* **2024**, *14*, 100171. [CrossRef]
12. Golmohamadi, H.; Keypour, R.; Bak-Jensen, B.; Pillai, J.R.; Khooban, M.H. Robust Self-Scheduling of Operational Processes for Industrial Demand Response Aggregators. *IEEE Trans. Ind. Electron.* **2020**, *67*, 1387–1395. [CrossRef]
13. Gregory, A.T.M.; Homan, K.; Aghajanzadeh, A. *Opportunities for Automated Demand Response in California's Dairy Processing Industry*; Lawrence Berkeley National Laboratory: Berkeley, CA, USA, 2015.

14. Olsen, A.; Goli, D.; Faulkner, S.; McKane, D. *Opportunities for Energy Efficiency and Demand Response in the California Cement Industry*; Lawrence Berkeley National Laboratory: Berkeley, CA, USA, 2010.
15. Bao, P.; Zhang, W.; Cheng, D.; Liu, M. Hierarchical control of aluminum smelter loads for primary frequency support considering control cost. *Int. J. Electr. Power Energy Syst.* **2020**, *122*, 106202. [CrossRef]
16. Helin, K.; Käki, A.; Zakeri, B.; Lahdelma, R.; Syri, S. Economic potential of industrial demand side management in pulp and paper industry. *Energy* **2017**, *141*, 1681–1694. [CrossRef]
17. Alarfaj, O.; Bhattacharya, K. Material Flow Based Power Demand Modeling of an Oil Refinery Process for Optimal Energy Management. *IEEE Trans. Power Syst.* **2019**, *34*, 2312–2321. [CrossRef]
18. Wang, J.; You, S.; Zong, Y.; Træholt, C.; Dong, Z.Y.; Zhou, Y. Flexibility of combined heat and power plants: A review of technologies and operation strategies. *Appl. Energy* **2019**, *252*, 113445. [CrossRef]
19. Golmohamadi, H.; Asadi, A. A multi-stage stochastic energy management of responsive irrigation pumps in dynamic electricity markets. *Appl. Energy* **2020**, *265*, 114804. [CrossRef]
20. Aghajanzadeh, A.; Therkelsen, P. Agricultural demand response for decarbonizing the electricity grid. *J. Clean. Prod.* **2019**, *220*, 827–835. [CrossRef]
21. Loschan, C.; Schwabeneder, D.; Lettner, G.; Auer, H. Flexibility potential of aggregated electric vehicle fleets to reduce transmission congestions and redispatch needs: A case study in Austria. *Int. J. Electr. Power Energy Syst.* **2023**, *146*, 108802. [CrossRef]
22. Li, Z.; Lei, X.; Shang, Y.; Jia, Y.; Jian, L. A genuine V2V market mechanism aiming for maximum revenue of each EV owner based on non-cooperative game model. *J. Clean. Prod.* **2023**, *414*, 137586. [CrossRef]
23. Heydarian-Forushani, E.; Golshan, M.; Shafie-Khah, M. Flexible interaction of plug-in electric vehicle parking lots for efficient wind integration. *Appl. Energy* **2016**, *179*, 338–349. [CrossRef]
24. Daryabari, M.K.; Keypour, R.; Golmohamadi, H. Stochastic energy management of responsive plug-in electric vehicles characterizing parking lot aggregators. *Appl. Energy* **2020**, *279*, 115751. [CrossRef]
25. Maouris, G.; Escriva, E.J.S.; Acha, S.; Shah, N.; Markides, C.N. CO_2 refrigeration system heat recovery and thermal storage modelling for space heating provision in supermarkets: An integrated approach. *Appl. Energy* **2020**, *264*, 114722. [CrossRef]
26. U.S. Energy Information Administration (EIA). Available online: https://www.eia.gov/tools/glossary/?id=electricity (accessed on 12 June 2024).
27. Golmohamadi, H. Flexible Operation of Power-To-X Energy Systems in Transportation Networks. In *Interconnected Modern Multi-Energy Networks and Intelligent Transportation Systems*; John Wiley & Sons: Hoboken, NJ, USA, 2024; pp. 117–164. [CrossRef]
28. IEA. World Electricity Final Consumption by Sector, 1974–2019. Available online: https://www.iea.org/data-and-statistics/charts/world-electricity-final-consumption-by-sector-1974-2018 (accessed on 10 June 2024).
29. Mugnini, A.; Ramallo-González, A.P.; Parreño, A.; Molina-Garcia, A.; Skarmeta, A.F.; Arteconi, A. Dynamic building thermal mass clustering for energy flexibility assessment: An application to demand response events. *Energy Build.* **2024**, *308*, 114011. [CrossRef]
30. Sadeghianpourhamami, N.; Demeester, T.; Benoit, D.; Strobbe, M.; Develder, C. Modeling and analysis of residential flexibility: Timing of white good usage. *Appl. Energy* **2016**, *179*, 790–805. [CrossRef]
31. Cibin, N.; Tibo, A.; Golmohamadi, H.; Skou, A.; Albano, M. Machine learning-based algorithms to estimate thermal dynamics of residential buildings with energy flexibility. *J. Build. Eng.* **2023**, *65*, 105683. [CrossRef]
32. Hasrat, I.R.; Jensen, P.G.; Larsen, K.G.; Srba, J. End-to-End Heat-Pump Control Using Continuous Time Stochastic Modelling and Uppaal Stratego. In *Theoretical Aspects of Software Engineering*; Aït-Ameur, Y., Crăciun, F., Eds.; Springer International Publishing: Cham, Switzerland, 2022; pp. 363–380.
33. Evens, M.; Arteconi, A. Design energy flexibility within a comfort and climate box–An experimental evaluation of the internal heat pump control effects. *Appl. Therm. Eng.* **2024**, *254*, 123842. [CrossRef]
34. Farzamkia, S.; Ranjbar, H.; Hatami, A.; Iman-Eini, H. A novel PSO (Particle Swarm Optimization)-based approach for optimal schedule of refrigerators using experimental models. *Energy* **2016**, *107*, 707–715. [CrossRef]
35. Vellei, M.; Le Dréau, J.; Abdelouadoud, S.Y. Predicting the demand flexibility of wet appliances at national level: The case of France. *Energy Build.* **2020**, *214*, 109900. [CrossRef]
36. Golmohamadi, H.; Keypour, R.; Bak-Jensen, B.; Pillai, J.R. Optimization of household energy consumption towards day-ahead retail electricity price in home energy management systems. *Sustain. Cities Soc.* **2019**, *47*, 101468. [CrossRef]
37. Tuomela, S.; Tomé, M.d.C.; Iivari, N.; Svento, R. Impacts of home energy management systems on electricity consumption. *Appl. Energy* **2021**, *299*, 117310. [CrossRef]
38. Khan, A.A.; Razzaq, S.; Khan, A.; Khursheed, F. Owais HEMSs and enabled demand response in electricity market: An overview. *Renew. Sustain. Energy Rev.* **2015**, *42*, 773–785. [CrossRef]
39. Prum, P.; Charoen, P.; Khan, M.A.; Bayati, N.; Charoenlarpnopparut, C. Energy Management Scheme for Optimizing Multiple Smart Homes Equipped with Electric Vehicles. *Energies* **2024**, *17*, 254. [CrossRef]
40. Golmohamadi, H. Stochastic energy optimization of residential heat pumps in uncertain electricity markets. *Appl. Energy* **2021**, *303*, 117629. [CrossRef]
41. Wang, Q.; Zhang, X.; Geng, X.; Chen, X.; Xing, M. Experiments on the characteristics of a sewage water source heat pump system for heat recovery from bath waste. *Appl. Therm. Eng.* **2022**, *204*, 117956. [CrossRef]

42. Kheiri, S.Z.; Mirzaei, M.A.; Parvania, M. Integrating the Energy Flexibility of Cold Climate Air Source Heat Pump in Home Energy Management Systems. In Proceedings of the 2024 IEEE Power & Energy Society Innovative Smart Grid Technologies Conference (ISGT), Washington, DC, USA, 19–22 February 2024; pp. 1–5. [CrossRef]
43. Juhl, R.; Kristensen, N.R.; Bacher, P.; Kloppenborg, J.; Madsen, H. Grey-box Modeling of the Heat Dynamics of a Building with CTSM-R. 2020. Available online: http://ctsm.info/building2.pdf (accessed on 15 June 2024).
44. Hasrat, I.R.; Jensen, P.G.; Larsen, K.G.; Srba, J. A toolchain for domestic heat-pump control using Uppaal Stratego. *Sci. Comput. Program.* **2023**, *230*, 102987. [CrossRef]
45. Hou, J.; Li, H.; Nord, N.; Huang, G. Model predictive control under weather forecast uncertainty for HVAC systems in university buildings. *Energy Build.* **2022**, *257*, 111793. [CrossRef]
46. Sawant, P.; Mier, O.V.; Schmidt, M.; Pfafferott, J. Demonstration of Optimal Scheduling for a Building Heat Pump System Using Economic-MPC. *Energies* **2021**, *14*, 7953. [CrossRef]
47. Vivian, J.; Prataviera, E.; Cunsolo, F.; Pau, M. Demand Side Management of a pool of air source heat pumps for space heating and domestic hot water production in a residential district. *Energy Convers. Manag.* **2020**, *225*, 113457. [CrossRef]
48. Nunna, A.C.; Zong, Y.; Georges, L.; You, S. Demand response with active phase change material based thermal energy storage in buildings. *Energy Rep.* **2023**, *9*, 227–235. [CrossRef]
49. Pakere, I.; Feofilovs, M.; Lepiksaar, K.; Vītoliņš, V.; Blumberga, D. Multi-source district heating system full decarbonization strategies: Technical, economic, and environmental assessment. *Energy* **2023**, *285*, 129296. [CrossRef]
50. Sorknæs, P. Hybrid energy networks and electrification of district heating under different energy system conditions. *Energy Rep.* **2021**, *7*, 222–236. [CrossRef]
51. Korpela, T.; Kaivosoja, J.; Majanne, Y.; Laakkonen, L.; Nurmoranta, M.; Vilkko, M. Utilization of District Heating Networks to Provide Flexibility in CHP Production. *Energy Procedia* **2017**, *116*, 310–319. [CrossRef]
52. Wahlroos, M.; Pärssinen, M.; Manner, J.; Syri, S. Utilizing data center waste heat in district heating–Impacts on energy efficiency and prospects for low-temperature district heating networks. *Energy* **2017**, *140*, 1228–1238. [CrossRef]
53. Fambri, G.; Marocco, P.; Badami, M.; Tsagkrasoulis, D. The flexibility of virtual energy storage based on the thermal inertia of buildings in renewable energy communities: A techno-economic analysis and comparison with the electric battery solution. *J. Energy Storage* **2023**, *73*, 109083. [CrossRef]
54. Todorov, O.; Alanne, K.; Virtanen, M.; Kosonen, R. A method and analysis of aquifer thermal energy storage (ATES) system for district heating and cooling: A case study in Finland. *Sustain. Cities Soc.* **2020**, *53*, 101977. [CrossRef]
55. Vandermeulen, A.; Van Der Heijde, B.; Helsen, L. Controlling district heating and cooling networks to unlock flexibility: A review. *Energy* **2018**, *151*, 103–115. [CrossRef]
56. Visakh, A.; Selvan, M.P. Analysis and mitigation of the impact of electric vehicle charging on service disruption of distribution transformers. *Sustain. Energy Grids Netw.* **2023**, *35*, 101096. [CrossRef]
57. Sharma, S.; Jain, P. Risk-averse integrated demand response and dynamic G2V charge scheduling of an electric vehicle aggregator to support grid stability. *Int. Trans. Electr. Energy Syst.* **2021**, *31*, e12867. [CrossRef]
58. Alfaverh, F.; Denaï, M.; Sun, Y. Optimal vehicle-to-grid control for supplementary frequency regulation using deep reinforcement learning. *Electr. Power Syst. Res.* **2023**, *214*, 108949. [CrossRef]
59. Sørensen, Å.L.; Lindberg, K.B.; Sartori, I.; Andresen, I. Analysis of residential EV energy flexibility potential based on real-world charging reports and smart meter data. *Energy Build.* **2021**, *241*, 110923. [CrossRef]
60. Karimi-Arpanahi, S.; Jooshaki, M.; Pourmousavi, S.A.; Lehtonen, M. Leveraging the flexibility of electric vehicle parking lots in distribution networks with high renewable penetration. *Int. J. Electr. Power Energy Syst.* **2022**, *142*, 108366. [CrossRef]
61. Daryabari, M.K.; Keypour, R.; Golmohamadi, H. Robust self-scheduling of parking lot microgrids leveraging responsive electric vehicles. *Appl. Energy* **2021**, *290*, 116802. [CrossRef]
62. Moshövel, J.; Kairies, K.-P.; Magnor, D.; Leuthold, M.; Bost, M.; Gährs, S.; Szczechowicz, E.; Cramer, M.; Sauer, D.U. Analysis of the maximal possible grid relief from PV-peak-power impacts by using storage systems for increased self-consumption. *Appl. Energy* **2015**, *137*, 567–575. [CrossRef]
63. Cerna, F.V. A hybrid PV scheme as support to relieve congestion in the domestic supply network. *Int. J. Electr. Power Energy Syst.* **2022**, *134*, 107413. [CrossRef]
64. Holweger, J.; Pena-Bello, A.; Jeannin, N.; Ballif, C.; Wyrsch, N. Distributed flexibility as a cost-effective alternative to grid reinforcement. *Sustain. Energy Grids Netw.* **2023**, *34*, 101041. [CrossRef]
65. Ali, A.O.; Hamed, A.M.; Abdelsalam, M.M.; Sabry, M.N.; Elmarghany, M.R. Energy management of photovoltaic-battery system connected with the grid. *J. Energy Storage* **2022**, *55*, 105865. [CrossRef]
66. Wang, Y.; Liu, L.; Wennersten, R.; Sun, Q. Peak shaving and valley filling potential of energy management system in high-rise residential building. *Energy Procedia* **2019**, *158*, 6201–6207. [CrossRef]
67. Li, Z.; Cheng, Z.; Si, J.; Zhang, S.; Dong, L.; Li, S.; Gao, Y. Adaptive Power Point Tracking Control of PV System for Primary Frequency Regulation of AC Microgrid With High PV Integration. *IEEE Trans. Power Syst.* **2021**, *36*, 3129–3141. [CrossRef]
68. Chaudhary, P.; Rizwan, M. Voltage regulation mitigation techniques in distribution system with high PV penetration: A review. *Renew. Sustain. Energy Rev.* **2018**, *82*, 3279–3287. [CrossRef]
69. Salim, O.M.; Dorrah, H.T.; Hassan, M.A. Wind speed estimation based on a novel multivariate Weibull distribution. *IET Renew. Power Gener.* **2019**, *13*, 2762–2773. [CrossRef]

70. Birge, J.R.; Louveaux, F. *Introduction to Stochastic Programming*, 2nd ed.; Springer Series in Operations Research and Financial Engineering; Springer: New York, NY, USA, 2011; Volume 31. [CrossRef]
71. Jordehi, A.R.; Tabar, V.S.; Mansouri, S.; Sheidaei, F.; Ahmarinejad, A.; Pirouzi, S. Two-stage stochastic programming for scheduling microgrids with high wind penetration including fast demand response providers and fast-start generators. *Sustain. Energy, Grids Netw.* **2022**, *31*, 100694. [CrossRef]
72. Zatti, M.; Martelli, E.; Amaldi, E. A three-stage stochastic optimization model for the design of smart energy districts under uncertainty. In *Computer Aided Chemical Engineering*; Espuña, A., Graells, M., Puigjaner, L., Eds.; Elsevier: Amsterdam, The Netherlands, 2017; Volume 40, pp. 2389–2394. [CrossRef]
73. Flores-Quiroz, A.; Strunz, K. A distributed computing framework for multi-stage stochastic planning of renewable power systems with energy storage as flexibility option. *Appl. Energy* **2021**, *291*, 116736. [CrossRef]
74. Aljohani, T.M. Multilayer Iterative Stochastic Dynamic Programming for Optimal Energy Management of Residential Loads with Electric Vehicles. *Int. J. Energy Res.* **2024**, *2024*, 6842580. [CrossRef]
75. Ben-Tal, A.; Nemirovski, A.; El Ghaoui, L. *Robust Optimization*; Princeton University Press: Princeton, NJ, USA, 2009.
76. Chen, X.; Li, N. Leveraging Two-Stage Adaptive Robust Optimization for Power Flexibility Aggregation. *IEEE Trans. Smart Grid* **2021**, *12*, 3954–3965. [CrossRef]
77. Pirouzi, S.; Aghaei, J.; Niknam, T.; Shafie-Khah, M.; Vahidinasab, V.; Catalão, J.P. Two alternative robust optimization models for flexible power management of electric vehicles in distribution networks. *Energy* **2017**, *141*, 635–651. [CrossRef]
78. Bahramara, S. Robust Optimization of the Flexibility-constrained Energy Management Problem for a Smart Home with Rooftop Photovoltaic and an Energy Storage. *J. Energy Storage* **2021**, *36*, 102358. [CrossRef]
79. Ding, Y.; Bai, Y.; Tian, Z.; Wang, Q.; Su, H. Coordinated optimization of robustness and flexibility of building heating systems for demand response control considering prediction uncertainty. *Appl. Therm. Eng.* **2023**, *223*, 120024. [CrossRef]
80. Tostado-Véliz, M.; Hasanien, H.M.; Turky, R.A.; Jordehi, A.R.; Mansouri, S.A.; Jurado, F. A fully robust home energy management model considering real time price and on-board vehicle batteries. *J. Energy Storage* **2023**, *72*, 108531. [CrossRef]
81. Ben-Haim, Y. Chapter 3-Robustness and Opportuneness. In *Info-Gap Decision Theory*, 2nd ed.; Ben-Haim, Y., Ed.; Academic Press: Oxford, UK, 2006; pp. 37–114. [CrossRef]
82. Nikoobakht, A.; Aghaei, J.; Mardaneh, M. Managing the risk of uncertain wind power generation in flexible power systems using information gap decision theory. *Energy* **2016**, *114*, 846–861. [CrossRef]
83. Gazijahani, F.S.; Salehi, J. IGDT-Based Complementarity Approach for Dealing With Strategic Decision Making of Price-Maker VPP Considering Demand Flexibility. *IEEE Trans. Ind. Inform.* **2020**, *16*, 2212–2220. [CrossRef]
84. Chen, Q.; Wang, W.; Wang, H.; Dong, Y.; He, S. Information gap-based coordination scheme for active distribution network considering charging/discharging optimization for electric vehicles and demand response. *Int. J. Electr. Power Energy Syst.* **2023**, *145*, 108652. [CrossRef]
85. Zeng, L.; Xu, J.; Wang, Y.; Liu, Y.; Tang, J.; Wen, M.; Chen, Z. Day-ahead interval scheduling strategy of power systems based on improved adaptive diffusion kernel density estimation. *Int. J. Electr. Power Energy Syst.* **2023**, *147*, 108850. [CrossRef]
86. Wang, D.; Zhang, C.; Jia, W.; Liu, Q.; Cheng, L.; Yang, H.; Luo, Y.; Kuang, N. A Novel Interval Programming Method and Its Application in Power System Optimization Considering Uncertainties in Load Demands and Renewable Power Generation. *Energies* **2022**, *15*, 7565. [CrossRef]
87. Morales, J.M.; Conejo, A.J.; Madsen, H.; Pinson, P.; Zugno, M. Clearing the Day-Ahead Market with a High Penetration of Stochastic Production. In *Integrating Renewables in Electricity Markets: Operational Problems*; Morales, J.M., Conejo, A.J., Madsen, H., Pinson, P., Zugno, M., Eds.; Springer: Boston, MA, USA, 2014; pp. 57–100. [CrossRef]
88. Conejo, A.J.; Carrión, M.; Morales, J.M. Pool Trading for Wind Power Producers. In *Decision Making Under Uncertainty in Electricity Markets*; Conejo, A.J., Carrión, M.J., Morales, M., Eds.; Springer: Boston, MA, USA, 2010; pp. 195–251. [CrossRef]
89. Morales, J.M.; Conejo, A.J.; Madsen, H.; Pinson, P.; Zugno, M. Balancing Markets. In *Integrating Renewables in Electricity Markets: Operational Problems*; Morales, J.M., Conejo, A.J., Madsen, H., Pinson, P., Zugno, M., Eds.; Springer: Boston, MA, USA, 2014; pp. 101–136. [CrossRef]
90. Yao, M.; Hu, Z.; Zhang, N.; Duan, W.; Zhang, J. Low-carbon benefits analysis of energy-intensive industrial demand response resources for ancillary services. *J. Mod. Power Syst. Clean Energy* **2015**, *3*, 131–138. [CrossRef]
91. Golmohamadi, H. Demand-side management in industrial sector: A review of heavy industries. *Renew. Sustain. Energy Rev.* **2022**, *156*, 111963. [CrossRef]
92. Golmohamadi, H.; Larsen, K.G.; Jensen, P.G.; Hasrat, I.R. Hierarchical flexibility potentials of residential buildings with responsive heat pumps: A case study of Denmark. *J. Build. Eng.* **2021**, *41*, 102425. [CrossRef]
93. Bahlawan, H.; Castorino, G.A.M.; Losi, E.; Manservigi, L.; Spina, P.R.; Venturini, M. Optimal management with demand response program for a multi-generation energy system. *Energy Convers. Manag. X* **2022**, *16*, 100311. [CrossRef]
94. Rana, A.; Gróf, G. Assessment of local energy trading in a residential energy hub with demand management. *Energy Rep.* **2024**, *11*, 1642–1658. [CrossRef]
95. Al Hadi, A.; Silva, C.A.S.; Hossain, E.; Challoo, R. Algorithm for Demand Response to Maximize the Penetration of Renewable Energy. *IEEE Access* **2020**, *8*, 55279–55288. [CrossRef]
96. Zeng, H.; Shao, B.; Dai, H.; Tian, N.; Zhao, W. Incentive-based demand response strategies for natural gas considering carbon emissions and load volatility. *Appl. Energy* **2023**, *348*, 121541. [CrossRef]

97. Hua, H.; Chen, X.; Gan, L.; Sun, J.; Dong, N.; Liu, D.; Qin, Z.; Li, K.; Hu, S. Demand-Side Joint Electricity and Carbon Trading Mechanism. *IEEE Trans. Ind. Cyber-Physical Syst.* **2024**, *2*, 14–25. [CrossRef]
98. Yang, H.; Zhang, X.; Chu, Y.; Ma, Y.; Zhang, D.; Guerrero, J.M. Multi-objective based demand response strategy optimization considering differential demand on reliability of power system. *Int. J. Electr. Power Energy Syst.* **2023**, *152*, 109202. [CrossRef]
99. Chandran, C.V.; Basu, M.; Sunderland, K.; Pukhrem, S.; Catalão, J.P. Application of demand response to improve voltage regulation with high DG penetration. *Electr. Power Syst. Res.* **2020**, *189*, 106722. [CrossRef]
100. Miow, X.C.; Lim, Y.S.; Hau, L.C.; Wong, J.; Patsios, H. Demand response for frequency regulation with neural network load controller under high intermittency photovoltaic systems. *Energy Rep.* **2023**, *9*, 2869–2880. [CrossRef]
101. Devarapalli, H.P.; Dhanikonda, V.S.S.S.S.; Gunturi, S.B. Demand-Side Management for Improvement of the Power Quality in Smart Homes Using Non-Intrusive Identification of Appliance Usage Patterns with the True Power Factor. *Energies* **2021**, *14*, 4837. [CrossRef]
102. Stawska, A.; Romero, N.; de Weerdt, M.; Verzijlbergh, R. Demand response: For congestion management or for grid balancing? *Energy Policy* **2021**, *148*, 111920. [CrossRef]
103. Golmohamadi, H. Operational scheduling of responsive prosumer farms for day-ahead peak shaving by agricultural demand response aggregators. *Int. J. Energy Res.* **2021**, *45*, 938–960. [CrossRef]
104. Wohlfarth, K.; Worrell, E.; Eichhammer, W. Energy efficiency and demand response–Two sides of the same coin? *Energy Policy* **2020**, *137*, 111070. [CrossRef]
105. Morales, J.M.; Conejo, A.J.; Madsen, H.; Pinson, P.; Zugno, M. Facilitating Renewable Integration by Demand Re-sponseDemand response. In *Integrating Renewables in Electricity Markets: Operational Problems*; Morales, J.M., Conejo, A.J., Madsen, H., Pinson, P., Zugno, M., Eds.; Springer: Boston, MA, USA, 2014; pp. 289–329. [CrossRef]
106. Freire, V.A.; Marquez, J.J.; Bordons, C.; Zafra-Cabeza, A.; de Arruda, L.V.R. Energy Management System for Microgrid Considering Operational Faults in Power Supply. In Proceedings of the 2020 International Conference on Smart Energy Systems and Technologies (SEST), Istanbul, Turkey, 7–9 September 2020; pp. 1–6. [CrossRef]
107. Morato, M.M.; Mendes, P.R.; Normey-Rico, J.E.; Bordons, C. LPV-MPC fault-tolerant energy management strategy for renewable microgrids. *Int. J. Electr. Power Energy Syst.* **2020**, *117*, 105644. [CrossRef]
108. Lu, R.; Bai, R.; Huang, Y.; Li, Y.; Jiang, J.; Ding, Y. Data-driven real-time price-based demand response for industrial facilities energy management. *Appl. Energy* **2021**, *283*, 116291. [CrossRef]
109. Kadlec, M.; Buhnova, B.; Tomsik, J.; Herman, J.; Druzbikova, K. Weather Forecast Based Scheduling for Demand Response Optimization in Smart Grids. In Proceedings of the 2017 Smart City Symposium Prague (SCSP), Prague, Czech Republic, 25–26 May 2017; pp. 1–6. [CrossRef]
110. Golmohamadi, H.; Larsen, K.G. Economic heat control of mixing loop for residential buildings supplied by low-temperature district heating. *J. Build. Eng.* **2021**, *46*, 103286. [CrossRef]
111. Ebeid, E.; Rotger-Griful, S.; Mikkelsen, S.A.; Jacobsen, R.H. A methodology to evaluate demand response communication protocols for the Smart Grid. In Proceedings of the 2015 IEEE International Conference on Communications Workshops (ICCW), London, UK, 8–12 June 2015; pp. 2012–2017. [CrossRef]
112. Hasrat, I.R.; Jensen, P.G.; Larsen, K.G.; Srba, J. Modelling of Hot Water Buffer Tank and Mixing Loop for an Intelligent Heat Pump Control. In *Formal Methods for Industrial Critical Systems*; Cimatti, A., Titolo, L., Eds.; Springer Nature: Cham, Switzerland, 2023; pp. 113–130.
113. Golmohamadi, H.; Larsen, K.G.; Jensen, P.G.; Hasrat, I.R. Optimization of power-to-heat flexibility for residential buildings in response to day-ahead electricity price. *Energy Build.* **2021**, *232*, 110665. [CrossRef]
114. UPPAAL Real Time Simulator. Available online: http://www.uppaal.org/ (accessed on 11 June 2024).
115. EL Zerk, A.; Ouassaid, M.; Zidani, Y. Development of a real-time framework between MATLAB and PLC through OPC-UA: A case study of a microgrid energy management system. *Sci. Afr.* **2023**, *21*, e01846. [CrossRef]
116. Suresh, V.; Janik, P.; Guerrero, J.M.; Leonowicz, Z.; Sikorski, T. Microgrid Energy Management System With Embedded Deep Learning Forecaster and Combined Optimizer. *IEEE Access* **2020**, *8*, 202225–202239. [CrossRef]
117. Akbari, E.; Shabestari, S.F.M.; Pirouzi, S.; Jadidoleslam, M. Network flexibility regulation by renewable energy hubs using flexibility pricing-based energy management. *Renew. Energy* **2023**, *206*, 295–308. [CrossRef]
118. Bhattarai, B.P.; Myers, K.S.; Turk, R.J.; Bak-Jensen, B. 21-Multi-time-scale energy management of distributed energy resources in active distribution grids. In *Smart Power Distribution Systems*; Yang, Q., Yang, T., Li, W., Eds.; Academic Press: Cambridge, MA, USA, 2019; pp. 503–528. [CrossRef]
119. Al Faruque, M.A.; Ahourai, F. GridMat: Matlab toolbox for GridLAB-D to analyze grid impact and validate residential microgrid level energy management algorithms. In Proceedings of the 2014 IEEE Power & Energy Society Innovative Smart Grid Technologies Conference (ISGT), Washington, DC, USA, 19–22 February 2014; pp. 1–5. [CrossRef]
120. Subramanya, S.A.; Parathodiyil, M.; Nagashree, A. Implementation of Peak Demand Reduction on a Distribution Feeder Using Python-OpenDSS Co-Simulation. In Proceedings of the 2021 IEEE International Conference on Electronics, Computing and Communication Technologies (CONECCT), Bangalore, India, 9–11 July 2021; pp. 1–6. [CrossRef]
121. Elhefny, A.; Jiang, Z.; Cai, J. Co-simulation and energy management of photovoltaic-rich residential communities for improved distribution voltage support with flexible loads. *Sol. Energy* **2022**, *231*, 516–526. [CrossRef]

122. Andersen, A.N.; Østergaard, P.A. A method for assessing support schemes promoting flexibility at district energy plants. *Appl. Energy* **2018**, *225*, 448–459. [CrossRef]
123. Wang, J.; Blonsky, M.; Ding, F.; Drew, S.C.; Padullaparti, H.; Ghosh, S.; Mendoza, I.; Tiwari, S.; Martinez, J.E.; Dahdah, J.J.D.; et al. Performance Evaluation of Distributed Energy Resource Management via Advanced Hardware-in-the-Loop Simulation. In Proceedings of the 2020 IEEE Power & Energy Society Innovative Smart Grid Technologies Conference (ISGT), Washington, DC, USA, 17–20 February 2020; pp. 1–5. [CrossRef]
124. Lipari, G.; Del Rosario, G.; Corchero, C.; Ponci, F.; Monti, A. A real-time commercial aggregator for distributed energy resources flexibility management. *Sustain. Energy, Grids Netw.* **2018**, *15*, 63–75. [CrossRef]
125. Rodriguez, M.; Arcos-Aviles, D.; Martinez, W. Fuzzy logic-based energy management for isolated microgrid using meta-heuristic optimization algorithms. *Appl. Energy* **2023**, *335*, 120771. [CrossRef]
126. Aguilar, M.; Riffo, S.; Veliz, A.; González-Castaño, C.; Restrepo, C. RT Box card for studying the control communication impacts on microgrid performance and stability. *HardwareX* **2022**, *12*, e00322. [CrossRef] [PubMed]
127. Harmouch, F.Z.; Ebrahim, A.F.; Esfahani, M.M.; Krami, N.; Hmina, N.; Mohammed, O.A. An Optimal Energy Management System for Real-Time Operation of Multiagent-Based Microgrids Using a T-Cell Algorithm. *Energies* **2019**, *12*, 3004. [CrossRef]
128. Maturana, F.; Staron, R.; Loparo, K.; Ambre, R.; Carnahan, D. Simulation-Based Environment for Modeling Distributed Agents for Smart Grid Energy Management. In Proceedings of the Factory Automation (ETFA 2011), Toulouse, France, 5–9 September 2011; pp. 1–7. [CrossRef]
129. Luo, Z.; Peng, J.; Cao, J.; Yin, R.; Zou, B.; Tan, Y.; Yan, J. Demand Flexibility of Residential Buildings: Definitions, Flexible Loads, and Quantification Methods. *Engineering* **2022**, *16*, 123–140. [CrossRef]
130. Golmohamadi, H. Demand-Side Flexibility in Power Systems: A Survey of Residential, Industrial, Commercial, and Agricultural Sectors. *Sustainability* **2022**, *14*, 7916. [CrossRef]
131. Söder, L.; Lund, P.D.; Koduvere, H.; Bolkesjø, T.F.; Rossebø, G.H.; Rosenlund-Soysal, E.; Skytte, K.; Katz, J.; Blumberga, D. A review of demand side flexibility potential in Northern Europe. *Renew. Sustain. Energy Rev.* **2018**, *91*, 654–664. [CrossRef]
132. Golmohamadi, H.; Larsen, K.G.; Jensen, P.G.; Hasrat, I.R. Integration of flexibility potentials of district heating systems into electricity markets: A review. *Renew. Sustain. Energy Rev.* **2022**, *159*, 112200. [CrossRef]
133. Iqbal, S.; Sarfraz, M.; Ayyub, M.; Tariq, M.; Chakrabortty, R.K.; Ryan, M.J.; Alamri, B. A Comprehensive Review on Residential Demand Side Management Strategies in Smart Grid Environment. *Sustainability* **2021**, *13*, 7170. [CrossRef]

energies

MDPI

Article

Optimal Configuration of Electricity-Heat Integrated Energy Storage Supplier and Multi-Microgrid System Scheduling Strategy Considering Demand Response

Yuchen Liu [1], Zhenhai Dou [1,*], Zheng Wang [1], Jiaming Guo [1], Jingwei Zhao [2] and Wenliang Yin [1]

[1] School of Electrical and Electronic Engineering, Shandong University of Technology, Zibo 255000, China; 13706383106@163.com (Y.L.); 22504040047@stumail.sdut.edu.cn (Z.W.); gjm000629@163.com (J.G.); yinwenliang@sdut.edu.cn (W.Y.)

[2] State Power Investment Corporation Haiyang Offshore Wind Power Co., Ltd., Guangzhou 510710, China; zhaojingwei98@163.com

* Correspondence: douzhenhai@sdut.edu.cn

Abstract: Shared energy storage system provides an attractive solution to the high configuration cost and low utilization rate of multi-microgrid energy storage system. In this paper, an electricity-heat integrated energy storage supplier (EHIESS) containing electricity and heat storage devices is proposed to provide shared energy storage services for multi-microgrid system in order to realize mutual profits for different subjects. To this end, electric boiler (EB) is introduced into EHIESS to realize the electricity-heat coupling of EHIESS and improve the energy utilization rate of electricity and heat storage equipment. Secondly, due to the problem of the uncertainty in user-side operation of multi-microgrid system, a price-based demand response (DR) mechanism is proposed to further optimize the resource allocation of shared electricity and heat energy storage devices. On this basis, a bi-level optimization model considering the capacity configuration of EHIESS and the optimal scheduling of multi-microgrid system is proposed, with the objectives of maximizing the profits of energy storage suppliers in upper-level and minimizing the operation costs of the multi-microgrid system in lower-level, and solved based on the Karush-Kuhn-Tucker (KKT) condition and Big-M method. The simulation results show that in case of demand response, the total operation cost of multi-microgrid system and the total operation profit of EHIESS are 51,687.73 and 11,983.88 CNY, respectively; and the corresponding electricity storage unit capacity is 9730.80 kWh. The proposed model realizes the mutual profits of EHIESS and multi-microgrid system.

Keywords: multi-microgrid system; electricity-heat integrated energy storage supplier; demand response; bi-level optimization model

1. Introduction

In recent years, with the overuse of fossil energy and the increase of greenhouse gas emissions, the energy crisis as well as the climate warming is intensified. The development of traditional electricity industry is limited by conditions such as the environment and resources, and is in urgent need of transformation [1,2]. Meanwhile, renewable energy generation such as wind power and photovoltaic are developing rapidly, and the penetration rate of renewable energy is increasing, which brings certain challenges to the stable operation of power system [3,4]. Microgrid stand to meet these challenges by virtue of their renewable energy consumption capacity and flexibility. With the growing market size of microgrid, the demand for microgrid construction and renovation has gradually increased. Individual microgrid faces the problem of high operation cost. To this end, the concept of multi-microgrid system has emerged [5,6]. Multi-microgrid system can optimize the distribution of resources and benefits among different subjects by interconnecting with the upper-level grid, effectively compensating for the impact of randomness and volatility of distributed energy sources on the system, and further improving the energy utilization rate and the stability of system operation [7,8]. Ref. [9] constructs a multi-microgrid system to address load demand variations due to the stochastic character of renewable energy generation. Ref. [10] achieves multi-energy complementary co-optimization in multi-microgrid systems and proposes a benefit distribution strategy balancing fairness and stability. Ref. [11] uses a two-layer optimization structure for multi-microgrid system day-ahead and real-time energy scheduling, respectively, where energy interactions between adjacent microgrids reduce the total operation cost. In Ref. [12], the planning problem of multi-microgrid system coupled with electricity, heat and hydrogen is studied, which significantly reduces carbon emissions and environmental pollution. Ref. [13] specifically addresses the optimization of the design and operation of islanded multi-microgrid system, using a column and constraint generation based approach to address both the design and operation phases. Energy storage equipment can be used to solve the problem of time inconsistency between renewable energy generation and load demand [14]. However, configuring separate energy storage system for each microgrid can significantly increase the construction cost, and the utilization rate of the corresponding energy storage is low. Accordingly, shared energy storage has become a hot research topic for the past few years.

Shared energy storage has become a more attractive way of energy storage configuration in multi-microgrid system by virtue of its flexibility and economic advantages, which makes up for the regulation needs on the microgrid side and ensures the stable operation of renewable energy power system [15,16]. Currently, the research on shared energy storage is mainly distributed in the planning of energy storage capacity and the optimization of operation mode [17]. Ref. [18] presents a shared energy storage capacity planning model for multi-microgrid system considering PV generation accommodation and loading capacity, and obtains a better economic return. Ref. [19] proposes a market game model for peer-to-peer (P2P) energy trading between residential user side and shared energy storage configuration, which allocates the optimal energy storage capacity with considering the competition between users, and the obtained results achieve a further reduction of energy costs. Ref. [20] proposes a business model of shared energy storage in data center cluster and considers the uncertainty of renewable energy output, which effectively reduce the investment of energy storage unit, and promote the consumption of renewable energy. Ref. [21] considers shared energy storage to provide storage capacity leasing services for large-scale photovoltaic base stations, and solves a two-stage joint optimization problem for capacity planning and operation cost optimization. The optimization results achieve coordinated operation and cost sharing between shared energy storage and base station. Studies mentioned above have thoroughly studied the configuration and operation of shared electricity energy storage services. However, users not only have electricity load demand, but their demand for heat, especially in winter, is still a non-negligible part. The high-frequency use of heating equipment in winter further increases the users' need for thermal energy storage. Refs. [22–24] focus on the application of shared thermal energy storage. Ref. [22] combines the advantages of distributed and centralized structures of energy storage system, and proposes an optimal scheduling model for an integrated energy microgrid system containing electricity and thermal storage unit, which minimizes the waste of electricity and thermal energy and obtains the optimal economic benefits. In Ref. [23], three different configurations of combined heat and power units are set up with multiple energy storage units, which improve energy utilization while reducing production costs and carbon emissions. Ref. [24] set up a hybrid power system containing wind-photovoltaic-battery-thermal energy storage, which utilizes the economic benefits and working flexibility of thermal storage units and batteries to improve the intermittently output of renewable energy. Shared electricity and thermal storage units are considered as independent devices in the above literatures, but the coupling characteristic of electricity and thermal energy in them is neglected. For shared energy storage system, the realization of the coupling of electricity and thermal energy can further enhance their flexibility and economic efficiency.

Most studies mentioned above focus on the optimization of the energy supply side problem, while the independence of users on the load side is rarely included. Load fluctuations during the actual electricity consumption by users can also affect the stable operation of the multi-microgrid system [25]. In order to maximize the profits of microgrid users, Ref. [26] proposes a novel demand response model considering the different electricity consumption behaviors of different customers during peak and valley periods, and a new algorithm called hybrid hybrid crow search algorithm—jaya algorithm (CSAJAYA) is proposed, which verifies that the demand response has a positive effect on the reduction of the system's power generation cost. Ref. [27] addresses the problem of sudden energy interruptions in an islanded microgrid by proposing a novel energy management tool for isolating the microgrid and incorporating demand response program, which ultimately reduces the system operation cost by 3% and increases the allowed cumulative faults by 13 h. Ref. [28] proposes an optimal scheduling model with combined cooling heating and power (CCHP) and carbon capture devices considering demand response, which reduces the carbon emissions of the system as well as the energy purchases of users through the dual optimization of the demand response mechanism and the carbon trading mechanism, thus contributing significantly to the protection of the environment and the reduction of costs. The user-side uncertainties make the actual operation of multi-microgrid system difficult to predict, which usually leads to difficulties in optimizing the configuration of shared energy storage system as expected, and in many cases with low utilization of the storage equipment.

Based on the summary of existing literature in Table 1, we note that: (1) Multi-microgrid system is acknowledged for its ability to consume renewable energy and flexibility of operation. Energy storage device can solve the problem of temporal inconsistency between renewable energy generation and load demand, but configuring separate storage systems for each microgrid leads to the high cost and low utilization problem. (2) Although various studies have been conducted on the configuration and optimization of shared energy storage, there is still a lack of research on the synergistic utilization of shared electricity and thermal storage unit, especially ignoring the coupling of electricity and thermal energy therein. (3) Uncertainty on the user side makes the actual operation of multi-microgrid system difficult to predict, which usually leads to difficulties in optimizing the configuration of shared energy storage system as expected, and in many cases the utilization of energy storage devices is low. In response to the research gaps presented above, this paper makes the following significant contributions:

Table 1. Details of the proposed problem of this paper compared to other studies.

Reference	Multi-Microgrid System	Shared Energy Storage Electricity/Heat/Coupling	Demand Response	Objective Storage Configuration/Operation Optimization
[11]	√	×/×/×	√	×/√
[20]	×	√/×/×	×	×/√
[21]	×	√/×/×	×	√/√
[22]	√	√/√/×	×	×/√
[24]	×	√/√/×	×	×/√
[26]	√	×/×/×	√	×/√

(1) An electricity-heat integrated energy storage supplier (EHIESS) specialized in providing electricity and heat storage services for multi-microgrid system is proposed. A bi-level planning model considering EHIESS capacity configuration and optimal operation of multi-microgrid system is established, and solved based on Karush-Kuhn-Tucker (KKT) condition and Big-M method.

(2) Electric boiler (EB) is introduced into EHIESS and leverages its electricity-heat coupling characteristics to realize the enhancement of the integrated electricity-heat efficiency of EHIESS. As well, the economic benefits of multi-microgrid system and EHIESS are maximized.

(3) Aiming at the volatility of user-side loads, a price-based load demand response mechanism is introduced into multi-microgrid system to dynamically regulate the loads and improve the utilization rate of shared energy storage equipment.

The rest of the paper is organized as follows. Section 2 introduces the multi-microgrid system model considering EHIESS. Section 3 introduces the price-based DR model. In Section 4, the simulation analysis is carried out. Finally, the main conclusions are summarized in Section 5.

2. Multi-Microgrid System Model Sharing EHIESS

2.1. Electricity-Heat-Gas Coupled Microgrid Model

The microgrid model constructed in this paper is shown in Figure 1, which consists of photovoltaic, wind turbine, CCHP, electric refrigerator, and electricity, heat and cooling loads. The microgrid model is connected to the upper energy grid (upstream network and natural gas grid). CCHP unit consists of gas turbine, gas boiler, waste heat boiler (WHB), heat exchange and lithium bromide absorption chiller (LBAC). In this section, the mathematical model of each device in the microgrid is presented as follows.

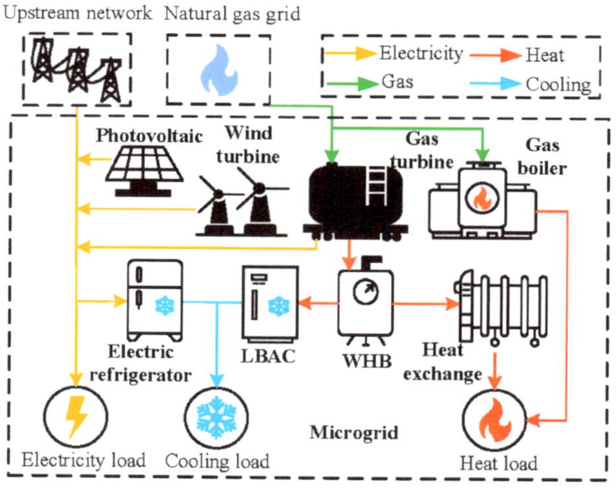

Figure 1. Topology of electricity-heat-gas coupled microgrid.

2.1.1. Combined Cooling Heating and Power

CCHP refers to the joint production of three different forms of energy for electricity, heat and cooling. CCHP provides a cost-effective way to increase energy utilization and improve environmental issues [29].

In CCHP, gas turbine generates electricity and heat by consuming natural gas. The electricity generated is used to supply electricity loads and the heat is processed by WHB and ultimately delivered to LBAC and heat exchange. The electricity and heat generated by gas turbine can be expressed as:

$$P_{\mathrm{GT}}^{t,i} = \eta_{\mathrm{GT}} \cdot L_{\mathrm{gas}} \cdot V_{\mathrm{GT}}^{t,i} \tag{1}$$

$$Q_{\mathrm{GT}}^{t,i} = \gamma_{\mathrm{GT}} \cdot P_{\mathrm{GT}}^{t,i} \tag{2}$$

where $P_{\mathrm{GT}}^{t,i}$ and $Q_{\mathrm{GT}}^{t,i}$ are the electricity and heat generated by gas turbine in microgrid i (MG i) during hour t, respectively; η_{GT} and γ_{GT} are the electricity and heat generation efficiency of gas turbine, respectively; L_{gas} is the heat value of natural gas, which is taken as 9.7 kWh/m^3; $V_{\mathrm{GT}}^{t,i}$ is the gas consumption volume of gas turbine in MG i during hour t.

Gas boiler can be used as a heat generation device in conjunction with gas turbine, and all heat it generates is used to supply the heat load. The heat generated by gas boiler in MG i during hour t, $Q_{\mathrm{GB}}^{t,i}$, can be calculated according to the following equation:

$$Q_{\mathrm{GB}}^{t,i} = \eta_{\mathrm{GB}} \cdot L_{\mathrm{gas}} \cdot V_{\mathrm{GB}}^{t,i} \tag{3}$$

where η_{GB} is the heat generation efficiency of gas boiler; $V_{\mathrm{GB}}^{t,i}$ is the gas consumption volume of gas boiler in MG i during hour t.

In CCHP, WHB can recycle the waste heat generated by gas turbine, and then deliver the resulting heat to heat exchange and LBAC, which can be expressed as:

$$Q_{\mathrm{WHB}}^{t,i} = \eta_{\mathrm{WHB}} \cdot Q_{\mathrm{GT}}^{t,i} \tag{4}$$

where $Q_{\mathrm{WHB}}^{t,i}$ is the heat absorbed by WHB in MG i during hour t; η_{WHB} is the heat absorption efficiency of WHB.

Heat exchange and LBAC can convert heat energy delivered by WHB to generate heat energy and cooling energy for users, respectively. The model of heat exchange and LBAC can be formulated as follows:

$$Q_{\mathrm{HE}}^{t,i} = \eta_{\mathrm{HE}} \cdot \alpha \cdot Q_{\mathrm{WHB}}^{t,i} \tag{5}$$

$$Q_{\mathrm{AC}}^{t,i} = \eta_{\mathrm{AC}} \cdot (1-\alpha) \cdot Q_{\mathrm{WHB}}^{t,i} \tag{6}$$

where $Q_{\mathrm{HE}}^{t,i}$ is the heat output of heat exchange in MG i during hour t; $Q_{\mathrm{AC}}^{t,i}$ is the cooling output of LBAC in MG i during hour t; η_{HE} and η_{AC} are the energy utilization rates of heat exchange and LBAC, respectively; α is the proportion of heat energy supplied to heat exchange by WHB.

2.1.2. Electric Refrigerator

Electric refrigerator, as a commonly used refrigeration equipment, supplies cooling load by consuming electricity. The cooling power generated by electric refrigerator in MG i during hour t, $Q_{\mathrm{ER}}^{t,i}$, can be calculated according to the following equation:

$$Q_{\mathrm{ER}}^{t,i} = \eta_{\mathrm{ER}} \cdot P_{\mathrm{ER}}^{t,i} \tag{7}$$

where $P_{\mathrm{ER}}^{t,i}$ is the power consumption of electric refrigerator in MG i during hour t; η_{ER} is the cooling efficiency of electric refrigerator.

2.2. *Electricity-Heat Integrated Energy Storage Supplier Model*

EHIESS forms a bi-level structure with multi-microgrid system, which can calculate the required capacity of energy storage equipments and the charging and discharging electricity and heat according to the energy consumption of multi-microgrid users, and carry out the construction and maintenance of energy storage equipments, thus providing shared energy storage service for multi-microgrid system, and earning profits from the energy trading and service fees. Specifically, the input of EB enables EHIESS more fully to utilize the complementarities and differences of different microgrid user loads, so that the capacity configuration of electricity and thermal storage unit is more reasonable, and to realize satisfying user's demand for energy storage capacity with less investment, and the economy of EHIESS can be further improved. The electricity and thermal storage units in EHIESS are connected to the multi-microgrid system through buses, and the electricity and heat interactions between microgrids are carried out through the buses, as shown in Figure 2 EHIESS is connected to MG 1 to MG n (n = 1, 2, . . . , N), and each microgrid is connected to upstream network and natural gas grid, respectively. Microgrid users who have excess electricity and heat use the energy storage service of EHIESS to store excess electricity and heat, and the microgrid users who are short of electricity and heat are supplied with electricity and heat by EHIESS.

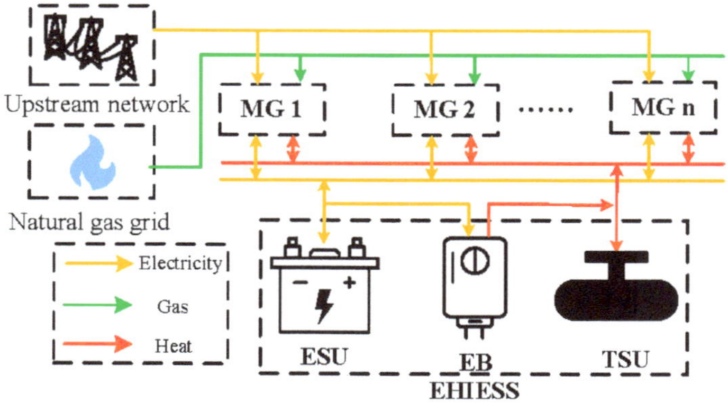

Figure 2. Multi-microgrid system shared EHIESS model.

The electricity stored in ESU during hour t, E^t, can be calculated as follows:

$$E^t = E^{t-1} + \eta_{\text{abs}} \cdot P_{\text{ch}}^{t-1} - \frac{1}{\eta_{\text{re}}} \cdot P_{\text{dis}}^{t-1} - P_{\text{EB}}^{t-1} \tag{8}$$

where η_{abs} and η_{re} are the charging and discharging efficiencies of ESU, respectively; P_{ch}^{t-1} and P_{dis}^{t-1} are the charging and discharging electricity of EHIESS during hour $t-1$, respectively; P_{EB}^{t-1} is the power consumption of EB during hour $t-1$.

The heat stored in TSU during hour t, H^t, can be calculated as follows:

$$H^t = H^{t-1} + \omega_{\text{abs}} \cdot Q_{\text{ch}}^{t-1} - \frac{1}{\omega_{\text{re}}} \cdot Q_{\text{dis}}^{t-1} + Q_{\text{EB}}^{t-1} \tag{9}$$

where ω_{abs} and ω_{re} are the heat charging and discharging efficiencies of TSU, respectively; Q_{ch}^{t-1} and Q_{dis}^{t-1} are the charging and discharging heat of EHIESS during hour $t-1$, respectively; Q_{EB}^{t-1} is the heat produced by EB during hour $t-1$.

In this paper, the initial storage capacity of the energy storage equipment is set to be 20% of the total capacity, which can be expressed as:

$$E^0 = 20\% \cdot E^{\max}, H^0 = 20\% \cdot H^{\max} \tag{10}$$

where E^0 and H^0 are the capacities of ESU and TSU at the initial time, respectively; $E^{\max}$ and $H^{\max}$ are the maximum capacities of ESU and TSU, respectively.

EB can connect ESU and TSU to convert heat into electricity when the heat load is high or ESU reaches its capacity limit. By incorporating EB equipment, the flexibility of EHIESS has been improved, resulting in higher profit for EHIESS. The model of EB can be expressed as follows:

$$Q_{\text{EB}}^t = \eta_{\text{EB}} \cdot P_{\text{EB}}^t \tag{11}$$

where η_{EB} is the conversion efficiency of EB.

2.3. *Price-Based Demand Response*

Different kinds of electricity loads have different sensitivities to the electricity price in practical work for multi-microgrid system, and the operation risk of microgrid and cost of users will increase without considering the different sensitivities of loads. Therefore, we considered the price-based DR in the load side. Price-based DR means that the upstream network guides users to use electricity reasonably by setting different electricity prices in different time periods, so as to achieve the purpose of smoothing the load curve, reducing the system operation risk, and decreasing the user's energy cost [30].

Currently, the ways that can be modeled to reflect the user's response behavior to the price are price elasticity coefficient matrix, consumer psychology and so on. Among them, the method of price elasticity coefficient matrix can accurately reflect the response behavior of users to different electricity prices [31]. Price elasticity coefficient is an index that measures the influence of price changes to load demands. When the price changes in a certain time period, it will not only affect the demand in this time period, but also indirectly change the demand in other time periods. The price elasticity coefficient includes the self-demand elasticity coefficient and cross-elasticity coefficient.

The elements of the price elasticity matrix, $e_{t,j}$, represent the elasticity coefficients of the electricity load during hour i with respect to the price of electricity during hour j, can be calculated as follows [32]:

$$e_{t,j} = \frac{\Delta P_{\mathrm{L}}^{t}}{P_{\mathrm{L},0}^{t}} \frac{\rho_0^j}{\Delta \rho^j} \tag{12}$$

where $\Delta P_{\mathrm{L}}^{t}$ is the electricity load changes after DR during hour t; $P_{\mathrm{L},0}^{t}$ is the initial electricity load; $\Delta\rho^j$ is the electricity price changes after DR during hour j; ρ_0^j is the initial electricity price during hour j. When t equals j, $e_{t,j}$ is the self-demand elasticity coefficient, when t and j are not equal, $e_{t,j}$ is the cross-elasticity coefficient.

2.3.1. Curtailable Load

Loads are categorized into curtailable load (CL) and shiftable load (SL) based on differences in sensitivity to prices. CL can choose whether or not to engage in load curtailment by comparing the price of electricity before and after considering DR for a given time period. The changes of CL after DR during hour t is given as follows:

$$\Delta P_{\mathrm{CL}}^{t} = P_{\mathrm{CL}}^{0}\left[\sum_{j=1}^{24} E_{\mathrm{CL}}(t,j)\frac{\rho^j - \rho_0^j}{\rho_0^j}\right] \tag{13}$$

where P_{CL}^{0} is the initial CL quantity; $E_{\mathrm{CL}}(t,j)$ is the price elasticity matrix with respect of CL; ρ^j is the electricity price during hour j.

2.3.2. Shiftable Load

SL can compare the prices before and after DR and flexibly choose whether to adjust the working hours or not. Customers will shift the load from peak hours into valley hours based on the time-of-use price. The change of SL after DR during hour t is given as follows:

$$\Delta P_{\mathrm{SL}}^{t} = P_{\mathrm{SL}}^{0}\left[\sum_{j=1}^{24} E_{\mathrm{SL}}(t,j)\frac{\rho^j - \rho_0^j}{\rho_0^j}\right] \tag{14}$$

where P_{SL}^{0} is the initial SL quantity; $E_{\mathrm{SL}}(t,j)$ is the price elasticity matrix with respect of SL.

2.4. Energy Balance in Microgrid

The three types of energy flow in the microgrid, electricity, heat and cooling, should satisfy the corresponding energy balance relationship. The expression for the electrical balance of the microgrid is as follows:

$$P_{\mathrm{GT}}^{t,i} + P_{\mathrm{WT}}^{t,i} + P_{\mathrm{PV}}^{t,i} + P_{\mathrm{grid}}^{t,i} + P_{\mathrm{mg,b}}^{t,i} = P_{\mathrm{mg,s}}^{t,i} + P_{\mathrm{ER}}^{t,i} + P_{\mathrm{el}}^{t,i} \tag{15}$$

where $P_{\mathrm{WT}}^{t,i}$ and $P_{\mathrm{PV}}^{t,i}$ are the electricity generated by WT and PV of MG i during hour t, respectively; $P_{\mathrm{grid}}^{t,i}$ is the electricity purchased by the MG i from the upstream network;

$P_{\mathrm{mg,b}}^{t,i}$ and $P_{\mathrm{mg,s}}^{t,i}$ are the electricity purchased and sold by MG i, respectively; $P_{\mathrm{el}}^{t,i}$ is the electricity load of MG i.

The cooling balance expression for the microgrid is given below:

$$Q_{\mathrm{ER}}^{t,i} + Q_{\mathrm{AC}}^{t,i} = Q_{\mathrm{cl}}^{t,i} \tag{16}$$

where $Q_{\mathrm{cl}}^{t,i}$ is the cooling load of MG i during hour t.

The heat balance expression for the microgrid is given below:

$$Q_{\mathrm{GB}}^{t,i} + Q_{\mathrm{HE}}^{t,i} + Q_{\mathrm{mg,b}}^{t,i} = Q_{\mathrm{hl}}^{t,i} + Q_{\mathrm{mg,s}}^{t,i} \tag{17}$$

where $Q_{\mathrm{hl}}^{t,i}$ is the heat load of MG i during hour t; $Q_{\mathrm{mg,b}}^{t,i}$ and $Q_{\mathrm{mg,s}}^{t,i}$ are the heat purchased and sold by MG i, respectively.

3. A Bi-Level Optimizing Model Considering Capacity Configuration of EHIESS and Optimal Operation of Multi-Microgrid System

In order to realize the win-win situation of multi-microgrid system and EHIESS, A bi-level optimization model is established considering capacity configuration of EHIESS and optimal operation of multi-microgrid system. The upper-level model is used to solve the optimal configuration problem of EHIESS, and the lower-level model is used to solve the optimal operation problem of multi-microgrid system. The upper-level model transfers the optimized EHEISS configuration information to the lower-level model, and then, in lower-level model, each microgrid interacts with EHIESS to optimize its own energy scheduling based on its own energy consumption and EHIESS optimal configuration information. Finally, EHIESS updates its configuration based on the microgrid's scheduling data, which ultimately satisfies the bi-level optimization model and obtains the optimal results.

3.1. *Upper Level Optimization Model*

3.1.1. Objective Function

The objective function of the upper-level model is to maximize the profits of EHIESS over the scheduling period, the optimization variables include the capacity configurations of ESU and TSU as well as the maximum charging and discharging power. The objective function for the upper-level model is as follows:

$$\max C_{\mathrm{upper}} = C_{\mathrm{buy}} - C_{\mathrm{sale}} - C_{\mathrm{inv}} + C_{\mathrm{serve}} \tag{18}$$

where C_{upper} is the total profits of EHIESS; C_{buy} is the cost of purchasing energy from the multi-microgrid system; C_{sale} is the profit of selling energy to the multi-microgrid system; C_{inv} is the investment and maintenance costs of the energy storage equipment in EHIESS; C_{serve} is the service fees charged by EHIESS.

The cost of purchasing energy from the microgrid includes the costs of purchasing electricity and heat, which can be expressed as:

$$C_{\mathrm{buy}} = \sum_{i=1}^{N}\sum_{t=1}^{N_T}(\lambda^t \cdot P_{\mathrm{mg,b}}^{t,i} + \varphi^t \cdot Q_{\mathrm{mg,b}}^{t,i}) \tag{19}$$

where N denotes the number of microgrids; N_T is the length of the scheduling period; λ^t is the price of electricity purchased from ESU during hour t; φ^t is the price of heat purchased from TSU during hour t.

Multi-microgrid system sells electricity and heat to EHIESS to generate profits, which can be expressed as follows:

$$C_{\mathrm{sale}} = \sum_{i=1}^{N}\sum_{t=1}^{N_T}(\delta^t \cdot P_{\mathrm{mg,s}}^{t,i} + \gamma^t \cdot Q_{\mathrm{mg,s}}^{t,i}) \tag{20}$$

where δ^t is the price of electricity sold to ESU during time t; γ^t is the price of heat sold to TSU during hour t.

The investment and maintenance costs of ESU and TSU can be calculated as:

$$C_{\text{inv}} = \frac{\eta_{\text{p}} P_{\text{es}}^{\max} + \eta_{\text{e}} E^{\max}}{T_{\text{ESU}}} - C_{\text{ESU}} + \frac{\eta_{\text{q}} Q_{\text{es}}^{\max} + \eta_{\text{h}} H^{\max}}{T_{\text{TSU}}} + C_{\text{TSU}} \tag{21}$$

where η_{p} and η_{e} are the electricity power cost and capacity cost of ESU, respectively; $P_{\text{es}}^{\max}$ is the maximum charging and discharging power of ESU; η_{q} and η_{h} are the heat power cost and capacity cost of TSU, respectively; $Q_{\text{es}}^{\max}$ is the maximum charging and discharging heat power of TSU; T_{ESU} and T_{TSU} is the operating life of ESU and TSU, respectively; C_{ESU} and C_{TSU} is the daily maintenance cost of ESU and TSU, respectively.

The service fees gained by EHIESS from multi-microgrid system, C_{serve}, can be calculated as:

$$C_{\text{serve}} = \sum_{i=1}^{N} \sum_{t=1}^{N_T} \theta^t \cdot \left(P_{\text{mg,b}}^{t,i} + P_{\text{mg,s}}^{t,i} + Q_{\text{mg,b}}^{t,i} + Q_{\text{mg,s}}^{t,i} \right) \tag{22}$$

where θ^t is the price of the service fees received by EHIESS during hour t.

3.1.2. Constraints

In the upper-level, we consider the storage capacity and charging/discharging power of each scheduling period of ESU and TSU in EHIESS operation, and set the upper limit constraint of storage capacity, charging/discharging power constraint and EB power constraint. The model of ESU and TSU can be formulated as follows:

$$E^{\max} = \eta_{\text{ESU}} \cdot P_{\text{es}}^{\max} \tag{23}$$

$$H^{\max} = \eta_{\text{TSU}} \cdot Q_{\text{es}}^{\max} \tag{24}$$

The maximum capacity of ESU, $E^{\max}$, and the maximum capacity of TSU, $H^{\max}$, are given by Equations (23) and (24), where η_{ESU} and η_{TSU} is the energy multiplication factor of ESU and TSU, respectively.

Equations (25) and (26) are the storage capacity constraints for ESU and TSU. In order to ensure the stability, efficiency, and prolong the service life of the equipments, the state of charge (SOC) limits are set to 0.1 and 0.9, respectively.

$$10\% \cdot E^{\max} \leq E^t \leq 90\% \cdot E^{\max} \tag{25}$$

$$10\% \cdot H^{\max} \leq H^t \leq 90\% \cdot H^{\max} \tag{26}$$

The charging and discharging electricity of ESU, and the charging and discharging heat of TSU constraints during hour t can be expressed as follows:

$$0 \leq P_{\text{ch}}^t \leq U_{\text{abs}}^t \cdot P_{\text{es}}^{\max}, 0 \leq P_{\text{dis}}^t \leq U_{\text{re}}^t \cdot P_{\text{es}}^{\max} \tag{27}$$

$$0 \leq Q_{\text{ch}}^t \leq V_{\text{abs}}^t \cdot Q_{\text{es}}^{\max}, 0 \leq Q_{\text{dis}}^t \leq V_{\text{re}}^t \cdot Q_{\text{es}}^{\max} \tag{28}$$

where U_{abs}^t and U_{re}^t are the charging and discharging states of ESU, respectively. When ESU is charging during hour t, U_{abs}^t is taken as 1 and U_{re}^t is taken as 0. When ESU is discharging during hour t, U_{abs}^t is taken as 0 and U_{re}^t is taken as 1. V_{abs}^t and V_{re}^t are the charging and discharging heat states of TSU. When TSU is charging during hour t, V_{abs}^t is taken as 1 and V_{re}^t is taken as 0, when TSU is discharging during hour t, V_{abs}^t is taken as 0 and V_{re}^t is taken as 1. Charging and discharging states cannot be performed at the same time.

The electricity consumption of EB during hour t is constrained as follows:

$$P_{\mathrm{EB}}^{\min} \leq P_{\mathrm{EB}}^{t} \leq P_{\mathrm{EB}}^{\max} \tag{29}$$

where $P_{\mathrm{EB}}^{\min}$ and $P_{\mathrm{EB}}^{\max}$ are the minimum and maximum power consumption of EB during hour t, respectively.

3.2. Lower Level Optimization Model

3.2.1. Objective Function

The objective function of the lower-level model is to minimize the operation costs of the multi-microgrid system in scheduling period. The decision variables include the power purchased from the upstream network, the output power of the equipments in CCHP, the electricity consumption of electric refrigerator and the power of the microgrids trading with EHEISS. The objective function of the lower-level model can be expressed as:

$$\min C_{\mathrm{lower}} = C_{\mathrm{grid}} + C_{\mathrm{gas}} + C_{\mathrm{buy}} - C_{\mathrm{sale}} + C_{\mathrm{serve}} + C_{\mathrm{ca}}^{t,i} \tag{30}$$

where C_{grid} is the cost of electricity purchased from the upstream network; C_{gas} is the cost of purchasing natural gas, which can be expressed separately as follows:

$$C_{\mathrm{grid}} = \sum_{i=1}^{N} \sum_{t=1}^{N_T} \tau_{\mathrm{grid}}^{t} \cdot P_{\mathrm{grid}}^{t,i} \tag{31}$$

$$C_{\mathrm{gas}} = \sum_{i=1}^{N} \sum_{t=1}^{N_T} \tau_{\mathrm{gas}}^{t} \cdot (V_{\mathrm{GT}}^{t,i} + V_{\mathrm{GB}}^{t,i}) \tag{32}$$

where τ_{grid}^{t} is the electricity price of the upstream network; τ_{gas}^{t} is the price of natural gas.

3.2.2. Constraints

In the multi-microgrid system scheduling model, expect for the electricity, cooling, and heat power balance constraints, the output constraints of each device such as microgrid purchase and sale power constraints are also considered. The electricity and heat sold/purchased by multi-microgrid during hour t should be the sum of the excess/insufficient energy of individual microgrid, can be expressed as follows:

$$P_{\mathrm{dis}}^{t} - P_{\mathrm{ch}}^{t} = \sum_{i=1}^{N} (P_{\mathrm{mg,b}}^{t,i} - P_{\mathrm{mg,s}}^{t,i}) \tag{33}$$

$$Q_{\mathrm{dis}}^{t} - Q_{\mathrm{ch}}^{t} = \sum_{i=1}^{N} (Q_{\mathrm{mg,b}}^{t,i} - Q_{\mathrm{mg,s}}^{t,i}) \tag{34}$$

The output constraints for each device in the microgrid are as follows:

$$\begin{cases} P_{\mathrm{GT}}^{\min} \leq P_{\mathrm{GT}}^{t,i} \leq P_{\mathrm{GT}}^{\max} \\ Q_{\mathrm{AC}}^{\min} \leq Q_{\mathrm{AC}}^{t,i} \leq Q_{\mathrm{AC}}^{\max} \\ Q_{\mathrm{ER}}^{\min} \leq Q_{\mathrm{ER}}^{t,i} \leq Q_{\mathrm{ER}}^{\max} \\ Q_{\mathrm{GB}}^{\min} \leq Q_{\mathrm{GB}}^{t,i} \leq Q_{\mathrm{GB}}^{\max} \\ Q_{\mathrm{HE}}^{\min} \leq Q_{\mathrm{HE}}^{t,i} \leq Q_{\mathrm{HE}}^{\max} \end{cases} \tag{35}$$

The power constraint for energy transfer from the upstream network to MG i during hour t is given as follow:

$$0 \leq P_{\mathrm{grid}}^{t,i} \leq P_{\mathrm{grid}}^{\max} \tag{36}$$

where $P_{\mathrm{grid}}^{\max}$ is the maximum power purchased from the upstream network.

The energy trading constraints of EHIESS and multi-microgrid system are as follows:

$$\begin{cases} 0 \leq P_{\mathrm{mg,s}}^{t,i} \leq P_{\mathrm{mg}}^{\max} \cdot U_{\mathrm{sale}}^{t,i}, 0 \leq P_{\mathrm{mg,b}}^{t,i} \leq P_{\mathrm{mg}}^{\max} \cdot U_{\mathrm{buy}}^{t,i} \\ 0 \leq Q_{\mathrm{mg,s}}^{t,i} \leq Q_{\mathrm{mg}}^{\max} \cdot V_{\mathrm{sale}}^{t,i}, 0 \leq Q_{\mathrm{mg,b}}^{t,i} \leq Q_{\mathrm{mg}}^{\max} \cdot V_{\mathrm{buy}}^{t,i} \\ U_{\mathrm{sale}}^{t,i} + U_{\mathrm{buy}}^{t,i} \leq 1, V_{\mathrm{sale}}^{t,i} + V_{\mathrm{buy}}^{t,i} \leq 1 \end{cases} \tag{37}$$

where $U_{\mathrm{sale}}^{t,i}$ and $U_{\mathrm{buy}}^{t,i}$ are the electricity purchase state bits of the MG i. When the MG i sells electricity to ESU, $U_{\mathrm{sale}}^{t,i}$ is equal to 1 and $U_{\mathrm{buy}}^{t,i}$ is equal to 0. When the MG i purchases electricity from ESU, $U_{\mathrm{sale}}^{t,i}$ is equal to 0 and $U_{\mathrm{buy}}^{t,i}$ is equal to 1. $V_{\mathrm{sale}}^{t,i}$ and $V_{\mathrm{buy}}^{t,i}$ are the heat purchase state bits of the MG i. When the MG i sells heat to TSU, $V_{\mathrm{sale}}^{t,i}$ is equal to 1 and $V_{\mathrm{buy}}^{t,i}$ is equal to 0. When the MG i purchases electricity from ESU, $V_{\mathrm{sale}}^{t,i}$ is equal to 0 and $V_{\mathrm{buy}}^{t,i}$ is equal to 1.

3.3. Solution Process

The solution process of the bi-level optimization model constructed in this paper is shown in Figure 3. It is difficult to solve the bi-level optimization problem directly because of the interaction between the upper-level model and the lower-level model. In this paper, we first construct the Lagrange function according to the objective function and constraints of the lower-level model [33], and then transform the lower-level model into the constraints of the upper-level model through the Karush-Kuhn-Tucker (KKT) optimization condition, which transforms a mixed integer linear problem into a single level nonlinear problem. Then the bi-level optimization model changes into a single level mixed integer linear problem according to the Big-M method [34]. Finally, the model is constructed under the MATLAB2021a platform and solved by invoking CPLEX and YALMIP. The specific solution procedure is shown below:

Step 1: The Lagrange function for the lower model is constructed as follows [35]:

$$\begin{aligned}
&\min C = \sum_{i=1}^{N} \sum_{t=1}^{N_T} \Delta t \Big\{ \tau_{\mathrm{grid}}^{t} \cdot P_{\mathrm{grid}}^{t,i} + \tau_{\mathrm{gas}}^{t} \cdot \frac{P_{\mathrm{GT}}^{t,i}}{\eta_{\mathrm{GT}} L_{\mathrm{gas}}} + \tau_{\mathrm{gas}}^{t} \frac{Q_{\mathrm{GT}}^{t,i}}{\eta_{\mathrm{GB}} L_{\mathrm{gas}}} + \lambda^{t} \cdot P_{\mathrm{mg,b}}^{t,i} + \varphi^{t} \cdot Q_{\mathrm{mg,b}}^{t,i} \\
&- \delta^{t} \cdot P_{\mathrm{mg,s}}^{t,i} - \gamma^{t} \cdot Q_{\mathrm{mg,s}}^{t,i} + \theta^{t} \cdot (P_{\mathrm{mg,b}}^{t,i} + P_{\mathrm{mg,s}}^{t,i} + Q_{\mathrm{mg,b}}^{t,i} + Q_{\mathrm{mg,s}}^{t,i}) \Big\} + \lambda_{1}^{t,i} \Big[P_{\mathrm{GT}}^{t,i} + P_{\mathrm{WT}}^{t,i} + \\
&P_{\mathrm{PV}}^{t,i} + P_{\mathrm{grid}}^{t,i} + P_{\mathrm{mg,b}}^{t,i} - P_{\mathrm{mg,s}}^{t,i} - P_{\mathrm{ER}}^{t,i} - P_{\mathrm{el}}^{t,i} \Big] + \lambda_{2}^{t,i} \Big[Q_{\mathrm{ER}}^{t,i} + Q_{AC}^{t,i} - Q_{\mathrm{cl}}^{t,i} \Big] \\
&+ \lambda_{3}^{t,i} \Big[Q_{\mathrm{GB}}^{t,i} + Q_{\mathrm{HE}}^{t,i} + Q_{\mathrm{mg,b}}^{t,i} - Q_{\mathrm{hl}}^{t,i} - Q_{\mathrm{mg,s}}^{t,i} \Big] + \lambda_{4}^{t,i} \Big[\frac{Q_{\mathrm{HE}}^{t,i}}{\eta_{\mathrm{HE}}} + \frac{Q_{\mathrm{AC}}^{t,i}}{\eta_{\mathrm{AC}}} - \eta_{\mathrm{WHB}} \cdot Q_{\mathrm{GT}}^{t,i} \Big] \\
&+ \lambda_{5}^{t,i} \Big[P_{\mathrm{dis}}^{t} - P_{\mathrm{ch}}^{t} - \sum_{i=1}^{N} \Big(P_{\mathrm{mg,b}}^{t,i} - P_{\mathrm{mg,s}}^{t,i} \Big) \Big] + \lambda_{6}^{t,i} \Big[Q_{\mathrm{dis}}^{t} - Q_{\mathrm{ch}}^{t} - \sum_{i=1}^{N} \Big(Q_{\mathrm{mg,b}}^{t,i} - Q_{\mathrm{mg,s}}^{t,i} \Big) \Big] \\
&+ \mu_{1}^{\min} \Big[P_{\mathrm{GT}}^{\min} - P_{\mathrm{GT}}^{t,i} \Big] + \mu_{1}^{\max} \Big[P_{\mathrm{GT}}^{t,i} - P_{\mathrm{GT}}^{\max} \Big] + \mu_{2}^{\min} \Big[Q_{\mathrm{AC}}^{\min} - Q_{\mathrm{AC}}^{t,i} \Big] + \mu_{2}^{\max} \Big[Q_{\mathrm{AC}}^{t,i} - Q_{\mathrm{AC}}^{\max} \Big] \\
&+ \mu_{3}^{\min} \Big[Q_{\mathrm{ER}}^{\min} - Q_{\mathrm{ER}}^{t,i} \Big] + \mu_{3}^{\max} \Big[Q_{\mathrm{ER}}^{t,i} - Q_{\mathrm{ER}}^{\max} \Big] + \mu_{4}^{\min} \Big[Q_{\mathrm{GB}}^{\min} - Q_{\mathrm{GB}}^{t,i} \Big] + \mu_{4}^{\max} \Big[Q_{\mathrm{GB}}^{t,i} - Q_{\mathrm{GB}}^{\max} \Big] \\
&+ \mu_{5}^{\min} \Big[Q_{\mathrm{HE}}^{\min} - Q_{\mathrm{HE}}^{t,i} \Big] + \mu_{5}^{\max} \Big[Q_{\mathrm{HE}}^{t,i} - Q_{\mathrm{HE}}^{\max} \Big] - \mu_{6}^{\min} \cdot P_{\mathrm{grid}}^{t,i} + \mu_{6}^{\max} \Big[P_{\mathrm{grid}}^{t,i} - P_{\mathrm{grid}}^{\max} \Big] \\
&- \mu_{7}^{\min} \cdot P_{\mathrm{mg,s}}^{t,i} + \mu_{7}^{\max} \Big[P_{\mathrm{mg,s}}^{t,i} - P_{\mathrm{mg}}^{\max} \cdot U_{\mathrm{sale}}^{t,i} \Big] + \mu_{8}^{\min} \cdot P_{\mathrm{mg,b}}^{t,i} + \mu_{8}^{\max} \Big[P_{\mathrm{mg,b}}^{t,i} - P_{\mathrm{mg}}^{\max} \cdot U_{\mathrm{buy}}^{t,i} \Big] \\
&+ \mu_{9}^{\max} \Big[V_{\mathrm{sale}}^{t,i} + V_{\mathrm{buy}}^{t,i} - 1 \Big] - \mu_{10}^{\min} \cdot Q_{\mathrm{mg,s}}^{t,i} + \mu_{10}^{\max} \Big[Q_{\mathrm{mg,s}}^{t,i} - Q_{\mathrm{mg}}^{\max} \cdot V_{\mathrm{sale}}^{t,i} \Big] + \mu_{11}^{\min} \cdot Q_{\mathrm{mg,b}}^{t,i} \\
&+ \mu_{11}^{\max} \Big[Q_{\mathrm{mg,b}}^{t,i} - Q_{\mathrm{mg}}^{\max} \cdot V_{\mathrm{buy}}^{t,i} \Big] + \mu_{12}^{\max} \Big[V_{\mathrm{sale}}^{t,i} + V_{\mathrm{buy}}^{t,i} - 1 \Big]
\end{aligned} \tag{38}$$

Step 2: Based on the constructed Lagrange function (38) and the complementary relaxation conditions of the lower-level model, the lower-level model can be transformed into an additional constraint for the upper model, given as follows:

$$\begin{cases} \tau_{\text{grid}}^{t} + \lambda_{1}^{t,i} + \mu_{6}^{\max} - \mu_{6}^{\min} = 0 \\ \frac{\tau_{\text{gas}}^{t}}{\eta_{\text{GT}} L_{\text{gas}}} + \lambda_{1}^{t,i} + \mu_{1}^{\max} - \mu_{1}^{\min} = 0 \\ \frac{\tau_{\text{gas}}^{t}}{\eta_{\text{GB}} L_{\text{gas}}} - \lambda_{4}^{t,i} \cdot \eta_{\text{WHB}} = 0 \\ \lambda^{t} + \theta^{t} + \lambda_{1}^{t,i} - \lambda_{5}^{t,i} + \mu_{8}^{\max} - \mu_{8}^{\min} = 0 \\ \varphi^{t} + \theta^{t} + \lambda_{3}^{t,i} - \lambda_{6}^{t,i} + \mu_{11}^{\max} - \mu_{11}^{\min} = 0 \\ -\delta^{t} + \theta^{t} - \lambda_{1}^{t,i} - \lambda_{5}^{t,i} + \mu_{7}^{\max} - \mu_{7}^{\min} = 0 \\ -\gamma^{t} + \theta^{t} - \lambda_{3}^{t,i} - \lambda_{6}^{t,i} + \mu_{10}^{\max} - \mu_{10}^{\min} = 0 \\ \lambda_{2}^{t,i} + \frac{\lambda_{4}^{t,i}}{\eta_{\text{AC}}} + \mu_{2}^{\max} - \mu_{2}^{\min} = 0 \\ \lambda_{2}^{t,i} + \mu_{3}^{\max} - \mu_{3}^{\min} = 0 \\ \lambda_{3}^{t,i} + \mu_{4}^{\max} - \mu_{4}^{\min} = 0 \\ \lambda_{3}^{t,i} + \frac{\lambda_{4}^{t,i}}{\eta_{\text{HE}}} + \mu_{5}^{\max} - \mu_{5}^{\min} = 0 \\ -\mu_{7}^{\max} \cdot P_{\text{mg}}^{\max} + \mu_{9}^{\max} = 0 \\ -\mu_{8}^{\max} \cdot P_{\text{mg}}^{\max} + \mu_{9}^{\max} = 0 \\ -\mu_{10}^{\max} \cdot Q_{\text{mg}}^{\max} + \mu_{12}^{\max} = 0 \\ -\mu_{11}^{\max} \cdot Q_{\text{mg}}^{\max} + \mu_{12}^{\max} = 0 \end{cases} \tag{39}$$

$$\begin{cases} 0 \le \mu_{1}^{\min} \perp \left(P_{\text{GT}}^{t,i} - P_{\text{GT}}^{\min}\right) \ge 0, 0 \le \mu_{1}^{\max} \perp \left(P_{\text{GT}}^{\max} - P_{\text{GT}}^{t,i}\right) \ge 0 \\ 0 \le \mu_{2}^{\min} \perp \left(Q_{\text{AC}}^{t,i} - Q_{\text{AC}}^{\min}\right) \ge 0, 0 \le \mu_{2}^{\max} \perp \left(Q_{\text{AC}}^{\max} - Q_{\text{AC}}^{t,i}\right) \ge 0 \\ 0 \le \mu_{3}^{\min} \perp \left(Q_{\text{ER}}^{t,i} - Q_{\text{ER}}^{\min}\right) \ge 0, 0 \le \mu_{3}^{\max} \perp \left(Q_{\text{ER}}^{\max} - Q_{\text{ER}}^{t,i}\right) \ge 0 \\ 0 \le \mu_{4}^{\min} \perp \left(Q_{\text{GB}}^{t,i} - Q_{\text{GB}}^{\min}\right) \ge 0, 0 \le \mu_{4}^{\max} \perp \left(Q_{\text{GB}}^{\max} - Q_{\text{GB}}^{t,i}\right) \ge 0 \\ 0 \le \mu_{5}^{\min} \perp \left(Q_{\text{HE}}^{t,i} - Q_{\text{HE}}^{\min}\right) \ge 0, 0 \le \mu_{5}^{\max} \perp \left(Q_{\text{HE}}^{\max} - Q_{\text{HE}}^{t,i}\right) \ge 0 \\ 0 \le \mu_{6}^{\min} \perp P_{\text{grid}}^{t,i} \ge 0, 0 \le \mu_{6}^{\max} \perp \left(P_{\text{grid}}^{\max} - P_{\text{grid}}^{t,i}\right) \ge 0 \\ 0 \le \mu_{7}^{\min} \perp P_{\text{mg,s}}^{t,i} \ge 0, 0 \le \mu_{7}^{\max} \perp \left(P_{\text{mg}}^{\max} \cdot U_{\text{sale}}^{t,i} - P_{\text{mg,s}}^{t,i}\right) \ge 0 \\ 0 \le \mu_{8}^{\min} \perp P_{\text{mg,b}}^{t,i} \ge 0, 0 \le \mu_{8}^{\max} \perp \left(P_{\text{mg}}^{\max} \cdot U_{\text{buy}}^{t,i} - P_{\text{mg,b}}^{t,i}\right) \ge 0 \\ 0 \le \mu_{9}^{\max} \perp \left(1 - U_{\text{sale}}^{t,i} - U_{\text{buy}}^{t,i}\right) \ge 0 \\ 0 \le \mu_{10}^{\min} \perp Q_{\text{mg,s}}^{t,i} \ge 0 \\ 0 \le \mu_{11}^{\min} \perp Q_{\text{mg,b}}^{t,i} \ge 0, 0 \le \mu_{11}^{\max} \perp \left(Q_{\text{mg}}^{\max} \cdot V_{\text{buy}}^{t,i} - Q_{\text{mg,b}}^{t,i}\right) \ge 0 \\ 0 \le \mu_{12}^{\max} \perp \left(1 - V_{\text{sale}}^{t,i} - V_{\text{buy}}^{t,i}\right) \ge 0 \end{cases} \tag{40}$$

Step 3: Based on the transformed single-level model, as shown in Equations (39) and (40), the Big-M method is used to introduce 0–1 variables to transform the nonlinear constraints in the model into mixed-integer linear constraints, for example:

$$0 \le \mu_{1}^{\min} \le M_{\mu}^{\min} v^{\min} \tag{41}$$

$$0 \le P_{\text{GT}}^{t,i} - P_{\text{GT}}^{\min} \le M_{\mu}^{\min}(1 - v^{\min}) \tag{42}$$

where $M_{\mu}^{\min}$ is a sufficiently large constant; $v^{\min}$ is a binary variable.

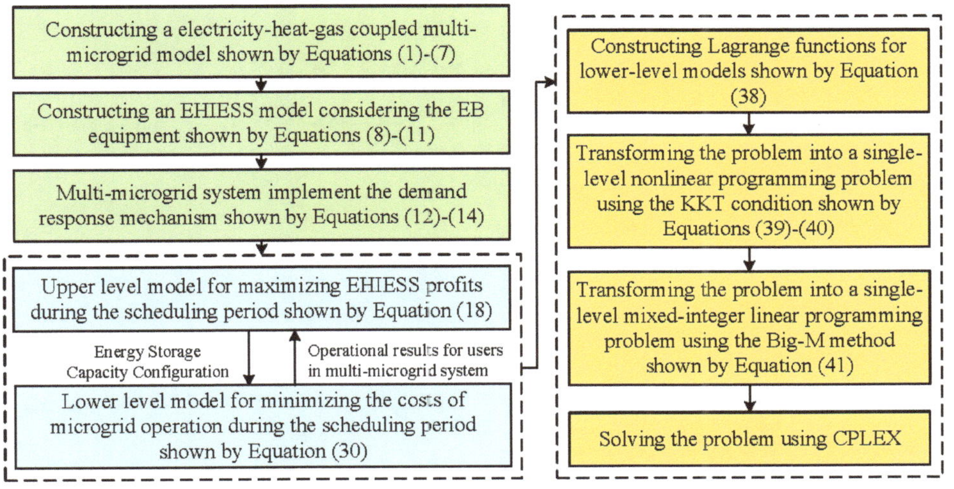

Figure 3. Schematic diagram of the solution process.

4. Simulation Analysis

4.1. Parameter Settings

The multi-microgrid system in this paper consists of three microgrids equipped with CCHP, wind turbine and photovoltaic, and is connected to EHIESS consisting of ESU, TSU and EB. The predicted output data of wind turbine and photovoltaic for the three microgrids in summer and winter, as well as three loads demands of electricity, heat and cooling are shown in Figure 4 [36], where MG 3 is not equipped with WT. The scheduling period is set to 24 h. The time-of-use price after DR and the transaction price between the multi-microgrid system and ESU are shown in Table 2 [36]. The detailed operation parameters of energy conversion equipment are shown in Table 3 [37]. The power cost of ESU and TSU is \$281.66/kW and \$70.40/kW (The original data used CNY as the unit, for reading convenience, this paper uses the RMB-USD exchange rate of 12 October 2024 to convert the unit to USD. It will not be separately noted later.). The capacity of ESU and TSU cost is \$267.11/kWh and \$26.753/kWh [37]. The service fees charged by EHIESS is \$0.0014/kW. The price of natural gas is taken as 0.31 $\$/m^3$. The price of heat sold to the TSU is \$0.014/kWh, and the price of heat purchased from the TSU is \$0.056/kWh.

Table 2. Time-of-use price parameters.

Time Period		Electricity Price/(\$/kWh)		
		Upstream Network Electricity Price	ESU's Selling Electricity Price	ESU's Purchasing Electricity Price
Peak period	08:00–13:00 16:00–21:00	0.19	0.16	0.13
Flat period	13:00–16:00 21:00–24:00	0.12	0.11	0.08
Valley period	00:00–08:00	0.05	0.06	0.03

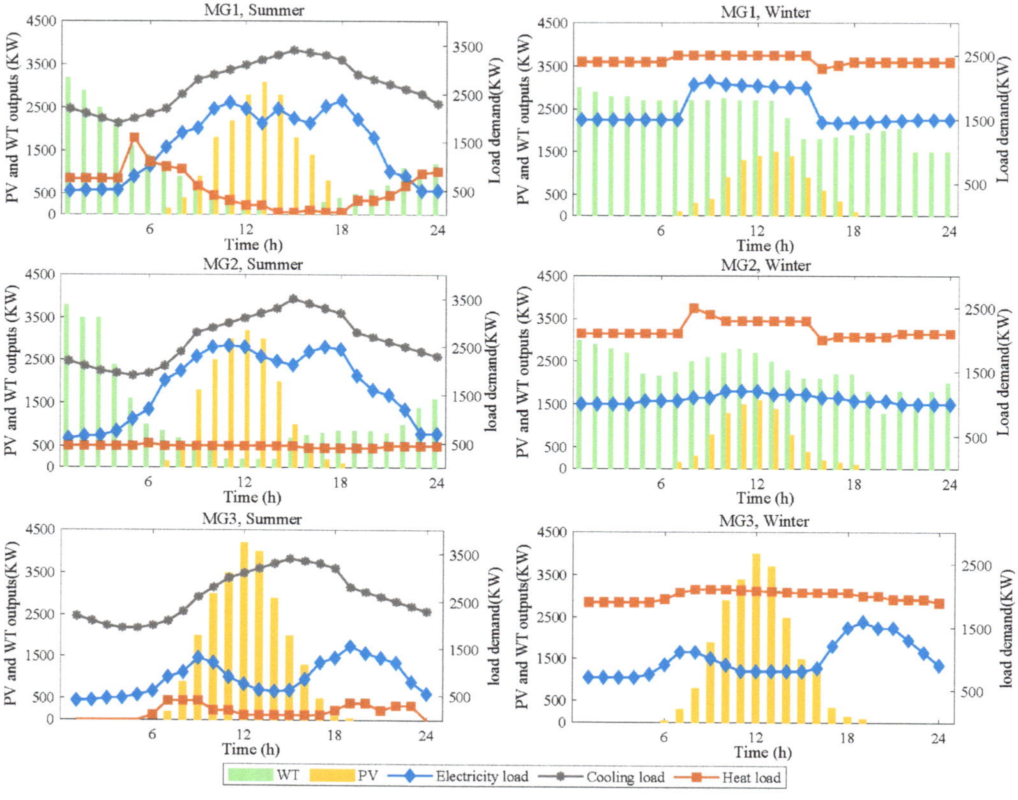

Figure 4. Loads demands and predicted power generation of PV and WT in multi-microgrid system.

Table 3. Equipment parameters of the microgrid.

Parameters	Numerical Value
Electricity generation efficiency of gas turbine, η_{GT}	0.3
Electricity-to-heat ratio of gas turbine, γ_{GT}	1.47
Heat generation efficiency of gas boiler, η_{GB}	0.9
Heat absorption efficiency of WHB, η_{WHB}	0.8
Cooling efficiency of electric refrigerator, η_{ER}	4
Energy utilization rates of heat exchange, η_{HE}	0.9
Energy utilization rates of LBAC, η_{AC}	1.2
Electricity charging and discharging efficiencies of ESU	0.95
Heat charging and discharging efficiencies of TSU	0.95
Electricity-heat conversion efficiency of EB, η_{EB}	3
Energy multiplication factor of ESU, η_{ESU}	2.7
Energy multiplication factor of TSU, η_{TSU}	0.2
Maximum power of gas turbine, kW	3000
Maximum power of LBAC, kW	4000
Maximum power of electric refrigerator, kW	4000
Maximum power of gas boiler, kW	4000
Maximum power of heat exchange, kW	4000
Maximum power transaction of multi-microgrid system, kW	4000

4.2. Analysis of the Operation and Economic Scheduling of EHIESS

In this paper, the impact of EHIESS on the multi-microgrid system operation is comparatively analyzed by setting up the following three cases.

Case 1: The multi-microgrid system is connected only to ESU, with ESU providing the electricity storage service.

Case 2: The multi-microgrid system is connected to ESU and TSU, with ESU and TSU providing electricity and heat storage service, respectively.

Case 3: The multi-microgrid system is connected to EHIESS, which includes ESU, TSU, and EB that can provide electricity and heat storage services, and realize electricity-heat coupling.

Table 4 shows the operation costs of the multi-microgrid system and the profits of EHIESS for three different cases. From Table 4, we can see that the profits of EHIESS keep increasing while the operation costs of the multi-microgrid system keep decreasing after adding TSU and EB devices in both summer and winter cases. Figure 5 reflects the charging and discharging energy of EHIESS under three cases in summer, from which we can see that the multi-microgrid system in Case 1 sells electricity to EHIESS during the lower load hours of 01:00–4:00 and 10:00–14:00, and purchases electricity from EHIESS during the higher load hours of 06:00–9:00 and 16:00–19:00. The addition of the TSU in Case 2 allows the multi-microgrid system to trade heat with EHIES. TSU can store excess heat to reduce heat waste, which reduces the heat production of CCHP in multi-microgrid system fairly, and reduces the operation costs of the multi-microgrid system in summer from $10,352.25 to $9884.21 in Case 1. Meanwhile, the operation costs of the multi-microgrid system in winter decreases from $36,745.27 to $35,291.77, and the profits of EHIESS in winter rises from $3754.61 to $6397.26. In Case 3, TSU is in heat discharging state for 23 h during the scheduling period, which is due to the input of EB equipment makes part of the electricity in ESU converted into heat. At this time, the profits of EHIESS in Case 3 are $1933.88 and $8825.53 in summer and winter, respectively, which are 12.25% and 37.96% higher compared to Case 2. The operation of EHIESS alleviates the heat production pressure and dependence on natural gas of gas turbine and gas boiler in multi-microgrid system, and realizes synergistic operation and complementary advantages between different microgrids.

Table 4. Profits of EHIESS and operation cost of multi-microgrid system in summer and winter.

Season	Case	Profits of EHIESS/$	Operation Costs of the Multi-Microgrid System/$
Summer	1	489.27	10,352.25
	2	1722.87	9884.21
	3	1933.88	7799.96
Winter	1	3754.61	36,745.27
	2	6397.26	35,291.77
	3	8825.53	26,628.18

Table 5 reflects the energy purchases from upper level energy grid by multi-microgrid system under Case 2 and Case 3. As can be seen from Table 5, the addition of the EB results in a significant increase of the purchased power of the multi-microgrid system both in summer and winter, whereas the purchased natural gas of the multi-microgrid system has a significant decrease in summer and winter by 6757.1 kWh and 41,766.6 kWh, respectively. The inclusion of EB realizes the coupled utilization of ESU and TSU so that the dependence of the multi-microgrid system on the energy storage has reduced. Besides, because of the higher construction cost of ESU relative to TSU, the configuration capacity of the ESU is reduced accordingly. Meanwhile, the multi-microgrid system in Case 3 has lower energy purchase costs in summer and winter compared to Case 2. The input of EB strengthens the energy interaction between different microgrids, and improves the utilization rate of the energy storage equipment in EHIESS, which significantly improves the economics of both the multi-microgrid system and EHIESS.

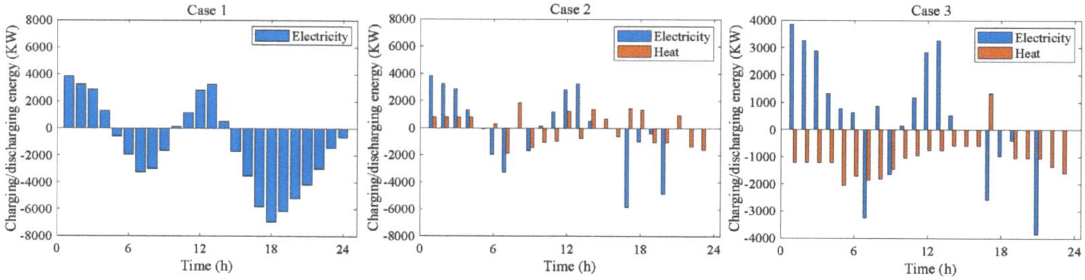

Figure 5. Charging and discharging energy EHIESS for Case 1 to Case 3 in summer.

Table 5. Energy purchase from upper-level energy grid for Case 2 and Case 3.

Season	Case2			Case3		
	Electricity Purchase /kWh	**Natural Gas Purchase /kWh**	**Cost of Energy Purchase /$**	**Electricity Purchase /kWh**	**Natural Gas Purchase /kWh**	**Cost of Energy Purchase /$**
Summer	8301.7	14,335.4	5609.60	14,059.3	7578.3	4880.98
Winter	57,091.5	62,568.8	27,420.73	79,857.7	20,802.2	13,092.99

In order to further evaluate the effectiveness of the shared energy storage optimal configuration model proposed in this paper, simulations are conducted in two scenarios of fixed energy storage capacity and optimal configuration of energy storage capacity in EHIESS. Summer load demands, wind and photovoltaic data of three microgrids are selected, and the optimal operation of EHIESS with the TSU capacity set to 10,000, 11,000, 12,000 kWh and TSU capacity set to 8000, 8500, 9000 kWh respectively is solved to compare with the optimal configuration scenarios of EHIESS, and the obtained results are shown in Table 6. When the ESU and TSU capacities are set to 10,000 kWh and 8000 kWh, respectively, EHIESS cannot guarantee the stable supply of electricity and heat to multi-microgrid systems, and the optimal solution cannot be obtained in this case. With the gradual increase of ESU and TSU capacity configurations, the energy storage construction and maintenance cost of EHIESS increases accordingly. Meanwhile, we find that as the storage capacity configuration increases, the storage capacity utilization rate gradually decreases, and the operation revenue of EHIESS during the scheduling period also gradually decreases.

Table 6. Operation results of EHIESS under fixed capacity configuration and optimal capacity configuration scenarios for energy storage unit.

Scenario	ESU Capacity/kWh	TSU Capacity/kWh	P_{es}^{max}/kW	Q_{es}^{max}/kW	Profit of EHIESS/$
Scenario 1 ine Scenario 2	10,000 11,000	8000 8500	- 4126.34	- 2833.33	- 1683.25
Scenario 3	12,000	9000	4501.46	3000	1654.28
Optimal configuration	10,263.33	8290.79	3850	2763.60	1933.88

4.3. Analysis of User-Side Price-Based Demand Response

Figure 6 shows the power balance and upstream network electricity price considering DR in Case 3. In the multi-microgrid system, the electricity is mainly supplied by PV, WT, and CCHP. The multi-microgrid system can trade energy with EHIESS, and each microgrid transmits energy to each other through EHIESS.

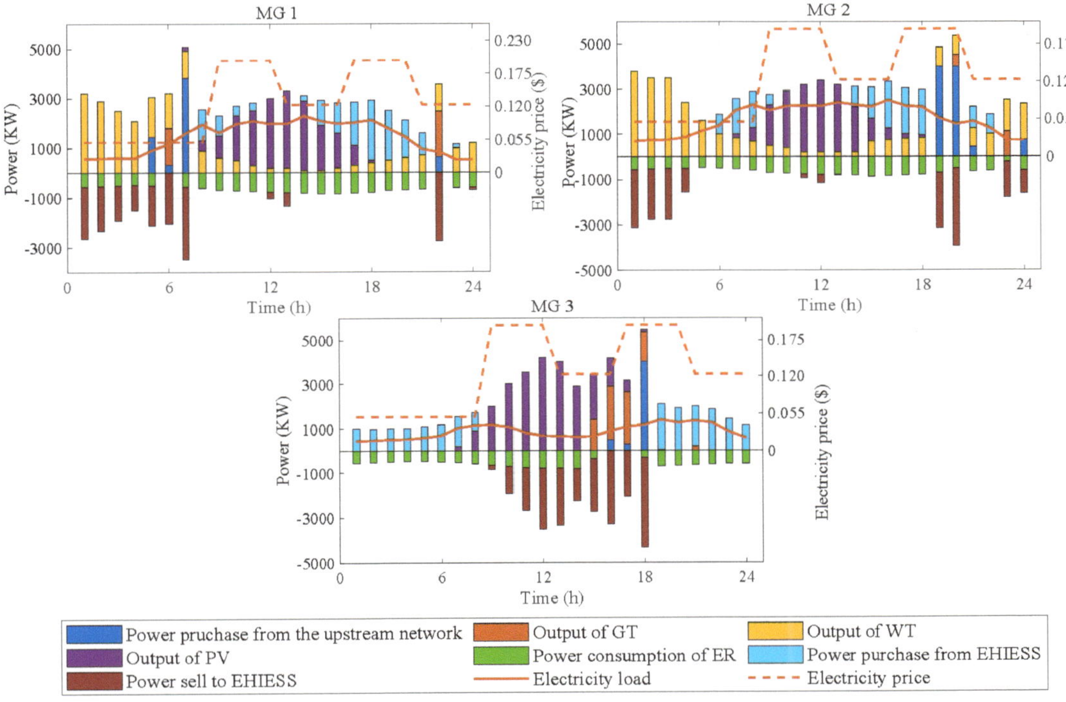

Figure 6. Power balance and upstream network electricity price considering DR in Case 3.

As we can see in Figure 6, after the introduction of renewable energy generation in conjunction with CCHP and the consideration of DR, the power generation of the multi-microgrid system is more diversified, which effectively improves the flexibility of the operation of the multi-microgrid system. Among them, electricity generation of PV and WT in each MG can be completely consumed, while the stochastic problem of WT and PV generation can be solved by EHIESS. In addition, the time-of-use price can guide the trading behavior of the multi-microgrid system with EHIESS. Besides, the sum of electricity generated by PV and WT and the output of CCHP is much larger than the user's demand from 01:00 to 08:00. In order to prevent the waste of energy, the time-of-use price can guide users to transfer their electricity consumption behavior to period 01:00–08:00. The three MGs transfer their loads from 01:00 to 08:00, which are 964.3 kW, 1198.1 kW and 655.8 kW, respectively, and the multi-microgrid system can sell surplus electricity to EHIESS to get profits. The 9:00–12:00 and 17:00–20:00 time periods are peak periods, during which DR can drive users to slash their electricity consumption behavior and prompt them to buy electricity from EHIESS, and the total load reduction of three MGs in peak periods are 120.5 kW, 131.6 kW, and 69.1 kW, respectively. The 13:00–16:00 and 21:00–24:00 time periods are the flat periods, and the multi-microgrid system can sell surplus electricity to EHIESS to gain profits. MG relies on the WT generation and electricity purchase from EHIESS to meet its own electricity load demands, and adjusts its electricity consumption behavior.

Table 7 reflects the operation situation of the multi-microgrid system before and after the implementation of DR in summer. From Table 6, it can be seen that the multi-microgrid system has better economic benefits after the implementation of DR, in which the operation cost of the multi-microgrid system is reduced by $522.05. In addition, the peak-to-valley difference of electric loads in the multi-microgrid system is reduced after considering DR in the user side, which leads to smoother load curves and more stable operation of the multi-microgrid system. This follows the fact that the implementation of DR prompts the users to shift the electricity load from the peak hours to other hours, which facilitates peak shaving and valley filling, thus reducing the value of the electricity load in peak periods. Finally, the capacity of ESU is reduced from 10,263.33 kWh to 9730.80 kWh after the implementation of DR in the user side. The implementation of DR makes the generation of each microgrid more compatible with the changes of electricity price, which reduces the generation of excess electricity in certain extent, so that the transactions of multi-microgrid system and EHIESS are reduced accordingly, and reduces the required capacity of the ESU.

Table 7. The operation situation of the multi-microgrid system before and after DR in summer.

Scenario	Operation Costs of Multiple Microgrids/$	Profits of EHIESS/$	Peak-to-Valley Difference in Electricity Loads/kW			ESU Capacity Configuration /kWh
			MG1	MG2	MG3	
Before DR	7799.96	1933.88	1830.00	1930.00	1150.00	10,263.33
After DR	7277.91	1687.40	1762.06	1804.60	927.48	9730.80

User satisfaction is an essential consideration in the operation of the multi-microgrid system, which is related to the changing magnitude of SL and CL. According to Ref. [38], the user satisfaction, S_u, can be expressed as follows:

$$S_{\mathrm{u}} = \frac{P_{\mathrm{L}}^{0} - \left|\Delta P_{\mathrm{CL}}^{t}\right| - \left|\Delta P_{\mathrm{SL}}^{t}\right|}{P_{\mathrm{L}}^{0}} \times 100\% \tag{43}$$

Simulations are carried out for different load shares of CL and SL in total load demands to analyze the impact on the system's operation. Figure 7 shows the operation costs of the multi-microgrid system and the customer satisfaction in the case of CL and SL each with a share of 5% to 25%. From Figure 7, it can be seen that as the share of CL and SL increases, the operation costs of the multi-microgrid system gradually decreases from $8558.33 to $7369.94, and the customer satisfaction gradually decreases from 98.44% to 92.22%. With the total microgrid load demand remaining unchanged, the DR behavior of multi-microgrid system increases as the CL and SL share gradually increases, prompting more load shifting from the peak hours to the valley hours and reducing the operation costs of the multi-microgrid system. Whereas, price-based DR forces users to shift the time of electricity consumption, which is contrary to the customer's willingness, and thus the increase of CL and SL share reduces the customer satisfaction.

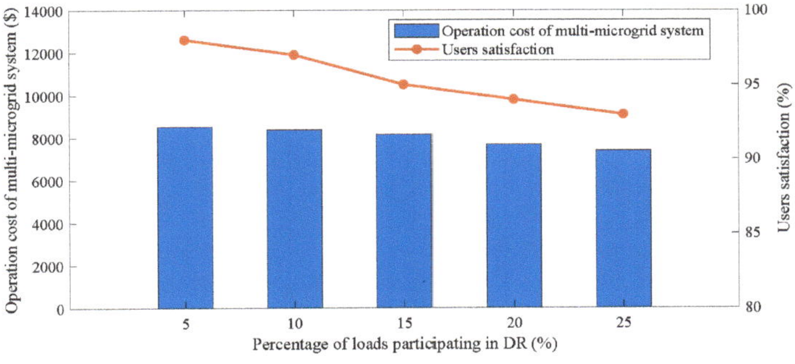

Figure 7. Operation costs and customer satisfaction of microgrids with different load shares after considering DR under Case3.

4.4. Analysis of Sensitivity

The variation range of variables (i.e., electricity price and load demands of multi-microgrid system, etc.) is set to −20–20%, based on Case 3 in summer, and the results of the sensitivity analysis of these variables on the results of the energy storage configuration, the EHIESS operation revenue, and the operation costs of multi-microgrid system are shown in Tables 8 and 9. As shown in Table 9, within the range of changes in electricity, with each 10% increase, EHIESS operation revenue and multi-microgrid system operation cost increase slightly, and ESU and TSU capacity configurations remain essentially unchanged. The sensitivity of electricity price to EHIESS operation revenue and operation cost of multi-microgrid system is relatively insignificant. For the load demand of multi-microgrid users, as shown in Table 9, with the loads changed by −20%, −10%, 10%, and 20%, the EHIESS operation revenue changed by −49.2%, −24.1%, 24.5%, and 46.9%, and the multi-microgrid system operation cost increased by −37.7%, −17.3%, 19.3% and 33.9%, respectively. As can be seen in Table 9, the sensitivity of the change in load demand to the EHIESS operation revenue and the operation cost of multi-microgrid system is more significant.

Table 8. Influence of electricity price changes on operation results.

Electricity Price	Profits of EHIESS/$	Operation Costs of Multi-Microgrid System/$	ESU Capacity/kWh	TSU Capacity/kWh
−20%	1930.89	7310.84	10,263.33	9090.23
−10%	1931.56	7484.73	10,263.33	8290.79
0%	1933.88	7799.96	10,263.33	8290.79
10%	1935.75	8537.65	10,263.33	8170.36
20%	1938.23	8620.57	10,263.33	8170.36

Table 9. Influence of load demand changes on operation results.

Electricity Price	Profits of EHIESS/$	Operation Costs of Multi-Microgrid System/$	ESU Capacity/kWh	TSU Capacity/kWh
−20%	981.87	4860.88	12,355.98	12,273.05
−10%	1468.40	6457.18	10,765.48	9090.23
0%	1933.88	7799.96	10,263.33	8290.79
10%	2408.21	9308.22	9863.46	6560.43
20%	2840.66	10,442.00	9463.59	6036.65

4.5. Analysis of Comparison of Solution Methods

In order to verify the effectiveness of the KKT condition and Big-M optimization method used in this paper to solve the bi-level planning problem of optimizing the shared energy storage configuration of multi-microgrid system, the computational efficiencies of the transformed single-level model and the original bi-level model are compared and analyzed. The computational results of the two solution methods are shown in Table 10. Since the original bi-level optimization model is a nonlinear optimization model, it is necessary to adopt corresponding intelligent algorithms to solve iteratively, which is easy to fall into the local optimal solution and consumes a large amount of computing time. It can be seen from the table that the computational time is significantly reduced after transforming into single-level optimization model, which also leads to a relative increase in the operation revenue of EHIESS and a decrease in the operation cost of multi-microgrid users.

Table 10. Calculation results of two solution methods.

Season	Calculation Time/s	Profits of EHIESS/$	Operation Costs of Multi-Microgrid System/$
Original bi-level model	319.6	1876.45	8209.87
Single-level model	8.6	1933.88	7799.96

5. Conclusions

In this paper, a bi-level optimization model considering EHIESS capacity configuration and optimal scheduling of multi-microgrid system is proposed to maximize the profits of EHIESS and minimize the operation costs of multi-microgrid system. The case studies demonstrate the effectiveness of the proposed methodology in solving the problem of high cost and low utilization of energy storage equipment configuration in multi-microgrid system. The main research results are as follows:

The introduction of EB in EHIESS makes the electricity-heat coupling in EHIESS more flexible, significantly improves the energy utilization rate of the storage equipment, and brings better economic benefits to the multi-microgrid system. Compared to the scenario considering only the separate configuration of electricity and heat energy storage, the addition of EB improves the total profits of EHIESS in one scheduling period by $211.01 and $2428.27 in summer and winter, respectively, and reduces the total operation costs of the multi-microgrid system by $2084.25 and $8663.59, respectively. In addition, DR provides an effective solution to the problem of load demand uncertainty on the user side, and users adjust their own electricity consumption behavior based on time-of-use price, which smooths out load fluctuations and effectively improves the flexibility of the operation of the multi-microgrid system. After the implementation of the price-based DR mechanism, the configuration of ESU and TSU in EHIESS is reduced by 532.53 kW, which further improves the economic efficiency of multi-microgrid system and EHIESS. In this case, the total operation cost of multi-microgrid system is reduced by $522.05.

This paper focuses on multi-microgrid shared energy storage systems, and future research can be extended to a wider range of application scenarios and consider different sizes and types of energy systems to provide broader application guidance and decision support. The main limitation of the proposed model is that it does not fully analyze the impact of renewable energy uncertainty on shared storage capacity allocation. In future research work, we will further consider the issue of evaluating the impact of multiple uncertainties, such as wind power and photovoltaic, on multi-microgrid shared energy storage systems. Besides, cost sharing between EHIESS and multi-microgrid system users will be further studied and explored, and game-based pricing mechanisms for service fees will be discussed.

Author Contributions: Conceptualization, Y.L.; methodology, Y.L.; software, Y.L., Z.W. and J.G.; validation, Y.L.; formal analysis, Z.D.; investigation, J.Z.; resources, W.Y.; writing—original draft preparation, Y.L.; writing—review and editing, Y.L. and Z.D.; visualization, Z.W. and J.G.; project administration, Z.D.; funding acquisition, Z.D. All authors have read and agreed to the published version of the manuscript.

Funding: This work was supported by National Natural Science Foundation of China under grant 52005306, Natural Science Foundation of Shandong Province under grant ZR2020QE220.

Institutional Review Board Statement: Not applicable.

Informed Consent Statement: Informed consent was collected from all participants in this study.

Data Availability Statement: Data is contained within the article.

Conflicts of Interest: Jingwei Zhao is from State Power Investment Corporation Haiyang Offshore Wind Power Co., Ltd. They declared no conflict of interest.

Nomenclature

Acronyms

CCHP	combined cooling heating and power
DR	demand response
EB	electric boiler
EHIESS	electricity-heat integrated energy storage supplier
ESU	electricity storage unit
KKT	Karush-Kuhn-Tucker
LBAC	lithium bromide absorption chiller
TSU	thermal storage unit
WHB	waste heat boiler

Variables

E^t	the electricity stored in ESU during hour t, MW
H^t	the heat stored in TSU during hour t, MW
P^t_{ch}	the electricity purchased by ESU during hour t, MW
P^t_{dis}	the electricity sold by ESU during hour t, MW
P^t_{EB}	the power consumption of EB during hour t, MW
$P^{t,i}_{\text{grid}}$	the electricity purchased of MG i from the upstream network during hour t, MW
$P^{t,i}_{\text{GT}}$	the electricity generated by gas turbine of MG i during hour t, MW
$P^{t,i}_{\text{mg,b}}$	the electricity purchased from EHIESS of MG i during hour t, MW
$P^{t,i}_{\text{mg,s}}$	the electricity sold to EHIESS of MG i during hour t, MW
$Q^{t,i}_{\text{AC}}$	the cooling output of LBAC in MG i during hour t, MW
Q^t_{ch}	the heat purchased by TSU during hour t, MW
Q^t_{dis}	the heat sold by TSU during hour t, MW
$Q^{t,i}_{\text{EB}}$	the heat produced by EB during hour t, MW
$Q^{t,i}_{\text{ER}}$	the cooling output of ER in MG i during hour t, MW
$Q^{t,i}_{\text{GB}}$	the heat generation of gas boiler in MG i during hour t, MW
$Q^{t,i}_{\text{GT}}$	the heat generation of gas turbine in MG i during hour t, MW
$Q^{t,i}_{\text{HE}}$	the heat output of heat exchange in MG i during hour t, MW
$Q^{t,i}_{\text{mg,b}}$	the heat purchased from EHIESS of MG i during hour t, MW
$Q^{t,i}_{\text{mg,s}}$	the heat sold to EHIESS of MG i during hour t, MW
$Q^{t,i}_{\text{WHB}}$	the heat absorbed by WHB of MG i during hour t, MW
$V^{t,i}_{\text{GB}}$	the gas consumption volume of gas boiler in MG i during hour t, MW
$V^{t,i}_{\text{GT}}$	the gas consumption volume of gas turbine in MG i during hour t, MW
ΔP^t_{CL}	the changes of curtailable load after DR during hour t, MW
ΔP^t_{SL}	the changes of shiftable load after DR during hour t, MW

Parameters

L_{gas}	the heat value of natural gas, kWh/m^3
η_{AC}	the heat utilization rate of LBAC
η_{EB}	the conversion efficiency of EB
η_{ER}	the cooling efficiency of electric refrigerator
η_{GB}	the heat generation efficiency of gas boiler
η_{GT}	the electricity generation efficiency of gas turbine
η_{HE}	the heat utilization rate of heat exchange
η_{WHB}	the heat absorption efficiency of WHB
η_{abs}	the charging efficiency of ESU
η_{re}	the discharging efficiency of ESU
ω_{abs}	the charging efficiency of TSU
ω_{re}	the discharging efficiency of TSU
γ_{GT}	the heat generation efficiency of gas turbine
λ^t	the price of electricity purchased from ESU during hour t, CNY/MW
φ^t	the price of heat purchased from TSU during hour t, CNY/MW
δ^t	the price of electricity sold to ESU during hour t, CNY/MW
γ^t	the price of heat sold to TSU during hour t, CNY/MW
θ^t	the price of service fees received by EHIESS during hour t, CNY/MW
τ^t_{grid}	the electricity price of the upstream network, CNY/MW
τ^t_{gas}	the price of natural gas, CNY/m^3

References

1. Dalala, Z.; Al-Omari, M.; Al-Addous, M.; Bdour, M.; Al-Khasawneh, Y.; Alkasrawi, M. Increased renewable energy penetration in national electrical grids constraints and solutions. *Energy* **2022**, *246*, 123361. [CrossRef]
2. Li, Y.; Wei, Y.; Zhu, F.; Du, J.; Zhao, Z.; Ouyang, M. The path enabling storage of renewable energy toward carbon neutralization in China. *Etransportation* **2023**, *16*, 100226. [CrossRef]
3. Ranjan, M.; Shankar, R. A literature survey on load frequency control considering renewable energy integration in power system: Recent trends and future prospects. *J. Energy Storage* **2022**, *45*, 103717. [CrossRef]
4. Thirunavukkarasu, M.; Sawle, Y.; Lala, H. A comprehensive review on optimization of hybrid renewable energy systems using various optimization techniques. *Renew. Sustain. Energy Rev.* **2023**, *176*, 113192. [CrossRef]
5. Zhong, X.; Zhong, W.; Liu, Y.; Yang, C.; Xie, S. Optimal energy management for multi-energy multi-microgrid networks considering carbon emission limitations. *Energy* **2022**, *246*, 123428. [CrossRef]
6. Zhou, B.; Zou, J.; Chung, C.Y.; Wang, H.; Liu, N.; Voropai, N.; Xu, D. Multi-microgrid energy management systems: Architecture, communication, and scheduling strategies. *J. Mod. Power Syst. Clean Energy* **2021**, *9*, 463–476. [CrossRef]
7. Xu, J.; Yi, Y. Multi-microgrid low-carbon economy operation strategy considering both source and load uncertainty: A Nash bargaining approach. *Energy* **2023**, *263*, 125712. [CrossRef]
8. Zhao, J.; Wang, W.; Guo, C. Hierarchical optimal configuration of multi-energy microgrids system considering energy management in electricity market environment. *Int. J. Electr. Power Energy Syst.* **2023**, *144*, 108572. [CrossRef]
9. Azizivahed, A.; Gholami, K.; Arefi, A.; Li, L.; Arif, M.; Haque, M. Stochastic scheduling of energy sharing in reconfigurable multi-microgrid systems in the presence of vehicle-to-grid technology. *Electr. Power Syst. Res.* **2024**, *231*, 110285. [CrossRef]
10. Lei, L.; Wu, N. An optimal scheduling strategy for electricity-thermal synergy and complementarity among multi-microgrid based on cooperative games. *Renew. Energy* **2024**, *237*, 121575. [CrossRef]
11. Jani, A.; Jadid, S. Two-stage energy scheduling framework for multi-microgrid system in market environment. *Appl. Energy* **2023**, *336*, 120683. [CrossRef]
12. Carvallo, C.; Jalil-Vega, F.; Moreno, R. A multi-energy multi-microgrid system planning model for decarbonisation and decontamination of isolated systems. *Appl. Energy* **2023**, *343*, 121143. [CrossRef]
13. Shi, Z.; Zhang, T.; Liu, Y.; Feng, Y.; Wang, R.; Huang, S. Optimal design and operation of islanded multi-microgrid system with distributionally robust optimization. *Electr. Power Syst. Res.* **2023**, *221*, 109437. [CrossRef]
14. Numan, M.; Baig, M.F.; Yousif, M. Reliability evaluation of energy storage systems combined with other grid flexibility options: A review. *J. Energy Storage* **2023**, *63*, 107022.
15. Li, L.; Peng, K.; Yang, X.; Liu, K. Coordinated design of multi-stakeholder community energy systems and shared energy storage under uncertain supply and demand: A game theoretical approach. *Sustain. Cities Soc.* **2024**, *100*, 105028. [CrossRef]
16. Hu, J.; Wang, Y.; Dong, L. Low carbon-oriented planning of shared energy storage station for multiple integrated energy systems considering energy-carbon flow and carbon emission reduction. *Energy* **2024**, *290*, 130139. [CrossRef]
17. Wang, Q.; Zhang, X.; Yi, C.; Li, Z.; Xu, D. A novel shared energy storage planning method considering the correlation of renewable uncertainties on the supply side. *IEEE Trans. Sustain. Energy* **2022**, *13*, 2051–2063. [CrossRef]

18. Han, J.; Fang, Y.; Li, Y.; Du, E.; Zhang, N. Optimal planning of multi-microgrid system with shared energy storage based on capacity leasing and energy sharing. In *IEEE Transactions on Smart Grid*; IEEE: Piscataway, NJ, USA, 2024.
19. Zheng, B.; Wei, W.; Chen, Y.; Wu, Q.; Mei, S. A peer-to-peer energy trading market embedded with residential shared energy storage units. *Appl. Energy* **2022**, *308*, 118400. [CrossRef]
20. Han, O.; Ding, T.; Zhang, X.; Mu, C.; He, X.; Zhang, H. A shared energy storage business model for data center clusters considering renewable energy uncertainties. *Renew. Energy* **2023**, *202*, 1273–1290. [CrossRef]
21. Zhang, X.; Wang, Z.; Liao, H.; Zhou, Z.; Ma, X.; Yin, X.; Wang, Z.; Liu, Y.; Lu, Z.; Lv, G. Optimal capacity planning and operation of shared energy storage system for large-scale photovoltaic integrated 5G base stations. *Int. J. Electr. Power Energy Syst.* **2023**, *147*, 108816. [CrossRef]
22. Dong, H.; Fu, Y.; Jia, Q.; Wen, X. Optimal dispatch of integrated energy microgrid considering hybrid structured electric-thermal energy storage. *Renew. Energy* **2022**, *199*, 628–639. [CrossRef]
23. Pradeep, N.; Reddy, K. Design and investigation of solar cogeneration system with packed bed thermal energy storage for ceramic industry. *Renew. Energy* **2022**, *192*, 243–263. [CrossRef]
24. He, Y.; Guo, S.; Zhou, J.; Ye, J.; Huang, J.; Zheng, K. Multi-objective planning-operation co-optimization of renewable energy system with hybrid energy storages. *Renew. Energy* **2022**, *184*, 776–790. [CrossRef]
25. Zeng, H.; Shao, B.; Dai, H.; Tian, N.; Zhao, W. Incentive-based demand response strategies for natural gas considering carbon emissions and load volatility. *Appl. Energy* **2023**, *348*, 121541. [CrossRef]
26. Dey, B.; Misra, S.; Marquez, F.P.G. Microgrid system energy management with demand response program for clean and economical operation. *Appl. Energy* **2023**, *334*, 120717. [CrossRef]
27. Tostado-Véliz, M.; Kamel, S.; Aymen, F.; Jordehi, A.R.; Jurado, F. A Stochastic-IGDT model for energy management in isolated microgrids considering failures and demand response. *Appl. Energy* **2022**, *317*, 119162. [CrossRef]
28. Yang, P.; Jiang, H.; Liu, C.; Kang, L.; Wang, C. Coordinated optimization scheduling operation of integrated energy system considering demand response and carbon trading mechanism. *Int. J. Electr. Power Energy Syst.* **2023**, *147*, 108902. [CrossRef]
29. Saberi-Beglar, K.; Zare, K.; Marzband, M.; Nojavan, S. Risk-embedded scheduling of a CCHP integrated with electric vehicle parking lot in a residential energy hub considering flexible thermal and electrical loads. *Appl. Energy* **2023**, *329*, 120265. [CrossRef]
30. Alfaverh, F.; Denai, M.; Sun, Y. A dynamic peer-to-peer electricity market model for a community microgrid with price-based demand response. *IEEE Trans. Smart Grid* **2023**, *14*, 3976–3991. [CrossRef]
31. Kong, X.; Wang, Z.; Liu, C.; Zhang, D.; Gao, H. Refined peak shaving potential assessment and differentiated decision-making method for user load in virtual power plants. *Appl. Energy* **2023**, *334*, 120609. [CrossRef]
32. Alikhani, M.; Moghaddam, M.P.; Moazzen, F.; Azadi, A. Optimal implementation of consumer demand response program with consideration of uncertain generation in a microgrid. *Electr. Power Syst. Res.* **2023**, *225*, 109859. [CrossRef]
33. Fang, X.; Li, F.; Wei, Y.; Cui, H. Strategic scheduling of energy storage for load serving entities in locational marginal pricing market. *IET Gener. Trans. Dis.* **2016**, *10*, 1258–1267. [CrossRef]
34. Li, L.; Xu, Q.; Wang, X.; Ling, J.; Sun, H. Optimal economic scheduling of industrial customers on the basis of sharing energy-storage station. *Electr. Power Constr.* **2020**, *41*, 100–107. (In Chinese)
35. Zhang, B.; Li, Q.; Wang, L.; Feng, W. Robust optimization for energy transactions in multi-microgrids under uncertainty. *Appl. Energy* **2018**, *217*, 346–360. [CrossRef]
36. Cao, W.; Xiao, J.W.; Cui, S.C.; Liu, X.K. An efficient and economical storage and energy sharing model for multiple multi-energy microgrids. *Energy* **2022**, *244*, 123124. [CrossRef]
37. Guo, Y.; Wang, C.; Shi, Y.; Guo, C.; Shang, J.; Yang, X. Comprehensive optimization configuration of electric and thermal cloud energy storage in regional integrated energy system. *Power Syst. Technol.* **2020**, *44*, 1611–1621. (In Chinese)
38. Hu, G.; Yan, Y.; Wu, H.; Jin, P. Emergency load control strategy and terminal implementation for distribution network based on collaboration of main grid and distribution network. *Autom. Electr. Power Syst.* **2022**, *46*, 180–187.

energies

MDPI

Article

An Equivalent Model for Frequency Dynamic Analysis of Large Power Grids Based on Regulation Performance Weighting Method

Junlong Zhang [1], Junchao Ma [2], Xuchen Yang [1], Jiajia Wang [1], Chenxu Wang [2] and Yiping Yu [1,*]

[1] School of Electrical and Power Engineering, Hohai University, Nanjing 210098, China; 231306080050@hhu.edu.cn (J.Z.); 231606030104@hhu.edu.cn (X.Y.); 231306080042@hhu.edu.cn (J.W.)

[2] State Grid Zhejiang Electric Power Co., Ltd., Research Institute, Hangzhou 310008, China; majunchao36@163.com (J.M.); wangchenxu@whu.edu.cn (C.W.)

* Correspondence: yyiping@hhu.edu.cn

Abstract: With the construction of the UHV (Ultra High Voltage) AC/DC hybrid power grid and the large-scale access to renewable energy such as wind power, frequency dynamic fluctuation has become a prominent problem affecting the safe and stable operation of large power grids. The expansion of the scale of the power system makes it impossible to use traditional fine modeling to analyze the power system. In order to reduce the calculation scale and storage capacity of power system frequency dynamic simulation, it is necessary to make appropriate equivalent simplification of the external system, so the appropriate dynamic equivalent method is of great significance. This paper mainly studies the equivalent model suitable for frequency dynamic analysis of large power grids. Firstly, the typical models of generator set and load are simplified, and the parameters that have a great influence on frequency in the simplified model are obtained through characteristic analysis. Then, a dynamic aggregation method of generator governor and prime mover parameters and load parameters based on regulation performance weighting (the parameters of the generator or load are weighted and summed according to its regulation ability on the system) is proposed. This method is applied to the simulation example of the East China Power Grid. The simulation proves that the frequency of the East China Power Grid before and after equivalence can be consistent under four different faults, which verifies the effectiveness of the equivalent method proposed in this paper in the frequency dynamic analysis of large power grids.

Keywords: frequency analysis; regulation performance weighting method; governor and prime mover equivalent; load equivalence

1. Introduction

One of the most important parameters in the power system is frequency, which can reflect the operation quality of the power grid [1]. To make the power grid safe and stable, it is necessary to maintain the frequency within a reasonable range [2]. Because the power system has a certain particularity, it is impossible to directly test the actual power grid to study the frequency problem, and it must be studied through modeling and simulation. The dynamic equivalence method of the power system is inseparable from the physical problems that need to be studied after equivalence. According to different research problems, dynamic equivalence can be divided into three methods: the coherency equivalence method [3,4], the mode equivalence method [5] and the online identification method [6].

There are four main models in the power system [7], including the generator model, excitation system model, prime mover-governor model and load model. So far, the first two of these four models have well-studied models because their models are mainly electrical quantities that can be obtained through actual measurements [8–10]. This paper mainly studies the latter two models. Because the power system is developing rapidly and the scale is expanding, a large number of additional governors, loads and other components must be taken into account. Therefore, the traditional comprehensive and refined modeling of the power system [11–13] is no longer applicable. It is necessary to make appropriate equivalent simplification of the external system. Therefore, the appropriate dynamic equivalent method is of great significance.

With the integration of large-scale renewable energy, such as wind power, into the power system, its instability will inevitably have an unpredictable impact on the grid frequency. This paper also studies the general model of wind power generation system, including four modules: wind turbine model, pitch angle controller, converter controller and generator model [14,15], so as to further study the frequency dynamic analysis of large power grids with new energy. The simulation results of the power system can be used as the reference data for the planning and construction of the power grid, but different simulation models will have different degrees of influence on the results of the simulation calculation [16–18]. In particular, the accuracy of the load model will play a crucial role in the final simulation results of the power grid [19]. It can be seen from references [20–22] that with the change in the power grid and power supply structure and load type, the frequency characteristics of the system will also change, and the security and stability of the power grid will also face new threats.

In this paper, a method of dynamic aggregation of generator governor-primitive motor parameters and load parameters based on adjustment performance weighting is proposed. This method overcomes the uncertainty of human experience values and has fast aggregation speed and high engineering accuracy. Table 1 shows the comparison between the regulation performance weighting method and other main equivalence methods. Using this method, the actual power grid can be equivalent to a single generator and a single load, which is convenient for system frequency simulation analysis. Finally, the method is applied to the East China Power Grid. The equivalent method of this paper is used to equalize the generator and load of the East China Power Grid. The simulation proves that the frequency of the East China Power Grid can be consistent before and after the equivalence under four different faults, which verifies the effectiveness of the equivalent method in the dynamic analysis of the large power grid frequency.

Table 1. The comparison of regulation performance weighting method and other methods.

	Regulation Performance Weighting Method	**Other Coherency-Based Equivalence Method**
Polymerization rates	Faster	Fast
Reliability	Overcome the uncertainty of human experience	Relying on human experience
Accuracy	High	Low

2. Equivalence Method for Governor and Prime Mover Based on Regulation Performance Weighting

2.1. Aggregation Equivalence Modeling for Governor and Prime Mover of Steam Turbine

Taking the governor and prime mover models commonly used in steam turbines as an example, the governor model is analyzed in detail and simplified reasonably. The governor and prime mover models commonly used in PSD-BPA (Power System Department-Bonneville Power Administration) are shown in Figures 1 and 2 [23].

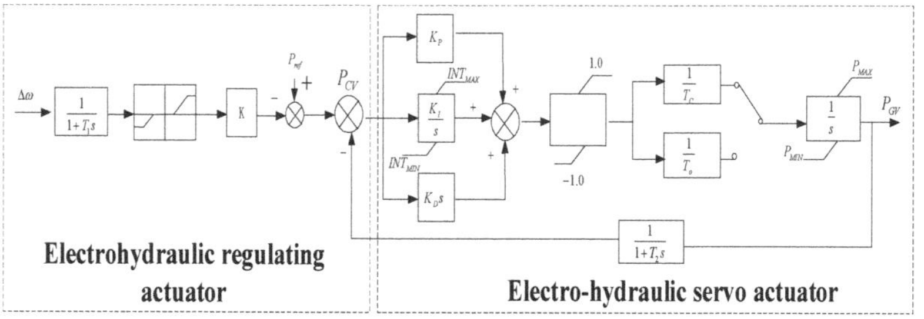

Figure 1. Structure model of steam turbine governing system.

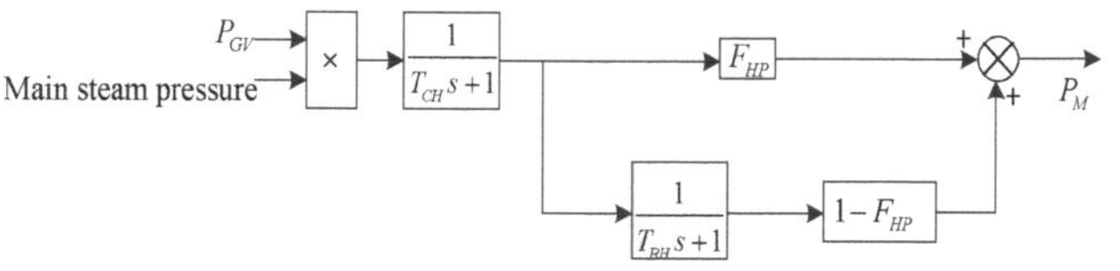

Figure 2. Simplified transfer function model of turbine prime mover.

In Figure 1, T_1 represents the time constant of rotational speed measurement, P_{ref} represents the given power reference value and T_C represents the opening time constant of the oil motor.

The transfer function of the electro-hydraulic servo actuator is as follows:

$$G_2 = \frac{P_{GV}}{P_{CV}} = \frac{\left(K_P + K_D s + \frac{K_I}{s}\right) . \frac{1}{T_O} \cdot \frac{1}{s}}{1 + \left(K_P + K_D s + \frac{K_I}{s}\right) . \frac{1}{T_O} \cdot \frac{1}{s} \cdot \frac{1}{1+T_2 s}} \tag{1}$$

T_2 represents the time of the oil motor travel feedback link, the typical value is 0.02 s; T_O represents the opening time constant of the oil motor, the typical value is 1–4 s; K_P, K_D and K_I are the proportional amplification link, differential link and integral link multiples of the PID module, respectively. In general, K_D and K_I are all 0, and K_P is about 10.

Under the typical parameters of the governor model, because P_{CV} is a step signal, take $P_{CV} = 1/s$ and the governor model is simplified to obtain the transfer function as shown in Equation (2).

$$P_{GV} = \frac{K_P . \frac{1}{T_O} \cdot \frac{1}{s^2}}{1 + K_P . \frac{1}{T_O} \cdot \frac{1}{s} \cdot \frac{1}{1+T_2 s}} \tag{2}$$

P_{GV} is a step signal, so when $P_{GV} = 1/s$, the transfer function of PM can be expressed as Equation (3).

$$P_M(s) = \frac{1}{s}\left[\frac{1}{T_{CH}s+1}F_{HP} + \frac{1}{T_{CH}s+1}\frac{1}{T_{RH}s+1}(1-F_{HP})\right] \tag{3}$$

Therefore, the steam turbine governor-primitive engine model has seven main parameters: speed deviation magnification K, oil motor opening time constant T_O, PID module proportional magnification multiple K_P, oil motor travel feedback link time T_2, steam volume time constant T_{CH}, reheater time constant T_{RH} and high-pressure cylinder mechanical power ratio coefficient F_{HP}. Therefore, the frequency characteristics of these parameters are analyzed to find out the parameters that have a greater impact on the frequency.

By analyzing Equation (1), taking $K_P = 7$, $T_2 = 0.02$, T_O = 1.15 and 4, the $P_{GV}(t)$ curve is obtained, as shown in Figure 3.

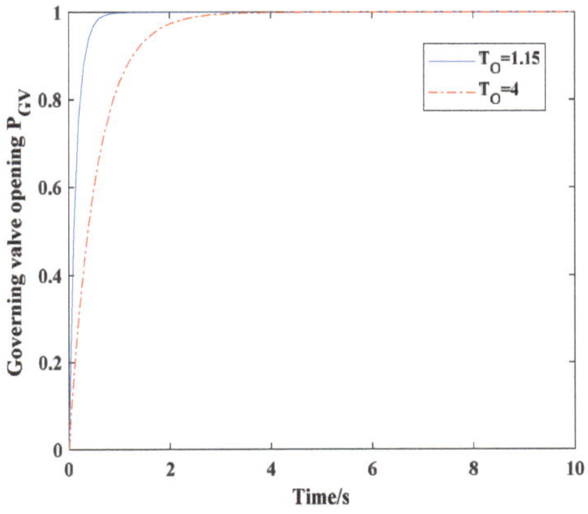

Figure 3. Governor valve opening under different T_O values.

From Figure 3, when T_O increases from 1.15 to 4, the $P_{GV}(t)$ curve changes greatly. It can be seen that the oil motor opening time constant T_O has a great influence on the characteristics of the steam turbine, and this parameter needs to be considered in the equivalent modeling.

The frequency characteristics of the remaining parameters are analyzed according to the same method, which is no longer shown here. Finally, it is concluded that the speed deviation amplification factor K, the oil motor opening time constant T_O, the reheater time constant T_{RH} and the high-pressure cylinder mechanical power proportional coefficient F_{HP} have a great influence on the frequency characteristics of the steam turbine. These four parameters are mainly considered when equivalent.

The steps of the dynamic aggregation method of turbine governor-prime mover parameters based on regulation performance weighting are shown in Figure 4:

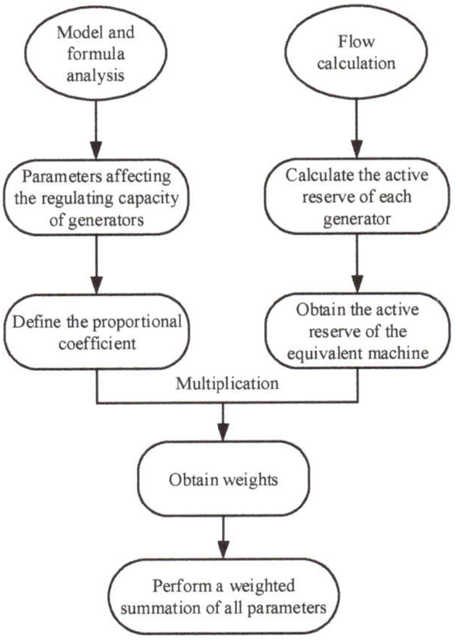

Figure 4. The flowchart of regulation performance weighting method.

(1) The system to be analyzed is divided into an internal system that should remain unchanged and an external system to be equivalent. The power flow calculation of the power grid is carried out through BPA, and the parameter variables (rated output of the generator prime mover P_{ei}) and state variables (actual active power output of the generator P_{GENi}) of each generator in the external system that needs to be equivalent are obtained.

(2) Through the calculation of the parameters of the M turbines to be aggregated, the spinning reserve of each turbine generator and the equivalent turbine generator are obtained.

According to (4), the spinning reserve of each turbine generator is obtained:

$$P_{SRi} = P_{ei}K_{MAXi} - P_{GENi} \tag{4}$$

In Equation (4), P_{ei} represents the rated output of the i-th turbine generator prime mover, K_{MAXi} represents the maximum regulating valve opening of the i-th turbine generator prime mover and P_{GENi} represents the actual active output of the i-th turbine generator.

According to (5), the spinning reserve of the equivalent turbine generator P_{SReq} is obtained:

$$P_{SReq} = \sum_{i=1}^{M} P_{SRi} \tag{5}$$

(3) Organize the body model and parameters of M governors to be aggregated in the external system and classify the governors of M steam turbine units according to the commonly used n governor models (G1/G2/G3/.../Gn). According to the different types of governors, the generators are classified, and the sum of the spins of turbo-generators under different types of governors is obtained.

$$\begin{cases} P_{G1} = \sum_{i=1}^{\mathrm{a}} P_{SRi} \\ P_{G2} = \sum_{i=1}^{\mathrm{b}} P_{SRi} \\ P_{G3} = \sum_{i=1}^{\mathrm{c}} P_{SRi} \\ \cdots \\ P_{Gn} = \sum_{i=1}^{\mathrm{n}} P_{SRi} \end{cases} \tag{6}$$

Comparing the sum of the spins of n governor models, if the proportional of the sum of the spins of a governor model to the sum of the spins of all models is less than 5%, the governor is ignored, and the others are equivalent according to different governor types.

(4) Through the superposition calculation of the rated output power of the prime mover of the M number of turbine generator sets to be aggregated, the rated output power of the equivalent turbine generator is obtained. Through the total rated output power, total actual active power and total spinning reserve of the unit to be aggregated, the maximum valve opening of the prime mover of the equivalent governor is obtained.

The rated output power of the equivalent turbine generator is as follows:

$$P_{eM} = \sum_{i=1}^{M} P_{ei} \tag{7}$$

The actual active power of the equivalent turbine generator is as follows:

$$P_{GENM} = \sum_{i=1}^{M} P_{GENi} \tag{8}$$

Maximum gate opening of prime mover of equivalent governor is as follows:

$$K_{MAXi} = \frac{P_{SReq} + P_{GENi}}{P_{eM}} \tag{9}$$

(5) The parameters of the equivalent steam turbine governor-primitive engine model are obtained by the method of dynamic aggregation of turbine governor-primitive engine parameters based on the weighted adjustment performance.

Define the proportional coefficient of turbine speed regulation K_{PMi1}:

$$K_{PMi1} = \frac{F_{HP}K}{T_{RH}T_O} \tag{10}$$

From the above analysis, it can be seen that the speed deviation amplification factor K, the oil motor opening time constant T_O, the reheater time constant T_{RH} and the high-pressure cylinder mechanical power proportional coefficient F_{HP} in the turbine governor-prime mover have a great influence on the characteristics of the turbine. Therefore, Equation (10) is expressed as the product of these four parameters, where T_O and T_{RH} takes the reciprocal form.

Define the turbine governor weight R_i:

$$R_i = P_{SRi}K_{PMi1} \tag{11}$$

The parameters in the turbine equivalent governor are aggregated as follows:

$$\varepsilon = \frac{\sum\limits_{i=1}^{M} \varepsilon_i R_i}{\sum\limits_{i=1}^{M} R_i} \tag{12}$$

In Equation (12), ε can be the amplification factor K of the equivalent machine speed deviation, the opening time constant T_O of the oil motor, the reheater time constant T_{RH} and the mechanical power proportional coefficient F_{HP} of the high-pressure cylinder, and ε_i are the corresponding parameters of the i-th generator governor.

2.2. Aggregation Equivalence Modeling for Governor and Prime Mover of Hydro-Turbine

The parameter aggregation method of the hydro-turbine governor and prime mover model is similar to that of the steam turbine. However, due to the large difference between the governor and prime mover model of hydro-turbine and steam turbine, the parameters are also different, so the aggregation steps are also different. The following describes the different steps of hydro-turbine parameter aggregation and steam turbine parameter aggregation.

In step 2 of the parameter aggregation method for steam turbine governors, the maximum governor valve opening K_{MAXi} of the steam turbine generator prime mover is required to obtain the spinning P_{SRi} of each generator. However, this parameter is not available in the typical governor model of the hydro-turbine. Therefore, the spinning P_{SRi} of each hydro-generator is obtained according to Equation (13):

$$P_{SRi} = P_{ei} - P_{GENi} \tag{13}$$

In Equation (13), P_{ei} represents the rated output of the prime mover of the *i*-th hydro-generator and P_{GENi} represents the actual active output of the *i*-th hydro-generator.

In step 4, only the rated output power of the equivalent hydro-generator and the actual active power output of the equivalent hydro-generator are required.

In step 5, because the parameters of the turbine governor are different from those of the steam turbine, the proportional coefficient K_{PMi2} of the turbine governor is defined:

$$K_{PMi2} = \frac{D_d}{RT_d T_G (T_W/2)} \tag{14}$$

According to the same method as the steam turbine, the frequency characteristics of hydro-turbine parameters are analyzed. It is found that the adjustment coefficient R, the soft feedback link coefficient D_d, the soft feedback time constant T_d, the governor response time T_G and the water hammer effect time constant $T_W/2$ have a great influence on the characteristics of the hydro-turbine. Therefore, K_{PMi2} in Equation (14) is expressed as the product of these five parameters, where R, T_d, T_G and $T_W/2$ are in the reciprocal form.

The other steps of the equivalent method of the hydro-turbine governor are the same as those of the steam turbine. Finally, the adjustment coefficient R, the soft feedback link coefficient D_d, the soft feedback time constant T_d, the governor response time T_G and the water hammer effect time constant $T_W/2$ in the hydraulic turbine equivalent governor can be obtained by Equation (12).

2.3. Aggregation Equivalence Modeling for Prime Mover and Controller of Wind Turbine

The general model of wind turbines mainly includes four modules: a wind turbine model, pitch angle controller, converter controller and generator model. The mechanical power captured by a wind turbine can be expressed as follows [14]:

$$P_{\mathrm{m}} = \frac{1}{2}\rho\pi R^2 C_p(\lambda,\beta)v^3 \tag{15}$$

In the formula, ρ is the air density, R is the radius of the wind wheel, v is the wind speed, λ is the tip speed ratio and C_p is the wind energy utilization coefficient.

In the general model, the wind turbine adopts the linear aerodynamic model, and the wind speed is assumed to be constant [14]:

$$P_m = P_{m0} - K_a\beta(\beta - \beta_0) \tag{16}$$

In the formula, β is the slurry pitch angle, β_0 is the initial pitch angle, P_{m0} is the initial mechanical power and K_a is the aerodynamic power coefficient.

The wind turbine drive shaft system model adopts a single mass model, and a rigid body is used to simulate the wind turbine blade, the drive shaft and the rotor shaft of the generator. Ignoring the internal differences of the shaft system, that is, the first-order inertia link is used to simulate the transmission process of the shaft torque, as shown in Equations (17) and (18):

$$T_J\frac{d\omega}{dt} = T_m - T_e \tag{17}$$

$$\frac{dT_m}{dt} = \frac{1}{t_h}(T_{ae} - T_m) \tag{18}$$

In the formulas, T_J is the total inertia time constant of the wind turbine and the generator; T_m, T_e and T_{ae} denote the mechanical torque of the rotor, electromagnetic torque of the generator and the torque of the shaft.

The pitch angle controller model and the converter-level controller model are shown in Figures 5 and 6.

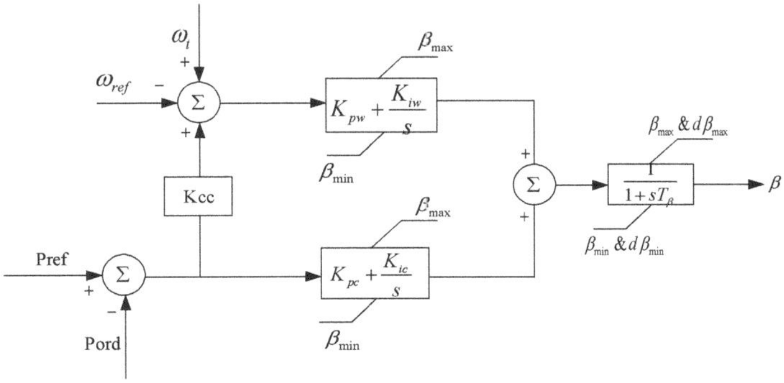

Figure 5. Simplified transfer function model of turbine prime mover. Where T_β is the reaction time constant of the blade; P_{ord} is the set value of power; P_{ref} is the reference power.

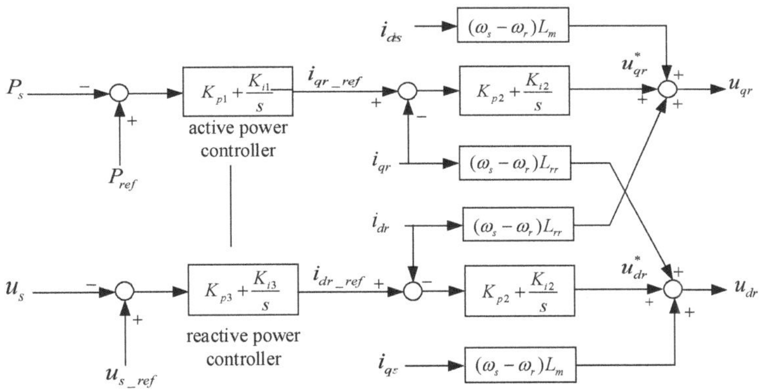

Figure 6. Simplified transfer function model of turbine prime mover. Where P_s is the value of input active power; u_s is the value of input voltage; P_{ref} is active power reference value; u_{s_ref} is the value of reference voltage; u_{qr} is the q-axis voltage of rotor and u_{dr} is the d-axis voltage of the rotor; u^*_{qr} is the q-axis voltage of rotor to be compensated and u^*_{dr} is the d-axis voltage of the rotor to be compensated.

Suppose the intermediate variables are x_1, x_2, x_3 and x_4. The transfer function expressions are shown as Equations (19)–(21):

$$\beta = \frac{1}{1+sT_\beta}[(K_{pc}+\frac{K_{ic}}{s})\Delta P + (\Delta PKcc + \Delta\omega)(K_{pw}+\frac{K_{iw}}{s})] \quad (19)$$

$$u_{qr} = K_{p2}(K_{p1}\Delta P + K_{i1}x_1 - i_{qr}) + K_{i2}x_2 + s\omega_s L_m i_{ds} + s\omega_s L_{rr} i_{qr} \quad (20)$$

$$u_{dr} = K_{p2}(K_{p3}\Delta u + K_{i3}x_3 - i_{dr}) + K_{i2}x_4 - s\omega_s L_m i_{qs} - s\omega_s L_{rr} i_{dr} \quad (21)$$

In the equation, T_β is the reaction time constant of the blade; L_{rr} is the inductance of the rotor winding; L_m is the mutual inductance between the stator and the rotor; ω_s is the electrical angular velocity of the stator. K_{p1} and K_{i1} are the proportional and integral coefficients of active power control; K_{p2} and K_{i2} are the proportional and integral coefficients of rotor-side current control;K_{p3} and K_{i3} are the proportional and integral coefficients of reactive power control.

Similarly, through the analysis of frequency characteristics, it is found that the inertia time constant T_J, the blade reaction time constant T_β, the active power control magnification K_1, the current control magnification K_2 and the reactive power control magnification K_3

have a great influence on the frequency. Therefore, these five parameters are mainly considered in the equivalence.

The parameter aggregation method of the wind turbine is similar to that of the hydro-turbine, but the parameters are different due to the large difference between the wind turbine and the hydro-turbine. The following describes the different steps of wind turbine parameter aggregation and water turbine parameter aggregation.

In step 2, the wind turbine is similar to the hydro-turbine in obtaining the spinning reserve of each unit, but the wind turbine must be grouped before aggregation. The working characteristics of the wind turbine are different under different wind speeds. The wind turbine is grouped according to the method based on critical wind speed and wind speed similarity.

In step 5, because the parameters of the wind turbine are different from those of the hydro-turbine, the proportional coefficient K_{PMi3} of wind turbine speed regulation is defined:

$$K_{PMi3} = \frac{K_1 K_2 K_3}{T_J T_\beta} \tag{22}$$

The other steps of the wind turbine equivalent method are the same as those of the hydro-turbine. Finally, the parameters of the wind turbine can be obtained by Equation (12).

3. Equivalence Method for Load Based on Regulation Performance Weighting

3.1. Static Load Model

The static load model is mainly represented by the polynomial model, and the LB (Load Balance) model is mainly used in BPA. The expression of the static load model is shown in Equation (23):

$$\begin{cases} P = P_0\left[P_1\left(\frac{V}{V_0}\right)^2 + P_2\left(\frac{V}{V_0}\right) + P_3\right](1 + \Delta f \cdot L_{DP}) \\ Q = Q_0\left[Q_1\left(\frac{V}{V_0}\right)^2 + Q_2\left(\frac{V}{V_0}\right) + Q_3\right](1 + \Delta f \cdot L_{DQ}) \end{cases} \tag{23}$$

where the second order term represents the constant impedance term, the first order term represents the constant current term, and the zeroth order term represents the constant power term. P_1, P_2 and P_3 are the active power proportional coefficients of each term, Q_1, Q_2 and Q_3 are the reactive power proportional coefficients of each term and $P_1 + P_2 + P_3 = 1$, $Q_1 + Q_2 + Q_3 = 1$. L_{DP} is the frequency response factor of active load, and L_{DQ} is the frequency response factor of reactive load.

The frequency characteristics of the load are characterized by term $(1 + \Delta f \cdot L_{DP})$ and $(1 + \Delta f \cdot L_{DQ})$, so in the static load model, the main factors affecting the frequency are the frequency response factor L_{DP} of the active load and the frequency response factor L_{DQ} of the reactive load.

Assuming that there are M static loads in the system when the static load is equivalent, the parameter aggregation steps of the equivalent model are as follows.

The capacity of the equivalent static load is the sum of the static load capacity (MVA) of each node:

$$S_M = \sum_{i=1}^{M} S_i \tag{24}$$

The parameter aggregation method based on the adjustment performance weighting method is also used to obtain the parameters of the equivalent static load.

Define the static load ratio coefficient:

$$K_{PMi3} = L_{DP} L_{DQ} \tag{25}$$

Considering the frequency characteristics of the static load model, the main influencing factors are the frequency response factor L_{DP} of the active load and the frequency response factor L_{DQ} of the reactive load. Therefore, K_{PMi3} in Equation (25) is expressed as the product of these two parameters.

Define the static load weight:

$$R_i = S_i K_{PMi3} \tag{26}$$

Then, the parameters in the equivalent static load model can also be aggregated according to Equation (12), where ε is the parameter of the equivalent static load model, which can be the frequency response factor L_{DP} of the active load and the frequency response factor L_{DQ} of the reactive load and ε_i is the corresponding parameters of the static load model of the i-th node.

3.2. Dynamic Load Model

The induction motor model is mainly considered in the dynamic load model, and its expressions are shown in Equations (27)–(29):

$$\begin{cases} \frac{dE'_d}{dt} = -\frac{1}{T'}\left[E'_d + (X - X')I_q\right] + \omega_0 sE'_q \\ \frac{dE'_q}{dt} = -\frac{1}{T'}\left[E'_q - (X - X')I_d\right] - \omega_0 sE'_d \end{cases} \tag{27}$$

$$\begin{cases} I_d = \frac{1}{R_s^2 + X'^2}\left[R_s(U_d - E'_d) + X'(U_q - E'_q)\right] \\ I_q = \frac{1}{R_s^2 + X'^2}\left[R_s(U_q - E'_q) + X'(U_d - E'_d)\right] \end{cases} \tag{28}$$

where s is the rotor slip; T' is the transient open circuit time constant; X is open-circuit impedance for the rotor and X' is short-circuit reactance for rotor stalling. I_d and I_q represent the d- and q-axis currents of the stator, respectively. R_s represent the stator resistance.

The rotor equation is as follows:

$$\begin{cases} T_J \frac{ds}{dt} = T_m - T_e \\ T_m = (A\omega_r^2 + B\omega_r + C)T_0 \end{cases} \tag{29}$$

where T_J is the inertia time constant of the rotor, ω_r is the angular velocity of the rotor, T_m is the mechanical torque of the motor and A, B and C are the torque coefficients of the mechanical load.

The induction motor model in BPA software (version number: FSDEdit 2.8) mainly adopts the ML model [24]. Among the main parameters, the inertia time constant T_J, the proportion of motor power to bus power P_{per}, the load rate K_L, the stator reactance X_s, the rotor reactance X_r and the torque equation constant A have a great influence on the frequency, which should be considered emphatically.

Assuming that there are M induction motors in the system, the following is the process of obtaining the parameters of the equivalent machine.

The capacity of the equivalent machine can be obtained by summing the capacity (MVA) of each induction motor:

$$S_M = \sum_{i=1}^{M} S_i \tag{30}$$

The parameters of the equivalent induction motor are also obtained by using the parameter aggregation method based on the adjustment performance weighting method.

Define the dynamic load ratio coefficient K_{PMi4}:

$$K_{PMi4} = T_j \cdot P_{per} \cdot K_L \cdot X_s \cdot X_r \cdot A \tag{31}$$

When studying the frequency characteristics of induction motors, the main influencing factors are the inertia time constant T_J, the ratio of motor power to bus power P_{per}, the load rate K_L, the stator reactance X_s, the rotor reactance X_r and the torque equation constant A. Therefore, K_{PMi4} in Equation (31) is expressed as the product of these six parameters.

Define the dynamic load weight R_i:

$$R_i = S_i K_{PMi4} \tag{32}$$

Then, the parameters in the equivalent induction motor model are also aggregated according to Equation (12), where ε is the parameter of the equivalent induction motor, which can be respectively the inertia time constant T_J, the ratio of motor power to bus power P_{per}, the load rate K_L, the stator reactance X_s, the rotor reactance X_r and the torque equation constant A. ε_i is the corresponding parameters of the i-th induction motor.

4. Case Analysis and Application of East China Power Grid

Using the equivalent method in this paper, Anhui, Shanghai and Fujian power grids are equivalent to a single generator (equivalent if there is a turbine and a steam turbine) and a single load (equivalent if there is a static load and a dynamic load). The Jiangsu and Zhejiang power grids remain unchanged to see whether the system frequency is consistent before and after the equivalence.

Among them, there are 31 generators in the Anhui power grid, all of which are turbine generators, 120 static load nodes and 102 induction motors. There are 20 generators in the Shanghai power grid, all of which are turbine generators, 64 static load nodes and 66 induction motors. Fujian power grid has 29 turbo-generators and 11 hydro-generators, 120 static load nodes and 125 induction motors. By using the governor-primitive motor and load equivalent method based on the weighted adjustment performance proposed in this paper, the generator and load models in the three power grids are equivalent. Each power grid is finally equivalent to a single generator (the Fujian power grid is a steam turbine and a turbine), a static load node and an induction motor. The following table shows some governors and load parameters of the Anhui power grid, Shanghai power grid and Fujian power grid. The following Tables 2–10 show some governor and load parameters of Anhui power grid, Shanghai power grid and Fujian power grid.

Table 2. Governor parameters of Anhui power grid (part).

Unit Name/Governor Parameters	P_{EI}/MW	K_{MAXI}	P_{GENI}/MW	P_{SRI}/MW	T_O	K	T_{RH}	F_{HP}
Wan Anqing_4	336.2	1.06	228.9	127.472	1.15	16.7	10	0.3
Wan Huaier_2	336.2	1.06	212.8	143.572	1.15	16.7	10	0.3
Wan Jiuhua_1	336.2	1.06	183.4	172.972	1.3	20	12	0.4
Wan Qiandian_2	151.2	1.06	87.25	73.022	1.3	20	12	0.4
Wan Tianmen_1	698.4	1.06	426.9	313.404	1.5	23	14	0.5
Wan Luohe_1	358.6	1.06	198.4	181.716	1.5	23	14	0.5
Wan Yongfeng_2	635.2	1.06	316.9	356.412	1.7	25	16	0.6
Wan Mayi_1	635.2	1.06	442.4	230.912	1.7	25	16	0.6
Wan Linhuan_3	336.2	1.06	164	192.372	1.9	28	18	0.7
Wan Hualiu_3	685.2	1.06	467.5	258.812	1.9	28	18	0.7
...	...	...	...	...	...	...	...	...
Equivalent generator	15,442.8	1.06	8755.21	7211.36	1.55	23.06	14.43	0.52

Table 3. Anhui power grid induction motor parameters (part).

Busbar Name	Capacity/MW	T_J	P_{PER}	K_L	X_S	X_R	A
Wan Kongdian21	−7.51652	3	1	0.8	0.067	0.17	1
Wan Huaibei22	1.456613	3	1	0.8	0.067	0.17	1
Wan Caishi22	64.64777	5	0.9	0.7	0.087	0.2	0.9
Wan Changlong22	105.7431	5	0.9	0.7	0.087	0.2	0.9
Wan Huizhou2E	87.15075	7	0.8	0.6	0.107	0.23	0.8
Wan Huigong22	96.08173	7	0.8	0.6	0.107	0.23	0.8
Wan Xishan22	84.51662	9	0.7	0.5	0.127	0.26	0.7
Wan Xiantong22	75.42928	9	0.7	0.5	0.127	0.26	0.7
...	...	...	...	...	...	...	...
Equivalent load	8489.95	6.87	0.81	0.61	0.106	0.23	0.81

Table 4. Static load parameters of security grid (part).

Busbar Name	Capacity/MW	L_{DP}	L_{DQ}
Wan Chenqiao22	40.5122	1.8	−2
Wan Linjiang22	100.8627	1.8	−2
Wan Shanmen21	26.27139	1.9	−2.1
Wan Puqing21	48.43346	1.9	−2.1
Wan Wenchang23	49.11798	2	−2.2
Wan Madian22	−0.48465	2	−2.2
Wan Qiaotou2A	63.72802	2.1	−2.5
Wan Qiaotou2_	67.3105	2.1	−2.5
...	...	...	...
Equivalent load	5365.01	1.88	−2.08

Table 5. Governor parameters of Shanghai power grid (part).

Unit Name/Governor Parameters	P_{EI}/MW	K_{MAXI}	P_{GENI}/MW	P_{SRI}/MW	T_O	K	T_{RH}	F_{HP}
Hu Jinshan_5	71.4	1.06	47.95	27.734	1.15	25	10	0.3
Hu Shidong_1	369.5	1.06	164	227.67	1.15	25	10	0.3
Hu Waigao_3	311.4	1.06	136.9	193.184	1.3	20	12	0.4
Hu Wure_1	336.2	1.06	230.1	126.272	1.5	16.7	14	0.5
Hu Waisan_1	1059	1.06	616.7	505.84	1.7	28	16	0.6
Hu Shangdian_2	1059	1.06	558.4	564.14	1.7	28	16	0.6
...	...	...	...	...	...	...	...	...
Equivalent generator	8370.3	1.06	4741.91	4130.68	1.51	24.68	14.06	0.50

Table 6. Induction motor parameters of Shanghai power grid (part).

Busbar Name	Capacity/MW	T_J	P_{PER}	K_L	X_S	X_R	A
Hu Hailu23	86.72697	3	1	0.8	0.067	0.17	1
Hu Haiyang22	57.55015	3	1	0.8	0.067	0.17	1
Hu Jingyi23	155.8081	5	0.9	0.7	0.087	0.2	0.9
Hu Jingfeng22	58.73741	5	0.9	0.7	0.087	0.2	0.9
Hu Shikou24	75.46927	7	0.8	0.6	0.107	0.23	0.8
Hu Shiran22	0.432161	9	0.7	0.5	0.127	0.26	0.7
...	...	...	...	...	...	...	...
Equivalent load	7303.38	6.75	0.81	0.61	0.104	0.23	0.81

Table 7. Static load parameters of Shanghai power grid (part).

Busbar Name	Capacity/MW	LDP	LDQ
Hu Zhangqiao23	155.7823	1.8	−2
Hu Dongting23	65.48597	1.9	−2.1
Hu Shenzhuan23	63.27155	2	−2.2
Hu Senlin23	71.13826	2	−2.2
Hu Hangji23	29.34815	2.1	−2.5
Hu Baobei2A	22.31632	2.1	−2.5
...	...	...	...
Equivalent load	5912.57	1.95	−2.22

Table 8. Governor parameters of Fujian power grid (part).

Unit Name/Governor Parameters	P_{EI}/MW	K_{MAXI}	P_{GENI}/MW	P_{SRI}/MW	K	F_{HP}	T_{RH}	T_O
Qingchuan Hong	1217.1	1.06	1079	211.126	25	0.3	10	1.15
Qingchuan Zhu	1217.1	1.06	1066	224.126	25	0.3	10	1.15
Houshi5	672.4	1.06	338.6	374.144	20	0.4	14	1.5
Houshi6	672.4	1.06	374.4	338.344	20	0.4	14	1.5
Jiangyin2	635.2	1.06	497.2	176.112	16.7	0.5	12	1.3
Gaoyu2	336.2	1.06	237.9	118.472	23	0.7	16	1.7
Gaoyu4	336.2	1.06	217	139.372	23	0.7	16	1.7
Zhangping6	336.2	1.06	219.7	136.672	28	0.6	18	1.9
Lianhua1	180.4	1.06	90.53	100.694	28	0.6	18	1.9
...	...	...	...	...	...	...	...	...
Equivalent generator	16,654.4	1.06	11,463.83	6189.834	21.7	0.5	13.6	1.47

Table 9. Induction motor parameters of Fujian power grid (part).

Busbar Name	Capacity/MW	T_J	P_{PER}	K_L	X_S	X_R	A
Min Qingchuan23	1.353096	3	1	0.8	0.067	0.17	1
Min Qingchuanhe	4.723641	3	1	0.8	0.067	0.17	1
Min Fuxing23	4.958399	3	1	0.8	0.067	0.17	1
Min Panlong22	146.6867	5	0.9	0.7	0.087	0.2	0.9
Min Shanfeng22	206.7169	5	0.9	0.7	0.087	0.2	0.9
Min Shangjing22	42.68303	5	0.9	0.7	0.087	0.2	0.9
Min Fuhuo24	3.775402	7	0.8	0.6	0.107	0.23	0.8
Min Gaolong23	8.835766	7	0.8	0.6	0.107	0.23	0.8
Min Gutian1A	−31.0459	7	0.8	0.6	0.107	0.23	0.8
Min Xiangshan23	73.66411	9	0.7	0.5	0.127	0.26	0.7
Min Xindian22	8.197	9	0.7	0.5	0.127	0.26	0.7
Min Xindu22	101.8102	9	0.7	0.5	0.127	0.26	0.7
...	...	...	...	...	...	...	...
Equivalent load	15,934.64	6.95	0.80	0.60	0.106	0.23	0.80

Table 10. Static load parameters of Fujian power grid (part).

Busbar Name	Capacity/MW	LDP	LDQ
Min Qingchuan23	1.353096	1.8	−2
Min Qingchuanhe	4.723641	1.8	−2
Min Fuxing 23	4.958399	1.8	−2
Min Huangli23	14.73251221	1.9	−2.1
Min Dongtai#2	−1.789323056	1.9	−2.1
Min Dongtai #1	189.0300536	1.9	−2.1
Min Fengban22	139.8962934	2	−2.2
Min Fuhuo24	3.7754017	2	−2.2
Min Gaolong23	8.835766464	2	−2.2
Min Shudou24	74.59382699	2.1	−2.3
Min Taqian22	158.8000653	2.1	−2.3
Min Tianbian23	44.59348724	2.1	−2.3
Min Tongcheng22	69.96865655	2.1	−2.3
...	...	...	...
Equivalent load	15,934.64112	1.965098481	−2.165098481

After the equivalence is completed according to the above method, the unipolar blocking fault of 'Jin-Su DC', the bipolar blocking fault of 'Jin-Su DC', the bipolar blocking fault of 'Jin-Su DC' and the bipolar blocking fault of 'Long-Zheng DC' and the load shedding fault are simulated. The frequency change of East China Power Grid with and without equivalence is monitored to verify the effectiveness of the equivalence method. The following is the graph of the frequency change of the power grid with and without the equivalence when four different faults occur in the system.

It can be seen from Figures 7–9 that under the three DC blocking conditions, the frequency changes of East China Power Grid with and without equivalence are consistent, which proves that the equivalence method has a good equivalence effect. And as shown in Figure 10, under the simulated load of −7200 MW, which is a high-frequency fault, the frequency difference of the whole network of the East China Power grid with and without equivalent is no more than 0.03 Hz, which shows that the overall equivalent effect of this equivalent method is good. It can be seen that the equivalent method of the governor-prime mover and load based on regulation performance weighting proposed in this paper has good effectiveness under small power shortage, high power shortage, low-frequency fault or high-frequency fault and can adapt to the frequency dynamic analysis of large power grid.

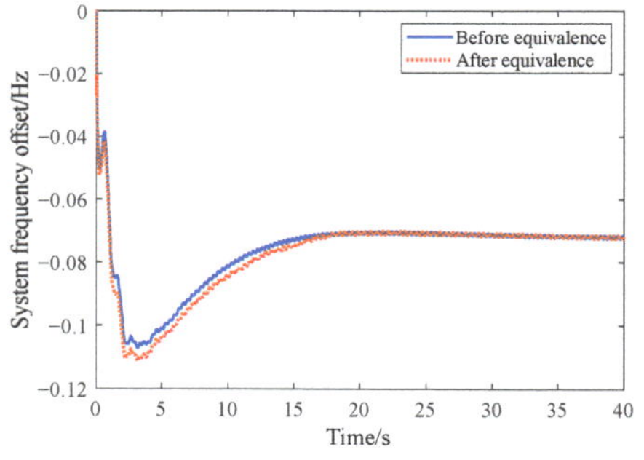

Figure 7. When simulating the unipolar blocking fault of 'Jinsu DC', the frequency change of East China Power Grid before and after equivalence is simulated.

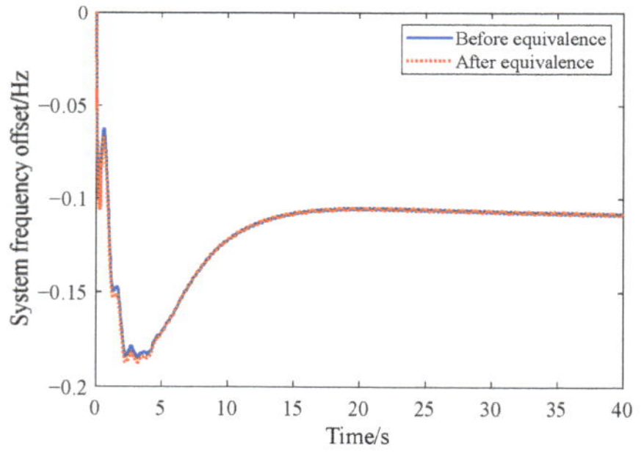

Figure 8. When simulating the bipolar blocking fault of 'Jinsu DC', the frequency change of East China Power Grid before and after equivalence is simulated.

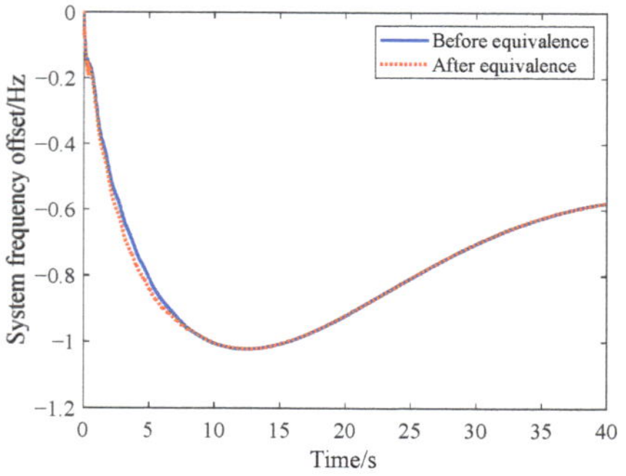

Figure 9. The frequency change of East China Power Grid before and after equivalence is simulated when two bipolar blocking faults of 'Jinsu DC' bipolar blocking and 'Longzheng DC' bipolar blocking are simulated.

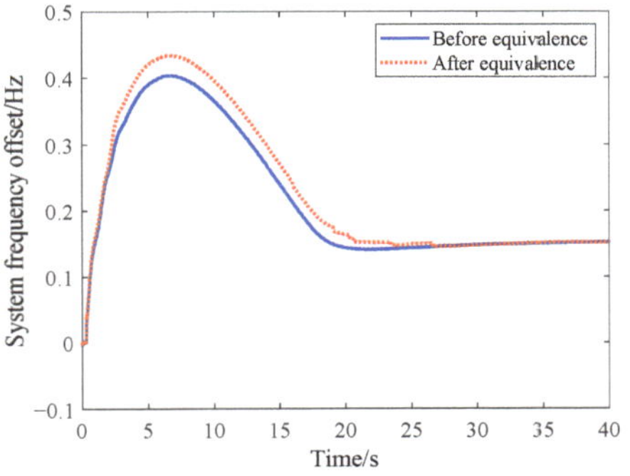

Figure 10. When the 7200 MW load fault is simulated, the frequency change of East China Power Grid before and after equivalence is simulated.

5. Conclusions

This paper studies the equivalent model suitable for large power grid frequency dynamic analysis. From the two aspects of generator and load, a method of equivalent aggregation of governor-primitive motor model and load model parameters based on regulation performance weighting is proposed. This method overcomes the uncertainty of artificial experience value, having fast aggregation speed and high engineering accuracy. Finally, the simulation verification is carried out in the example of the East China Power Grid. The results show that the equivalent method can meet the accuracy and safety requirements of the project in the dynamic analysis of large power grid frequency.

However, due to the complexity and particularity of the governor and load model, when the parameters of the governor-primitive motor are equivalently aggregated in this paper, there are fewer types of governors in the selected examples. Although the equivalent results are ideal, the case of more types of governors is not considered. In the follow-up study, a power grid with more types of governors can be selected as an example analysis to comprehensively verify the effect of the equivalent method.

Author Contributions: Conceptualization, J.Z. and J.W.; methodology, J.W.; software, J.M.; validation, J.M. and J.W.; formal analysis, C.W.; investigation, C.W.; data curation, J.M.; writing—original draft preparation, J.Z.; writing—review and editing, X.Y.; supervision, Y.Y. All authors have read and agreed to the published version of the manuscript.

Funding: This research was funded by the National Natural Science Foundation of China under grant 52077058 and The Science and Technology Project of State Grid Zhejiang Electric Power Co., Ltd. under grant 5211DS23000F.

Data Availability Statement: The original contributions presented in the study are included in the article, further inquiries can be directed to the corresponding author.

Conflicts of Interest: Authors Junchao Ma and Chenxu Wang were employed by the company State Grid Zhejiang Electric Power Co., Ltd. This employment had no influence on the study design, data analysis, results, or decision to publish. The remaining authors declare that the research was conducted in the absence of any commercial or financial relationships that could be construed as a potential conflict of interest. The authors declare that this study received funding from State Grid Zhejiang Electric Power Co., Ltd. The funder was not involved in the study design, collection, analysis, interpretation of data, the writing of this article or the decision to submit it for publication.

References

1. Milano, F.; Manjavacas, A.O. Frequency in Power Systems. In *Frequency Variations in Power Systems: Modeling, State Estimation, and Control*; John Wiley & Sons: Hoboken, NJ, USA, 2020; pp. 269–270.
2. Ju, P.; Qin, C.; Luo, J.Y.; Jiang, Y.F.; Xue, F.; Zhou, H.Q. *Frequency Security Analysis and Control of Modern Power System*; Science Press: Beijing, China, 2023.

3. Shabanikia, N.; Nia, A.A.; Tabesh, A.; Khajehoddin, S.A. Weighted Dynamic Aggregation Modeling of Induction Machine-Based Wind Farms. *IEEE Trans. Sustain. Energy* **2021**, *12*, 1604–1614. [CrossRef]
4. Di, S.; Yake, J.; Gengbo, W.; Lei, W. Research on transient modeling method and applicability of distributed wind farm. In Proceedings of the 16th IET International Conference on AC and DC Power Transmission (ACDC 2020), Online Conference, 2–3 July 2020; pp. 510–517. [CrossRef]
5. Dorile, P.O.; Jagessar, D.R.; Guardado, L.; McCann, R.A.; Ahanch, M. Grid Strength Assessment and Maximum Loadability of a Wind-Dominated Power System Through QV Modal Analysis. In Proceedings of the 2021 IEEE PES/IAS PowerAfrica, Nairobi, Kenya, 23–27 August 2021; pp. 1–5.
6. Han, M.; Zhao, H.; Cheng, J.; Niu, L.; Zheng, W. Real Time Equivalent Modeling Method for Wind Farms Based on Wind Speed Forecast. In Proceedings of the 10th Renewable Power Generation Conference (RPG 2021), Online Conference, 14–15 October 2021; pp. 1193–1202. [CrossRef]
7. Ni, Y.X.; Chen, S.S.; Zhang, B.L. *Theory and Analysis of Dynamic Power System*; Tsinghua University Press: Beijing, China, 2002; pp. 72–80.
8. Kan, W.M.; Tian, F.; Xiao, X.Q.; Deng, S.X.; Peng, B.; Wu, X.C.; Li, Y.X. Modeling and parameter identification of prime mover and governor. *China South. Power Grid Technol.* **2009**, *3*, 85–88.
9. Wang, S.B. *Research on Accurate Modeling and Parameter Identification of Generator Excitation System*; Huazhong University of Science and Technology: Wuhan, China, 2016.
10. Batchu, S.; Teeparthi, K. Transient Stability Assessment and Enhancement in the Realm of Power System Dynamic Security Analysis: A Review and Scope of Methods. In Proceedings of the 2023 4th IEEE Global Conference for Advancement in Technology (GCAT), Bangalore, India, 6–8 October 2023; pp. 1–8.
11. Marandi, V.J.; Dinavahi, V.; Strunz, K.; Martinez, J.A.; Ramirez, A. Interfacing techniques for transient stability and electromagnetic transient programs. *IEEE Trans Power Deliv.* **2009**, *24*, 2385–2395. [CrossRef]
12. Poon, K.K.; Emami, R.; Bose, A.; Abur, A. External data exchange issues for state estimation in power systems. *IEEE Trans. Power Syst.* **2012**, *27*, 849–856. [CrossRef]
13. Neto, A.C.; Rodrigues, A.B.; Prada, R.B.; da Silva, M.D.G. External equivalent for electric power distribution networks with radial topology. *IEEE Trans. Power Syst.* **2008**, *23*, 889–895. [CrossRef]
14. Ju, P.; Wu, F.; Jin, Y.Q.; Pan, X.P.; Yuan, Y.; Qin, C. *Modeling and Control of Renewable Energy Power Generation Systems*; Science Press: Beijing, China, 2014.
15. Gupta, A.P.; Mohapatra, A.; Singh, S.N. An Effective Dynamic Equivalent Model of Wind Farm Connected to Power System Network. In Proceedings of the 2019 IEEE PES Asia-Pacific Power and Energy Engineering Conference (APPEEC), Macao, China, 1–4 December 2019; pp. 1–5. [CrossRef]
16. Wang, J.; Yang, L.; Ma, Y.; Wang, J.; Tolbert, L.M.; Wang, F.; Tomsovic, K. Static and dynamic power system load emulation in a converter-based reconfigurable power grid emulator. *IEEE Trans. Power Electron.* **2016**, *31*, 3239–3251. [CrossRef]
17. Ma, Y.; Wang, Z.; Jiang, X.; Fan, B. Research on adaptive synthesis dynamic load model based on multiple model ideology. In Proceedings of the 2015 IEEE PES Asia-Pacific Power and Energy Engineering Conference (APPEEC), Brisbane, QLD, Australia, 15–18 November 2015; pp. 1–5. [CrossRef]
18. Ma, Y.T. *Parameter Identification of Complex Load Model and Its Influence on Power Grid Stability*; North China Electric Power University: Beijing, China, 2021. [CrossRef]
19. Hou, Y.; Xu, J.; Wang, J.; Wu, X. A Review on Dynamic Analysis Methods of Power System Frequency. In Proceedings of the 2019 IEEE Innovative Smart Grid Technologies—Asia (ISGT Asia), Chengdu, China, 21–24 May 2019; pp. 27–32. [CrossRef]
20. Sun, Y. *Analysis of Frequency Dynamic Characteristics of Power System Considering Wind Power Integrated Frequency Regulation*; Northeast Electric Power University: Jilin, China, 2023. [CrossRef]
21. Rainey, M.J.; Oyedokun, D.T.O. Load Modelling Effects on Power System Inertia Response. In Proceedings of the 2020 IEEE PES/IAS PowerAfrica, Nairobi, Kenya, 25–28 August 2020; pp. 1–5. [CrossRef]
22. Wang, L.; Wang, G.; Ma, G.; Yu, Q.; Li, L.; Li, X.; Gao, Y. Research on Response Characteristics of the Load. In Proceedings of the 2023 13th International Conference on Power, Energy and Electrical Engineering (CPEEE), Tokyo, Japan, 25–27 February 2023; pp. 260–264. [CrossRef]
23. China Electric Power Research Institute. *Chinese Version of BPA Transient Stability Program User Manual*; China Electric Power Research Institute: Beijingm China, 2004.
24. Guan, H.W. *Research on Load Model Parameter Identification and Grid Stability with Photovoltaic Power Generation System*; North China Electric Power University: Beijing, China, 2022. [CrossRef]

energies

MDPI

Article

Research on Safe-Economic Dispatch Strategy for Renewable Energy Power Stations Based on Game-Fairness Empowerment

Zhen Zhang [1], Wenjun Xian [1], Weijun Tan [1], Jinghua Li [1] and Xiaofeng Liu [2,*]

[1] Qinghai Electric Power Dispatching Control Center, Xining 810001, China; zhangzhen.qh@163.com (Z.Z.); 15110997757@139.com (W.X.); twjxing@163.com (W.T.); ljh-741741@163.com (J.L.)
[2] School of Electrical and Automation Engineering, Nanjing Normal University, Nanjing 210000, China
* Correspondence: liuxiaofeng@njnu.edu.cn

Abstract: The optimal dispatching of renewable energy power stations is particularly crucial in scenarios where the stations face energy rationing due to the large proportion of renewable energy integrated into the power system. In order to achieve safe, economical, and fair scheduling of renewable energy power stations, this paper proposes a two-stage scheduling framework. Specifically, in the initial stage, the maximum consumption space of renewable energy for the system can be optimized by optimizing the formulated safe-economic dispatch model. In the second stage, the fair allocation mechanism of renewable energy power stations is proposed based on the game-fairness empowerment approach. In order to obtain a comprehensive evaluation of renewable energy power stations, an evaluation index system is constructed considering equipment performance, output characteristics, reliability, flexibility, and economy. Subsequently, the cooperative game weighting method is proposed to rank the performance of renewable energy power stations as the basis for fair dispatching. Simulation results show that the proposed scheduling strategy can effectively ensure the priority of renewable energy power stations based on their comprehensive ranking, and improve the safety, economy, and fairness of power station participation in scheduling.

Keywords: renewable energy; fair dispatching; safe-economic dispatching; cooperative game; evaluation index

1. Introduction

With the gradual depletion of traditional energy resources and the escalating severity of environmental issues, the exploration and utilization of renewable energy have become a global priority [1,2]. Wind and solar energy, as important components of renewable energy, have witnessed rapid development in recent years [3,4]. According to the annual report "Renewable Capacity Statistics 2024" by the International Renewable Energy Agency, the cumulative installed capacity of global renewable energy has reached 3.87 TW. Notably, in China, by the end of July 2024, the installed capacity of photovoltaic (PV) and wind turbine (WT) power generation amounted to 1.206 billion kW. Based on the development plan for renewable energy, China anticipates that the proportion of renewable energy in power generation will surpass 50% by 2060, thereby becoming the primary pillar of the country's energy power supply.

However, the intermittency and uncertainty of renewable energy sources pose new challenges to the safe and stable operation of the power grid [5,6]. Effectively integrating and making full use of renewable energy has become an urgent problem to solve. It is particularly important to establish a scientific and efficient scheduling scheme. Currently, research on the dispatching of renewable energy power generation mainly focuses on exploring more flexible and effective dispatching methods, including economic dispatching [7–10], optimal dispatching considering the uncertainty of renewable energy output [11,12], and dispatching strategies aimed at maximizing the utilization of renewable energy [13,14]. For example, a study in [15] introduces and exemplifies a probabilistic methodology aimed at addressing the unit commitment challenge. This approach revolves around resolving the economic dispatch problem, incorporating the probabilistic characterization of variables such as thermal generator output power distributions, unmet energy demands, electricity surpluses, generation expenses, and spinning reserve capacities. The authors of [16] propose a new reliable power flow optimization tool to enhance the utilization level of WT, and a simulation based on the IEEE 39-bus system is conducted to verify that the proposed method maximizes the utilization level of WT without affecting system safety. The work presented in [17] advances a coordinated day-ahead and intraday optimal dispatch framework tailored specifically for PV and WT power generation, acknowledging the inherent uncertainty in energy sources and loads. Furthermore, the authors of [18] contribute a novel dispatch strategy labeled as modified predictive dispatch, which innovatively integrates proactive battery charging/discharging strategies and demand-side management tactics, premised on anticipated short-term electricity surplus profiles. The implementation of this strategy has been demonstrated to significantly enhance the self-utilization index of excess renewable energy, lifting it from under 25% to approximately 90%, thereby highlighting its potential to optimize energy utilization and reduce waste.

Although there have been a large number of studies on renewable energy dispatching, there is still a lack of systematic research on the fair scheduling of renewable energy power stations. It is very important to dispatch the output of each renewable energy station in a fair and reasonable manner as renewable energy stations emerge. When a large proportion of renewable energy is integrated into the energy system, the stations may face energy rationing in order to guarantee the stable operation of the system due to the fluctuation of renewable energy. Especially in China, there is a widespread phenomenon of energy rationing in renewable energy stations, and how these stations can participate in dispatching fairly has become a key issue. At present, fair dispatching mainly concentrates on ensuring that all power stations have relatively equal opportunities to meet their annual contract energy targets based on their installed capacities. For example, in [19], to address the fairness of consumption across various field groups and sections in PV and WT, a multi-level active power control strategy has been developed that takes into account the weighted consumption capacity of different sections to ensure equitable distribution and utilization of generated power. In [20], the authors develop an optimal dispatch model for a WT-PV thermal integrated system, focusing on economic efficiency and equity. This model is designed to tackle the challenges associated with large-scale PV and WT, ensuring the economic viability of system operations while achieving fair scheduling among renewable energy power stations. The authors of [21] formulate an optimal dispatch model incorporating WT, utilizing the Gini coefficient from economics as a fairness constraint, with power output serving as the metric. Nevertheless, this approach primarily addresses the fairness of generation stations for conventional units within a regional system and does not sufficiently explore dispatch fairness for renewable energy power stations.

However, although some scheduling models and fairness evaluation criteria have been proposed in the above studies, existing research still has certain limitations. Especially when facing multiple influencing factors, how to achieve priority dispatching of renewable energy power stations while ensuring fair competition among stations remains a complex and unsolved problem. Due to differences in resource conditions and power station facilities, there are significant variations in the actual power generation capacity for different renewable energy stations. Relying solely on the completion status of annual contract energy to measure the effectiveness of fair dispatching is clearly not comprehensive enough. Furthermore, it is critical for energy power systems to guarantee safe operation as well as economic and fair dispatching. Therefore, in this context, this paper mainly focuses on the optimal dispatch strategy for renewable energy power stations from the perspectives of security, economy, and fairness to better promote the development of renewable energy and ensure the stability of the power system. The research results not only contribute to improving the utilization rate of renewable energy but also provide strong support for achieving the country's long-term energy transition goals. Table 1 shows the comparison result of the proposed approach of this paper with recent works.

Table 1. Comparative result of the proposed approach with recent works.

Reference	Economy	Safety	Fairness	Main Concerns
[7]	Yes	Yes	No	Economic dispatching considering retrofitting thermal power plants
[11]	Yes	No	No	Dispatching strategy considering the uncertainty of renewable energy
[13]	Yes	No	No	Dispatching strategy aimed at maximizing the utilization of renewable energy
[15]	Yes	No	No	Economic dispatching considering the unit commitment challenge
[18]	Yes	Yes	No	Predictive dispatch considering charging/discharging of battery bank
[21]	Yes	No	Yes	Fairness dispatch considering generation stations for conventional units
This paper	Yes	Yes	Yes	Dispatching strategy considering the safety, economy, and fairness of renewable energy stations

In summary, the contributions of this paper are as follows:

(1) A two-stage scheduling framework is proposed for an energy power system with a high proportion of renewable energy, wherein the initial dispatching plan is optimized from the safe economic level and then the output of renewable energy power stations is fairly allocated based on game-fairness empowerment.

(2) An evaluation indicator system is established for the operation state of renewable energy power stations from the five dimensions of equipment performance, output characteristics, reliability, flexibility, and economy, which serves the fair dispatching needs of renewable energy stations.

(3) A bi-level evaluation approach is proposed for the ranking of renewable energy power stations in fair dispatching, in which a cooperative game method is formulated to combine the subjective weights determined by the anti-entropy weight (AEW) and objective weights determined by the attribute hierarchy model (AHM).

The rest of this paper is organized as follows. The scheduling framework is introduced in Section 2. The safe-economic dispatch strategy for energy power systems is formulated in Section 3. In Section 4, a fair dispatching strategy is proposed among renewable energy stations. A case study is presented in Section 5. Finally, conclusions are drawn in Section 6.

2. Two-Stage Scheduling Framework for Renewable Energy Power Stations

In the proposed scenario, we mainly focus on the dispatch problem of energy power systems when renewable energy stations face energy rationing. Accordingly, a two-stage dispatch strategy framework for the output of renewable energy power stations is designed to address how large-scale renewable energy power stations can participate in safe, economic, and fair system scheduling, as illustrated in Figure 1.

The initial stage of dispatching is conducted from the level of safety and economy. Based on the predicted output of PV-WT and load demand, economic scheduling is optimized with the goal of minimizing the overall system operating cost. Meanwhile, constraints such as system power flow and voltage are considered to achieve safe scheduling. By solving the optimization model, the optimal output of each power station can be obtained under the requirements of safety and economy. Accordingly, the maximum consumption space of renewable energy for the system can be calculated by accumulating the optimal output of all PV-WT stations.

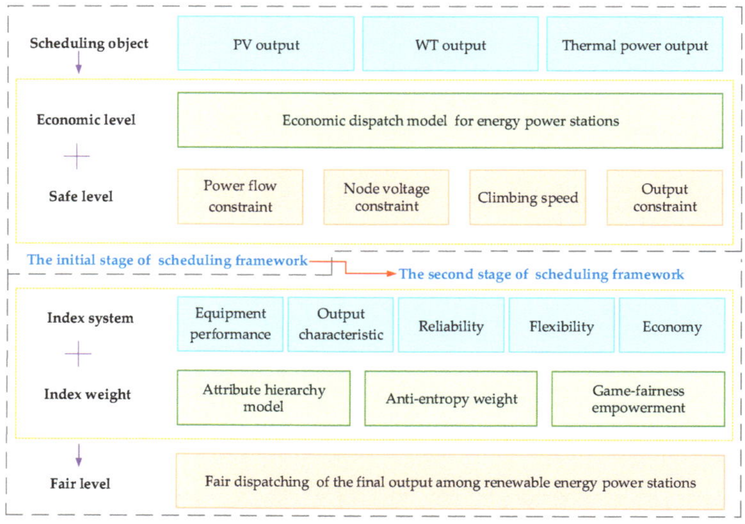

Figure 1. Two-stage scheduling framework for renewable energy power stations.

The second stage of dispatching is conducted from the level of fairness. With the high proportion of renewable energy integrated into the power system leading to energy rationing on PV-WT, the maximum consumption space of renewable energy will be less than the total predicted value of all stations. Therefore, the allocation of PV-WT's energy output is conducted. In order to realize fair allocation, an evaluation index system for renewable energy power stations is constructed from five dimensions: equipment performance, output characteristics, reliability, flexibility, and economy. The subjective and objective weights are determined using the AEW and AHM methods, respectively. Furthermore, the game-fairness empowerment approach is proposed to empower the comprehensive weighs by treating subjective and objective weights as the main players in the game. Here, a cooperative game is chosen to derive the weights for ranking each renewable energy power station. Through the two-stage optimal scheduling, the dispatching of renewable energy power stations can be ensured under the safe, economical, and fair mode.

3. Safe-Economic Dispatch Strategy for Energy Power Stations

For energy power systems with a high proportion of renewable energy sources, the goal of energy dispatching is to achieve optimal economic operation while ensuring system safety. At the economic aspect, in order to promote the utilization of PV and WT, the economic cost is mainly considered from the generation cost of thermal power units and the cost of abandoning PV and WT [7]. Accordingly, the dispatching objective function can be expressed as

$$\min\ C_{\text{total}} = \sum_{t=1}^{T} (C_{\text{TH},t} + C_{\text{PV},t} + C_{\text{WT},t}) \tag{1}$$

where T is the scheduling time period, and $C_{\text{TH},t}$, $C_{\text{PV},t}$, and $C_{\text{WT},t}$ are the generation cost of thermal power units, the cost of abandoning PV, and the cost of abandoning WT during time period t, respectively. The cost models are defined as [22]

$$C_{\text{TH},t} = \sum_{k_{\text{TH}}=1}^{N_{\text{TH}}} \left(a_{k_{\text{TH}}} \left(P_{\text{TH},t}^{k_{\text{TH}}} \right)^2 + b_{k_{\text{TH}}} P_{\text{TH},t}^{k_{\text{TH}}} + c_{k_{\text{TH}}} \right) \tag{2}$$

$$C_{\text{PV},t} = \sum_{k_{\text{PV}}=1}^{N_{\text{PV}}} \lambda_{\text{PV}} \left(\alpha_{\text{PV}} P_{\text{PVP},t}^{k_{\text{PV}}} - P_{\text{PV},t}^{k_{\text{PV}}} \right) \tag{3}$$

$$C_{\text{WT},t} = \sum_{k_{\text{WT}}=1}^{N_{\text{WT}}} \lambda_{\text{WT}} \left(\alpha_{\text{WT}} P_{\text{WTP},t}^{k_{\text{WT}}} - P_{\text{WT},t}^{k_{\text{WT}}} \right) \tag{4}$$

where N_{TH} refers to the number of thermal power units participating in the dispatch of the regional power grid; N_{PV} and N_{WT} refer to the number of WTs and PVs participating in the dispatch of the regional power grid; $a_{k_{\text{TH}}}$, $b_{k_{\text{TH}}}$, and $c_{k_{\text{TH}}}$ are the generation cost coefficients for the given thermal power unit k_{TH}; $P_{\text{PVP},t}^{k_{\text{PV}}}$ and $P_{\text{WTP},t}^{k_{\text{WT}}}$ are the predicted output values for PV k_{PV} and WT k_{TH} at time t; $P_{\text{PV},t}^{k_{\text{PV}}}$ and $P_{\text{WT},t}^{k_{\text{WT}}}$ are the dispatching plan for PV k_{PV} and WT k_{TH} at time t; λ_{PV} and λ_{WT} are the unit cost of abandoning PV and WT; α_{PV} and α_{WT} are the prediction error coefficients for the energy output of PV and WT; and $\alpha_{\text{PV}} P_{\text{PVP},t}^{k_{\text{PV}}}$ and $\alpha_{\text{WT}} P_{\text{WTP},t}^{k_{\text{WT}}}$ represent the actual energy output ability for PV and WT at time t. Here, the prediction error coefficients are mainly used to correct the prediction of PV-WT's energy output. Since the energy output of PV-WT is hard to predict precisely, the actual output ability for PV-WT needs to be regulated with the prediction error coefficient in order to more accurately calculate the abandoned amount of PT-WT.

At the safe level, in order to ensure the safe operation of the system, the optimization problem in (1) has to satisfy various constraints, such as power flow constraints, voltage security constraints, and safety operation constraints of generations [23,24]. Concretely,

(1) Equation constraint of power flow

The equation constraint of power flow is mainly used to describe the specific relationship between the voltage, active power, and reactive power of each node, in order to ensure the safe and stable operation of the system.

$$\begin{cases} P_{\text{TH},i,t} + P_{\text{PV},i,t} + P_{\text{WT},i,t} - P_{\text{LD},i,t} = \\ U_{i,t} \sum\limits_{j=1}^{n} U_{j,t} (G_{ij} \cos \delta_{ij,t} + B_{ij} \sin \delta_{ij,t}) \\ Q_{\text{TH},i,t} + Q_{\text{PV},i,t} + Q_{\text{WT},i,t} - Q_{\text{LD},i,t} = \\ U_{i,t} \sum\limits_{j=1}^{n} U_{j,t} (G_{ij} \sin \delta_{ij,t} - B_{ij} \cos \delta_{ij,t}) \end{cases} \tag{5}$$

where $P_{\text{TH},i,t}$, $P_{\text{PV},i,t}$, $P_{\text{WT},i,t}$ and $P_{\text{LD},i,t}$ are the active power of thermal power units, PV, WT, and load at the i-th node time t, $Q_{\text{TH},i,t}$, $Q_{\text{PV},i,t}$, $Q_{\text{WT},i,t}$, and $Q_{\text{LD},i,t}$ are the reactive power, G_{ij}, B_{ij}, and δ_{ij} are the conductance, reactance, and phase angle between nodes i and j.

(2) Safety constraint of node voltage

The safety constraint of node voltage refers to the requirement that the voltage of each node in the power system should be maintained within a certain allowable range that is mainly determined based on the stable operation requirements of the power system.

$$U_{i,\min} \leq U_i \leq U_{i,\max} \; i = 1, 2, \cdots N_{\text{ND}} \tag{6}$$

where $U_{i,\min}$ and $U_{i,\max}$ are the lower and upper limits of the voltage at node i, U_i is the voltage amplitude of node i, and N_{ND} is the number of system nodes. Generally, the lower/upper permissible voltages are set to 0.9 pu and 1.1 pu, respectively [25].

(3) Output constraint of renewable energy power station

In the dispatching of the power system, the dispatching plan for PV and WT should not be higher than the predicted value. That is,

$$\begin{cases} P_{\text{PV},t}^{k_{\text{PV}}} \leq P_{\text{PVP},t}^{k_{\text{PV}}} \\ P_{\text{WT},t}^{k_{\text{WT}}} \leq P_{\text{WTP},t}^{k_{\text{WT}}} \end{cases} \tag{7}$$

(4) Output constraint of thermal power units

Considering the capacity limitation of the generator units, the output of thermal power units must be within a certain range. That is,

$$P_{\mathrm{TH,min}}^{k_{\mathrm{TH}}} \le P_{\mathrm{TH},t}^{k_{\mathrm{TH}}} \le P_{\mathrm{TH,max}}^{k_{\mathrm{TH}}} \quad (8)$$

where $P_{\mathrm{TH,min}}^{k_{\mathrm{TH}}}$ and $P_{\mathrm{TH,max}}^{k_{\mathrm{TH}}}$ are the lower and upper limits of the output of thermal power unit k_{TH}, respectively.

(5) Climbing speed constraint of thermal power units

The climbing speed constraint mainly refers to the maximum output power that the generation unit can increase or decrease per unit of time, which is usually influenced by factors such as the physical design of the unit, fuel supply, etc.

$$\begin{cases} P_{\mathrm{TH},t}^{k_{\mathrm{TH}}} - P_{\mathrm{TH},t-1}^{k_{\mathrm{TH}}} \le \mathrm{UP}_{\mathrm{TH}}^{k_{\mathrm{TH}}} \\ P_{\mathrm{TH},t-1}^{k_{\mathrm{TH}}} - P_{\mathrm{TH},t}^{k_{\mathrm{TH}}} \le \mathrm{DO}_{\mathrm{TH}}^{k_{\mathrm{TH}}} \end{cases} \quad (9)$$

where $\mathrm{UP}_{\mathrm{TH}}^{k_{\mathrm{TH}}}$ and $\mathrm{DO}_{\mathrm{TH}}^{k_{\mathrm{TH}}}$ are the maximum upward and downward climbing power of thermal power unit k_{TH}, respectively.

4. Fair Dispatch Strategy Based on Cooperative Game Empowerment

4.1. Indicator System of Renewable Energy Power Stations

The evaluation objective of traditional fair scheduling only focuses on the completion rate of annual power generation plans, neglecting the execution process. Given the differences in equipment, output characteristics, and other factors among renewable energy stations, adhering to the principle of unified allocation may unfairly affect the rights of certain stations. Therefore, a comprehensive indicator system is proposed for the fair scheduling of renewable energy power stations, as shown in Figure 2 [20,26,27]. This indicator system can identify and quantify the differentiated impacts of PV and WT to ensure fair allocation of energy output.

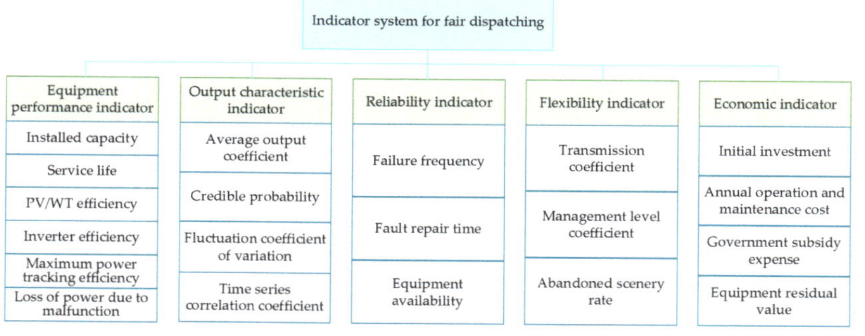

Figure 2. Indicator system of renewable energy power stations.

(1) Equipment performance indicator

The equipment performance indicator refers to the evaluation of renewable energy station status, mainly including installed capacity, unit performance, conversion efficiency, and related losses. They are quantified through device state coefficients, which indicate the significant impact of device performance changes on renewable energy generation capacity. In addition, they play a crucial role in determining the fairness of power grid scheduling. The comprehensive calculation model for equipment performance indicators is summarized as follows:

$$\begin{cases} I_{\mathrm{C}} = \dfrac{IP_{k_{\mathrm{PV/WT}}}}{\sum\limits_{k_{\mathrm{PV/WT}}=1}^{N_{\mathrm{PV/WT}}} IP_{k_{\mathrm{PV/WT}}}} \quad I_{\mathrm{L}} = \dfrac{T_{k_{\mathrm{PV/WT}}}}{\sum\limits_{k_{\mathrm{PV/WT}}=1}^{N_{\mathrm{PV/WT}}} T_{k_{\mathrm{PV/WT}}}} \\ I_{\mathrm{E}} = I_{\mathrm{C}} I_{\mathrm{L}} \eta_1 \eta_2 \eta_3 (1 - \eta_4) \end{cases} \tag{10}$$

where I_{C}, I_{L}, and I_{E} are the installed capacity coefficient, service life coefficient, and equipment performance; $IP_{k_{\mathrm{PV/WT}}}$ and $T_{k_{\mathrm{PV/WT}}}$ are the installed capacity and service life of the renewable energy power station $k_{\mathrm{PV/WT}}$ (here, $k_{\mathrm{PV/WT}}$ represents k_{PV} and k_{WT}, $N_{\mathrm{PV/WT}}$ represents N_{PV} and N_{WT}); η_1, η_2, and η_3 are the PV/WT efficiency, inverter efficiency, and maximum power tracking efficiency; and η_4 is the percentage of power reduction caused by fault losses.

(2) Output characteristic indicator

The output characteristic indicator refers to the description of the output characteristics of renewable energy stations, mainly including average output, credible probability, fluctuation degree, and temporal correlation. The higher the average output, the higher the effective power level. The credible probability can serve as a robust indicator of enhanced reliability. Similarly, the fluctuations in output directly correlate with an increase in stability. The intensity of temporal correlation directly influences the precision of predictive outcomes, with stronger correlations yielding higher accuracy in forecasting results. They are quantified through output characteristic indicators, which indicate the significant impact of renewable energy output characteristics on power generation capacity. Accordingly, the output characteristic indicator can be described as follows:

$$I_{\mathrm{O}} = \gamma_1 \gamma_2 (1 - \gamma_3) \gamma_4 \tag{11}$$

where I_{O} is the output characteristic coefficient, and γ_1, γ_2, γ_3, and γ_4 are the average output coefficient, credible probability, fluctuation coefficient of variation, and time series correlation coefficient, respectively.

(3) Reliability indicator

The reliability indicator refers to the evaluation of the reliability level of system operation, mainly including the number of failures, failure repair time, and availability. A reduction in the frequency of failures is directly proportional to a decrease in repair duration, which subsequently leads to an increase in system availability. In essence, minimizing failure occurrences not only expedites recovery processes but also fortifies the dependability of the system, thereby enhancing its overall performance and trustworthiness. Expressed in terms of the reliability coefficient, its calculation model is as follows:

$$I_{\mathrm{R}} = (1 - N_{\mathrm{FA}} T_{\mathrm{FA}}) \rho \tag{12}$$

where I_{R} is the reliability coefficient, N_{FA} is the number of malfunctions, T_{FA} is the fault repair time, and ρ is the equipment availability.

(4) Flexibility indicator

The flexibility indicator refers to the evaluation of the space for renewable energy consumption, mainly including network architecture, management level, and abandoned wind and solar power. The enhanced transmission capacity of the power grid is positively correlated with an increased proportion of flexible power sources, which in turn elevates the efficiency of dispatch and management. This synergistic enhancement contributes to a reduction in the rate of abandoned wind and solar power, thereby significantly improving the overall consumption scenario of renewable energy sources. Represented by a flexibility coefficient, its calculation model is as follows:

$$I_{\mathrm{F}} = \theta_{\mathrm{g}} \theta_{\mathrm{m}} (1 - \theta_{\mathrm{a}}) \tag{13}$$

where θ_g is the transmission coefficient for the power grid, θ_m is the management level coefficient, and θ_a is the abandoned scenery rate.

(5) Economic indicator

The economic indicator refers to the consideration of investment costs for renewable energy power stations, mainly including initial investment, annual operation and maintenance costs, government subsidy expenses, and equipment residual value. The economic indicator is a negative indicator, which means that the higher the cost of renewable energy stations, the lower the corresponding evaluation value. The calculation model is as follows:

$$I_E = E_I + (E_O - E_G)T_{k_{PV/WT}} - E_R \tag{14}$$

where E_I, E_O, E_G, and E_R represent initial investment, annual operation and maintenance costs, government subsidy expense, and equipment residual value.

4.2. Evaluation Method Based on Cooperative Game

Based on the above indicators, the weights of the indicators need to be calculated. In this paper, the objective weights of the indicators are determined using the anti-entropy weight (AEW) method [28,29] and the subjective weights are determined with the attribute hierarchy model (AHM) method [30]. The role of AEW is to optimize the weight distribution of each indicator, minimize entropy or uncertainty, and ensure that the weights reflect the relative importance of each indicator in the overall evaluation. Meanwhile, the AHM effectively incorporates expert evaluations into the decision-making process, ensuring that their subjective evaluations are consistently incorporated into the weighted system while maintaining objectivity through entropy minimization. Finally, the game-fairness empowerment approach is used to construct an optimization solution model to obtain the comprehensive weights.

(1) Objective weighting method based on AEW

Entropy is a concept used in thermodynamic systems to measure the degree of disorder in the system, which is later introduced into information theory. Assuming that there are M possible states in the system, and the probability of each possible state occurring is p_m (m = 1,2,. . .M), entropy can be defined as follows:

$$h = -\sum_{m=1}^{M} p_m \ln(p_m) \tag{15}$$

where $0 \leq p_m \leq 1$.

Under the AEW method, assuming that for the multi-attribute decision-making problem, the number of evaluation objects is M and the number of indicators is O, the indicator values can be expressed as $x_{o,m}$, and the corresponding decision matrix can be written as

$$\mathbf{X} = [x_{o,m}]_{O\times M} \tag{16}$$

Therefore, the inverse entropy values of each indicator can be expressed as

$$h_o = -\sum_{m=1}^{M} r_{o,m} \ln(1 - r_{o,m}) \tag{17}$$

where $r_{o,m} = x_{o,m}/\sum_m x_{o,m}$ represents the dimensionless value of the indicator value. By normalizing the inverse entropy value of the indicator, the AEW weight of the indicator can be obtained:

$$w_{1o} = h_o/\sum_{o=1}^{O} h_o \tag{18}$$

(2) Subjective weighting method based on the AHM

The AHM simplifies the weight calculation process without the need for consistency checks, making it easier and more practical to obtain subjective weights of indicators. The subjective weight calculation of indicators using the AHM is divided into the following three steps:

(a) Construct a comparative judgment matrix

Using a 1–9 scale method to represent the relative importance of indicators, a comparative judgment matrix $\boldsymbol{A} = [a_{o,o'}]_{O\times O}$ is obtained by the expert scoring method.

(b) Calculate the attribute judgment matrix

Convert the comparison judgment matrix $\boldsymbol{A} = [a_{o,o'}]_{O\times O}$ into an attribute judgment matrix $\boldsymbol{B} = [b_{o,o'}]_{O\times O}$ using the following equation

$$b_{o,o'} = \begin{cases} \frac{R}{2R+1} & a_{o,o'} = R, o \neq o' \\ \frac{1}{2R+1} & a_{o,o'} = \frac{1}{R}, o \neq o' \\ 0.5 & a_{o,o'} = 1, o \neq o' \\ 0 & a_{o,o'} = 1, o = o' \end{cases} \tag{19}$$

where R is a positive integer greater than or equal to 2.

(c) Determine subjective weights

The subjective weights of indicators are calculated as

$$w_{2o} = \frac{2}{O(O-1)} \sum_{o'=1}^{O} b_{o,o'} \tag{20}$$

(3) Comprehensive weights based on the game-fairness empowerment approach

In order to realize the equilibrium between the subjective weights and the objective weights, the assignment of evaluation index weights will be made more scientific and reasonable. Combine the indicator weight vectors $\boldsymbol{W}_1$ and $\boldsymbol{W}_2$ obtained by AEW and the AHM in any linear way

$$\boldsymbol{W} = \alpha_1 \boldsymbol{W}_1^{\mathrm{T}} + \alpha_2 \boldsymbol{W}_2^{\mathrm{T}} \tag{21}$$

where $\boldsymbol{W}$ is the combined weight vector, α_1 and α_2 are the combination coefficients.

Based on cooperative game theory [31–33], the game-fairness empowerment approach is used to determine the comprehensive weights by treating subjective and objective weights as the main players in the game. The combination coefficients α_1 and α_2 are optimized with the objective of minimizing the deviation between the combination weights $\boldsymbol{W}$ and $\boldsymbol{W}_1$ and $\boldsymbol{W}_2$. The objective function is

$$\min \left\| \sum_{s=1}^{2} \alpha_s \mathbf{W}_s^{\mathrm{T}} - \mathbf{W}_s \right\|_2 \tag{22}$$

If the equilibrium solution of the optimal problem is (α_1^*, α_2^*), then the combination weights based on a cooperative game can be expressed as:

$$\boldsymbol{W} = \frac{\alpha_1^*}{\alpha_1^* + \alpha_2^*} \boldsymbol{W}_1^{\mathrm{T}} + \frac{\alpha_2^*}{\alpha_1^* + \alpha_2^*} \boldsymbol{W}_2^{\mathrm{T}} \tag{23}$$

Since cooperative game theory emphasizes collective rationality and focuses on the fair distribution of benefits in decision-making schemes, the equilibrium solution of the formulated cooperative game model can achieve a balance between subjective and objective weights, thereby minimizing potential biases in weights. Therefore, the comprehensive weights determined by the game-fairness empowerment approach can aggregate the advantages of AEW and the AHM.

4.3. Fair Allocation of Renewable Energy Power Stations

In the economic dispatch model, the maximum consumption space of the renewable energy power stations is obtained, and then the dispatch plan is allocated to obtain the final energy output of the stations. Considering various indicators such as equipment performance and output characteristics, comprehensive weights based on game-fairness empowerment can ensure that power stations with superior performance receive more energy output, while power stations with poor performance receive fewer scheduling plans. Therefore, allocating the scheduling plan of renewable energy power stations based on comprehensive ranking priority can ensure fairness. The fair allocation process among renewable energy power stations is as follows.

(1) Based on the renewable energy consumption space obtained from safe-economic dispatch, the power station dispatch plan is preliminarily allocated according to the weights calculated by each power station.

$$G_{PV/WT,t}^{k_{PV/WT}} = \frac{W_{k_{PV/WT}} P_{PVP,t}^{k_{PV/WT}}}{\sum\limits_{k_{PV/WT}=1}^{N_{PV/WT}} W_{k_{PV/WT}} P_{PVP,t}^{k_{PV/WT}}} \sum_{k_{PV/WT}=1}^{N_{PV/WT}} P_{PV/WT,t}^{k_{PV/WT}} \tag{24}$$

where $G_{PV/WT,t}^{k_{PV/WT}}$ is the initial fair distribution plan calculated based on weights for the power station, and $W_{k_{PV/WT}}$ is the weight of PV k_{PV} or WT k_{WT}.

(2) Comparing the preliminary allocation plan of the power station with the predicted output value at time t, the power stations participating in scheduling at time t can be further divided into two categories:

(a) The first category is $G_{PV/WT,t}^{k_{PV/WT}} \geq P_{PVP,t}^{k_{PV/WT}}$, and the scheduling plan ultimately executed by the power station is $G_{PV/WT,t}^{k_{PV/WT}} = P_{PVP,t}^{k_{PV/WT}}$. Moreover, the remaining consumption plan of the power grid at time t is equal to

$$S_{PV/WT,t}^{k_{PV/WT}} = G_{PV/WT,t}^{k_{PV/WT}} - P_{PVP,t}^{k_{PV/WT}} \tag{25}$$

(b) The second category is $G_{PV/WT,t}^{k_{PV/WT}} < P_{PVP,t}^{k_{PV/WT}}$, and the power station needs to take $G_{PV/WT,t}^{k_{PV/WT}}$ as the intermediate scheduling plan. The remaining power generation space of the power station at time t can be described as

$$R_{PV/WT,t}^{k_{PV/WT}} = P_{PVP,t}^{k_{PV/WT}} - G_{PV/WT,t}^{k_{PV/WT}} \tag{26}$$

(3) According to the number of power stations in the second category, the remaining consumption plan is reallocated to obtain the adjustment quota

$$F_{PV/WT,t}^{k_{PV/WT}} = \frac{W_{k_{PV/WT}} R_{PVP,t}^{k_{PV/WT}}}{\sum\limits_{k_{PV/WT}=1}^{N_{PV/WT-2}} W_{k_{PV/WT}} R_{PVP,t}^{k_{PV/WT}}} \sum_{k_{PV/WT}=1}^{N_{PV/WT-1}} S_{PV/WT,t}^{k_{PV/WT}} \tag{27}$$

where $N_{PV/WT-1}$ and $N_{PV/WT-2}$ are the number of first and second-category power stations after initial allocation. The final fair scheduling plan for the second type of power station can be contained as follows:

$$GF_{PV/WT,t}^{k_{PV/WT}} = G_{PV/WT,t}^{k_{PV/WT}} + F_{PVP,t}^{k_{PV/WT}} \tag{28}$$

Based on the above allocation principle among renewable energy power stations, the final scheduling scheme can guarantee the dispatch strategy of stations in a fair manner.

5. Case Study

5.1. Simulation Parameters

Based on the IEEE 39-bus system, the high proportion of renewable energy power systems is taken as an example to verify the effectiveness of the proposed approach. The system contains seven renewable energy power stations and seven thermal power units, as shown in Figure 3. Tables 2–4 illustrate the parameters of renewable energy stations and thermal power units. PV_1~PV_4 are four PV stations and WT_1~WT_3 are three WT stations, with PV_1, PV_3, and WT_2 having been in operation for 5 years and the other stations having been in operation for 3 years. The parameters of thermal power units G_1~G_7 are presented in Table 4, containing the consumption characteristic coefficients, output upper and lower limits, and ramp rate parameters. It should be noted that the energy power system based on the IEEE 39-bus is taken as an example to show the effectiveness of the proposed approach. However, the proposed approach is not only limited to the systems with the IEEE 39-bus. It can still be applied to different regional grids or systems with higher variability in renewable energy types. If the system has other renewable energy types, the index system needs to be adjusted according to the new energy types, and other dispatching processes are the same.

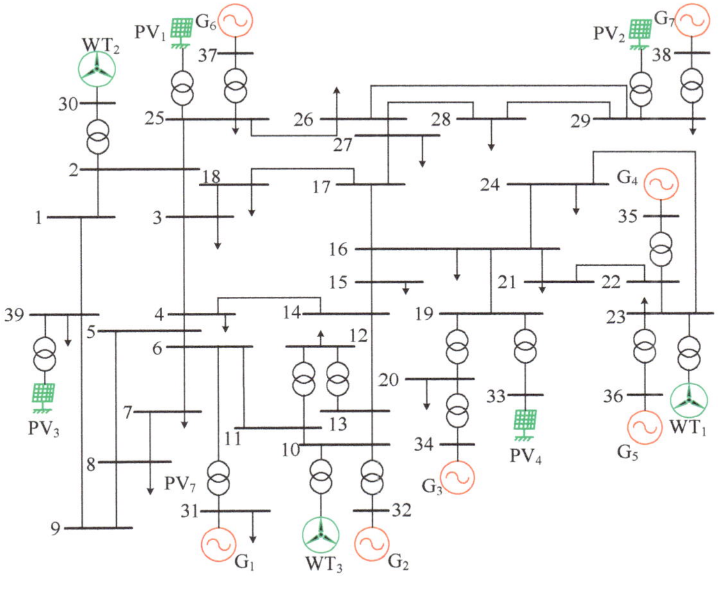

Figure 3. The energy power system based on IEEE 39-bus.

Table 2. Parameters of PV stations.

	Indicator Parameters	PV_1	PV_2	PV_3	PV_4
Equipment performance indicator	Installed capacity	830 MW	830 MW	950 MW	590 MW
	Installed capacity coefficient	0.1656	0.1656	0.1895	0.1177
	Service life	0.1290	0.1935	0.1613	0.1290
	PV/WT efficiency	15%	17.50%	17%	15%
	Inverter efficiency	96.50%	97%	96.80%	96%
	Maximum power tracking efficiency	98.00%	99.00%	99.00%	98.00%
	Loss of power due to malfunction	3%	1.20%	1.90%	2.60%
Output characteristic indicator	Average output coefficient	20.00%	28.00%	25.00%	21.00%
	Credible probability	7%	9%	8.50%	7.50%
	Fluctuation coefficient of variation	80.00%	90.00%	85.00%	85.00%
	Time series correlation coefficient	50%	65.00%	60.00%	55.00%
Reliability indicator	Failure frequency	19.1	13.2	15.4	18.8
	Fault repair time	0.025	0.026	0.027	0.024
	Equipment availability	55.00%	55.00%	60.00%	50.00%
Flexibility indicator	Transmission coefficient	20.00%	25.00%	25.00%	20.00%
	Management level coefficient	50.00%	50.00%	50.00%	30.00%
	Abandoned scenery rate	5%	7%	6%	7%

Table 2. *Cont.*

	Indicator Parameters	PV_1	PV_2	PV_3	PV_4
Economic indicator	Initial investment	500	480	550	330
	Annual operation and maintenance cost	7	8	8	6
	Government subsidy expense	15	15	16	14
	Equipment residual value	50	55	70	35

Table 3. Parameters of WT stations.

	Indicator Parameters	WT_1	WT_2	WT_3
Equipment performance indicator	Installed capacity	650 MW	312 MW	850 MW
	Installed capacity coefficient	0.1297	0.0623	0.1696
	Service life	0.1290	0.1290	0.1290
	PV/WT efficiency	45%	43.50%	44%
	Inverter efficiency	91%	91.30%	90%
	Maximum power tracking efficiency	97.00%	96.00%	97.00%
	Loss of power due to malfunction	2.50%	2.30%	2.60%
Output characteristic indicator	Average output coefficient	30.00%	32.00%	31.00%
	Credible probability	15%	13.70%	14.20%
	Fluctuation coefficient of variation	70.00%	65.00%	65.00%
	Time series correlation coefficient	80.00%	80.00%	80.00%
Reliability indicator	Failure frequency	15.6	17.8	15.9
	Fault repair time	0.028	0.027	0.027
	Equipment availability	60.00%	65.00%	60.00%
Flexibility indicator	Transmission coefficient	20.00%	22.00%	20.00%
	Management level coefficient	50.00%	30.00%	50.00%
	Abandoned scenery rate	12%	10%	10%
Economic indicator	Initial investment	530	260	700
	Annual operation and maintenance cost	20	18	22
	Government subsidy expense	12	10	15
	Equipment residual value	55	25	70

Table 4. Parameters of thermal power units.

Unit	$a_{k_{TH}}$/(\$/(MWh)^2	$b_{k_{TH}}$/(\$/MWh)	$c_{k_{TH}}$/(\$/h)	$P^{k_{TH}}_{TH,max}$/MW	$P^{k_{TH}}_{TH,min}$/MW	$DO^{k_{TH}}_{TH}$/(MW/h)	$UP^{k_{TH}}_{TH}$/(MW/h)
G1	0.056	35.095	905.105	347	140	200	200
G2	0.075	39.816	919.695	812	325	400	400
G3	0.069	36.832	724.645	635	255	350	350
G4	0.079	41.824	702.425	812	325	350	350
G5	0.058	35.675	1142.929	700	280	300	300
G6	0.087	37.111	757.327	675	270	300	300
G7	0.054	38.541	854.501	1037	415	500	500

5.2. Comprehensive Score of Renewable Energy Power Stations

Based on the parameters in Table 1, the indicator scores for each station can be calculated according to indicator models, which are presented in Table 5. According to expert ratings, the judgment matrix A for each indicator is shown in Table 6. Based on the AHM, the attribute judgment matrix B can be obtained, which is shown in Table 7.

Table 5. Normalization results of indicators for each station.

Indicator	PV_1	PV_2	PV_3	PV_4	WT_1	WT_2	WT_3
Performance indicator	0.089	0.162	0.149	0.063	0.197	0.091	0.249
Output characteristic indicator	0.034	0.039	0.046	0.031	0.259	0.295	0.296
Reliability indicator	0.125	0.158	0.153	0.120	0.147	0.147	0.149
Flexibility indicator	0.153	0.187	0.189	0.090	0.141	0.096	0.145
Economic indicator	0.149	0.153	0.150	0.158	0.128	0.142	0.119

Table 6. Judgment matrix A.

Indicator	Performance Indicator	Output Characteristic Indicator	Reliability Indicator	Flexibility Indicator	Economic Indicator
Performance indicator	1	4	2	6	3
Output characteristic indicator	1/4	1	1/6	3	1/5
Reliability indicator	1/2	6	1	5	2
Flexibility indicator	1/6	1/3	1/5	1	1/4
Economic indicator	1/3	5	1/2	4	1

Table 7. Attribute judgment matrix B.

Indicator	Performance Indicator	Output Characteristic Indicator	Reliability Indicator	Flexibility Indicator	Economic Indicator
Performance	0.00	0.44	0.40	0.46	0.43
Output characteristic	0.11	0.00	0.08	0.43	0.09
Reliability	0.20	0.46	0.00	0.45	0.40
Flexibility	0.08	0.14	0.09	0.00	0.11
Economy	0.14	0.45	0.20	0.44	0.00

The anti-entropy weight method is used to calculate the objective weights of each indicator corresponding to the comprehensive evaluation index system, and the AHM is used to calculate the subjective weights of each indicator corresponding to the comprehensive evaluation index system. The specific weight values are shown in Table 8. Based on the comprehensive weights, the corresponding scores of all renewable energy power stations can be calculated, which are presented in Table 9.

Table 8. Comprehensive weights of each indicator.

Indicator	Subjective Weight	Objective Weight	Comprehensive Weight
Performance indicator	0.31	0.200	0.254
Output characteristic indicator	0.13	0.189	0.157
Reliability indicator	0.27	0.204	0.237
Flexibility indicator	0.08	0.203	0.139
Economic indicator	0.22	0.204	0.212

Table 9. Comprehensive score of each renewable energy power station.

Indicator	WT_3	WT_1	WT_2	PV_2	PV_3	PV_1	PV_4
Performance indicator	0.063	0.050	0.023	0.041	0.038	0.023	0.016
Output characteristic indicator	0.047	0.041	0.046	0.006	0.007	0.005	0.005
Reliability indicator	0.035	0.035	0.035	0.037	0.036	0.030	0.028
Flexibility indicator	0.020	0.020	0.013	0.026	0.026	0.021	0.012
Economic indicator	0.025	0.027	0.030	0.033	0.032	0.032	0.034
Comprehensive ranking	0.191	0.173	0.148	0.143	0.139	0.111	0.096

5.3. Dispatching Plan of Renewable Energy Power Stations

In the proposed dispatching scheme, renewable energy power stations first need to broadcast the predictive value of energy output to the dispatch center of the power system. Then, the dispatch center makes the dispatching plan based on the predictive value of energy output from the perspective of safety, economy, and fairness. The predictive value of energy output for PV and WT stations are assumed in Figures 4 and 5 [34]. The load demand of the system is shown in Figure 6. Accordingly, the dispatching plan of renewable energy power stations can be obtained based on the scores of each station in Table 8. Note that the dispatching period of a whole day is divided into 96 time slots, and each time slot is 15 min.

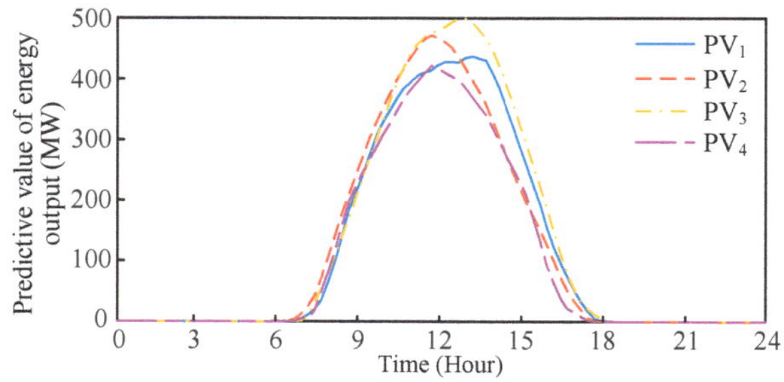

Figure 4. Predictive value of energy output for PV stations.

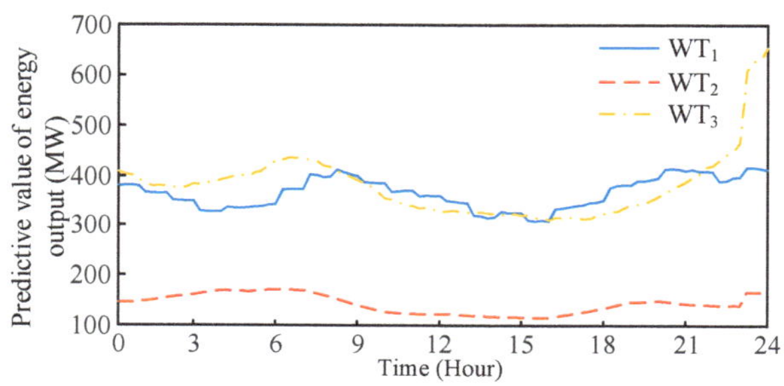

Figure 5. Predictive value of energy output for WT stations.

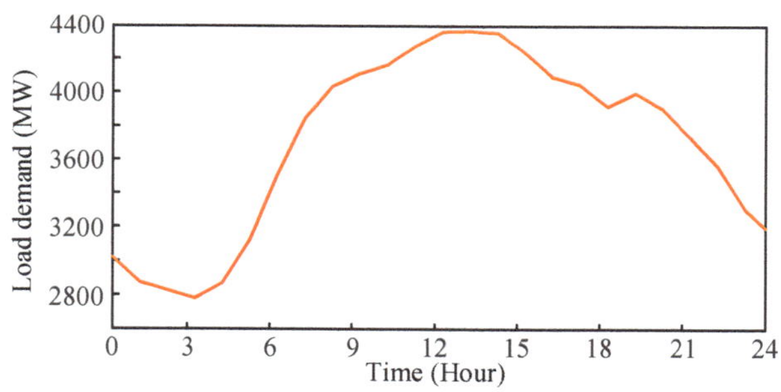

Figure 6. Load demand of energy power system.

Moreover, in order to more accurately calculate the abandoned amount of PT and WT, the parameter settings about the prediction error coefficients for PV and WT need to be closer to the actual situation. Through the statistical analysis of the collected data from PV and WT stations, the average prediction errors are taken as the prediction error coefficients. That is, $\alpha_{PV} = 1.2$ for PV and $\alpha_{WT} = 1.15$ for WT. The unit cost of abandoning PV is 703 yuan/MWh and the unit cost of abandoning WT is 1044 yuan/MWh. Note that the priority coordination scheduling method studied in this paper is applicable to the development of scheduling plans for renewable energy power stations to ensure fairness in participating in grid scheduling when there is a power restriction policy in the regional power grid.

By configuring generation parameters, load demand, and prediction of renewable energy power stations, an optimal scheduling plan for the renewable energy power station can be obtained. The safe-economical-fair scheduling results for PV stations are presented in Figure 7, while the results for WT stations are presented in Figure 8. Figures 7 and 8 describe the day-ahead power prediction curve and dispatching scheme formulated with the proposed fair dispatching approach. It can be seen that the optimal dispatching result is highly consistent with the prediction curve, which can basically achieve the maximum consumption of renewable energy power stations. The energy scheduling results of PV stations, WT stations, and thermal power units are presented in Figure 9. Moreover, Figure 10 shows the voltage amplitude of the renewable energy power stations during the scheduling periods. It depicts that the voltage of each station implementing the proposed scheduling scheme is all within the limit, which complies with the safety operation standards of the power grid.

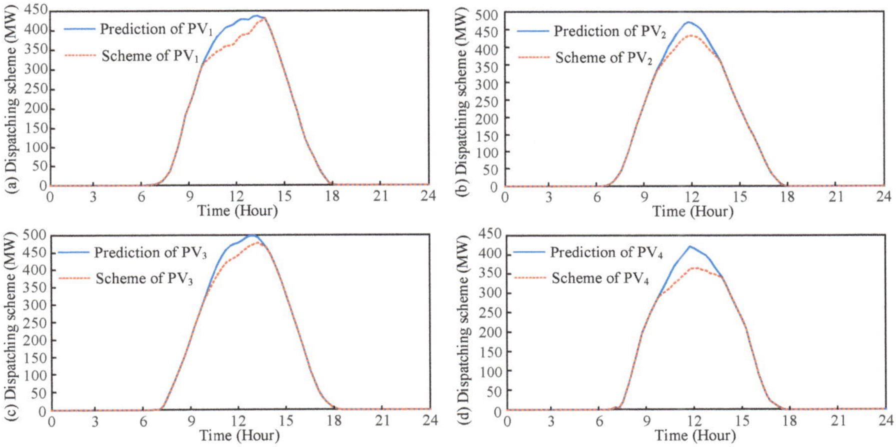

Figure 7. Optimal dispatch scheme of PV stations.

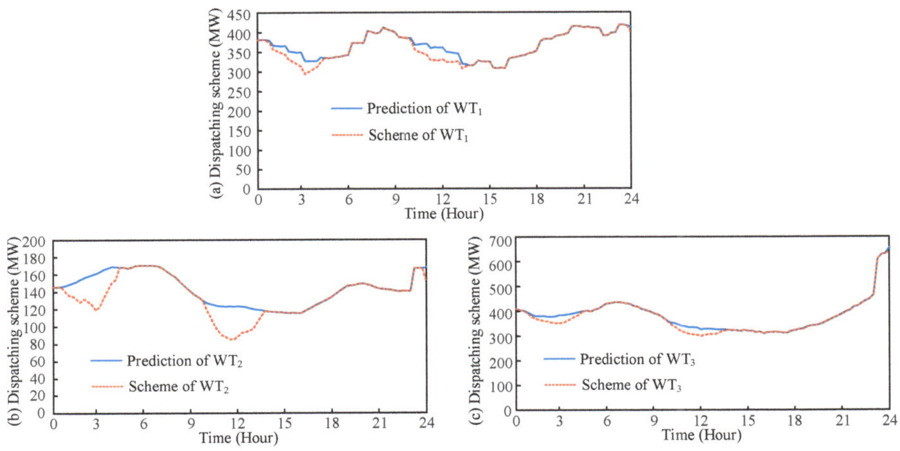

Figure 8. Optimal dispatch scheme of WT stations.

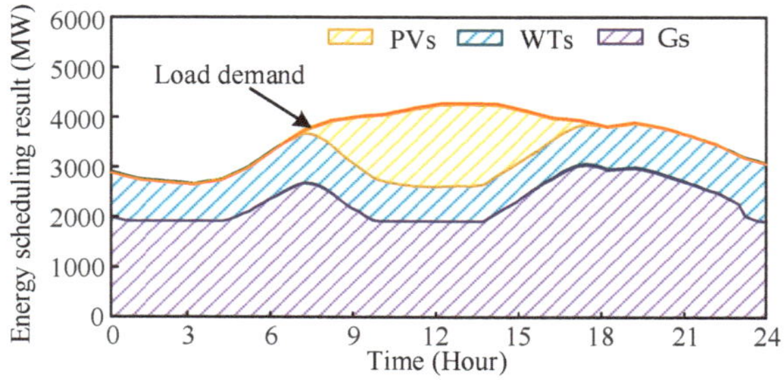

Figure 9. Energy scheduling results of the system.

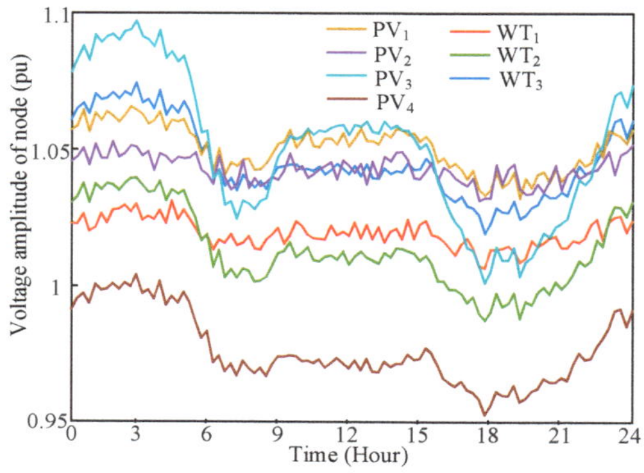

Figure 10. Voltage amplitude of renewable energy power stations.

In order to show the performance of the proposed approach (PA), it is compared with the traditional scheduling method (TSM), in which only the economic and safety aspects of scheduling are considered. The comparison results are shown in Table 10. Overall, the proposed scheduling approach has achieved utilization rates of over 94% for all renewable energy stations. However, the utilization rate of PV_3 stations under the traditional scheduling method is only 86.29%. The average utilization rate of renewable energy in the regional power grid has been increased from 95.89% to 96.26% through PA. Here, the effectiveness of the PA is verified by analyzing the Pearson correlation coefficient (PCC) [35]. The Pearson correlation coefficients between the scores and utilization rates of the four PV stations and three WT stations are 0.9636 and 0.9285, indicating a strong positive correlation between scores and utilization rates. It indicates that based on the comprehensive evaluation method, PA has achieved a good rating in renewable energy dispatch, thus realizing fair dispatch.

Table 10. Comprehensive results of scheduling scheme.

Station	PV_1	PV_2	PV_3	PV_4	WT_1	WT_2	WT_3
Total prediction power/MW	11,119	11,023	12,265	9860	35,190	13,721	36,632
Total scheduling power with TSM/MW	10,541	10,526	10,584	9714	34,881	13,441	36,275
Total scheduling power with PA/MW	10,608	10,628	11,859	9271	34,554	12,984	36,059
Utilization rate with TSM	94.80%	95.49%	86.29%	98.52%	99.12%	97.96%	99.03%
Utilization rate with PA	95.40%	96.42%	96.69%	94.03%	98.19%	94.63%	98.44%
PCC with TSM	0.3017				0.0748		
PCC with PA	0.9636				0.9285		

In summary, when the rating of renewable energy power stations is high, the stations can obtain larger scheduling plans to further improve their consumption capacity. However, renewable energy power stations with lower ratings need to continuously develop and improve their overall ranking by enhancing the performance in the evaluation to obtain more scheduling plans. Therefore, introducing renewable energy station ratings can achieve differentiation in scheduling schemes and ensure fairness in allocation.

6. Conclusions

In this paper, a two-stage scheduling framework is proposed for renewable energy power stations, aiming to address the safe, economical, and fair dispatching problem posed by the increasing integration of renewable energy into power systems. The safe and economic dispatching target is satisfied by formulating the economic optimization problem with the safe operation constraints. The fair allocation of renewable energy power stations is realized by evaluating the performance of stations. In order to obtain the comprehensive weights of indicators, game-fairness empowerment based on cooperative game theory is proposed to reach the balance between the subjective weight determined by AEW and the objective weight determined by the AHM. The simulation results show that the proposed approach can maintain the utilization rate of all renewable energy stations at over 94% and increase the utilization of renewable energy from 95.89% to 96.26%. In addition, the Pearson correlation coefficients of both PV and WT are greater than 0.9, indicating a strong positive correlation between the scores obtained by the evaluation method and the final utilization level of renewable energy power stations. This demonstrates that the proposed two-stage scheduling mechanisms have a good performance in achieving economic, safe, and fair dispatching.

Author Contributions: Conceptualization, Z.Z.; Data curation, W.T.; Formal analysis, J.L.; Funding acquisition, W.X.; Investigation, W.T.; Methodology, X.L.; Project administration, X.L.; Resources, J.L.; Supervision, X.L.; Validation, Z.Z.; Visualization, J.L.; Writing—original draft, W.X.; Writing—review and editing, X.L. All authors have read and agreed to the published version of the manuscript.

Funding: This work is supported by the Technology Project of Qinghai Electric Power Corporation "Research on the Technology of Improving the Utilization Efficiency of Renewable Energy Based on the Evaluation of the Whole Process of Power Generation" (Contract No. SGQH0000DKJS2400387).

Data Availability Statement: The original contributions presented in the study are included in the article. Further inquiries can be directed to the corresponding author.

Acknowledgments: We thank the Technology Project of Qinghai Electric Power Corporation for funding support.

Conflicts of Interest: The authors declare no conflicts of interest. The authors declare that this study received funding from Technology Project of Qinghai Electric Power Corporation. The funder was involved in the study design, collection, analysis, interpretation of data, the writing of this article and the decision to submit it for publication.

References

1. Ahmad, E.; Khan, D.; Anser, M.K. The influence of grid connectivity, electricity pricing, policy-driven power incentives, and carbon emissions on renewable energy adoption: Exploring key factors. *Renew. Energy* **2024**, *232*, 121108. [CrossRef]
2. Yang, Y.; Lo, K. China's renewable energy and energy efficiency policies toward carbon neutrality: A systematic cross-sectoral review. *Energy Environ.* **2024**, *35*, 491–509. [CrossRef]
3. Li, L.; Lin, N.; Meng, H. Study of a dynamic effect-based method for wind farm yaw control optimization. *Int. J. Green Energy* **2024**, *21*, 2142–2154. [CrossRef]
4. Zhou, L.; Qi, F.; Yan, X. A review of research on the passive effect of building photovoltaic systems and analysis of influencing factors. *Sol. Energy* **2024**, *278*, 112766. [CrossRef]
5. Luo, Y.; Yuan, H.; Hu, Z. Optimal scheduling of micro-energy grid based on pareto frontier under uncertainty and pollutant emissions. *IEEE Trans. Smart Grid* **2023**, *15*, 368–380. [CrossRef]
6. Pani, S.R.; Samal, R.K. Evaluation of power grid vulnerability indices accounting for wind power uncertainty. *Sustain. Energy Grids Netw.* **2024**, *38*, 101354. [CrossRef]

7. Zhang, X.; Ding, T.; Mu, C. Dual Stochastic Dual Dynamic Programming for Multi-Stage Economic Dispatch With Renewable Energy and Thermal Energy Storage. *IEEE Trans. Power Syst.* **2023**, *39*, 3725–3737. [CrossRef]
8. Li, Y.; Wu, F. Multistrategy Fusion Particle Swarm for Dynamic Economic Dispatch Optimization of Renewable Energy Sources. *Int. J. Energy Res.* **2024**, *2024*, 5992081. [CrossRef]
9. Huang, Y.; Wang, Y.; Liu, N. Low-carbon economic dispatch and energy sharing method of multiple Integrated Energy Systems from the perspective of System of Systems. *Energy* **2022**, *244*, 122717. [CrossRef]
10. Harasis, S.; Sozer, Y.; Elbuluk, M. Minimizing the Expected Energy Deficiency of a Distributed Generation System Using Dynamic Optimal Power Management. In Proceedings of the 2018 IEEE Energy Conversion Congress and Exposition (ECCE), Portland, OR, USA, 23–27 September 2018.
11. Ahmed, M.M.R.; Mirsaeidi, S.; Koondhar, M.A. Mitigating Uncertainty Problems of Renewable Energy Resources Through Efficient Integration of Hybrid Solar PV/Wind Systems Into Power Networks. *IEEE Access* **2024**, *12*, 30311–30328. [CrossRef]
12. Wu, G.; Yi, C.; Xiao, H.; Wu, Q.; Zeng, L.; Yan, Q.; Zhang, M. Multi-objective optimization of integrated energy systems considering renewable energy uncertainty and electric vehicles. *IEEE Trans. Smart Grid* **2023**, *14*, 4322–4332. [CrossRef]
13. Moosavian, S.F.; Noorollahi, Y.; Shoaei, M. Renewable energy resources utilization planning for sustainable energy system development on a stand-alone island. *J. Clean. Prod.* **2024**, *439*, 140892. [CrossRef]
14. Lee, J.; Jung, S.; Lee, Y. Energy storage mix optimization based on time sequence analysis methodology for surplus renewable energy utilization. *Energies* **2023**, *16*, 6031. [CrossRef]
15. Lujano-Rojas, J.M.; Osório, G.J.; Catalão, J.P.S. New probabilistic method for solving economic dispatch and unit commitment problems incorporating uncertainty due to renewable energy integration. *Int. J. Electr. Power Energy Syst.* **2016**, *78*, 61–71. [CrossRef]
16. Albatran, S.; Harasis, S.; Ialomoush, M.; Alsmadi, Y.; Awawdeh, M. Realistic Optimal Power Flow of a Wind-Connected Power System with Enhanced Wind Speed Model. *IEEE Access* **2020**, *8*, 176973–176985. [CrossRef]
17. Xu, M.; Li, W.; Feng, Z. Economic dispatch model of high proportional new energy grid-connected consumption considering source load uncertainty. *Energies* **2023**, *16*, 1696. [CrossRef]
18. Rad, M.A.V.; Kasaeian, A. Evaluation of stand-alone hybrid renewable energy system with excess electricity minimizer predictive dispatch strategy. *Energy Convers. Manag.* **2024**, *299*, 117898.
19. Shi, G.R.; Sun, R.F.; Xu, H.X.; Qiao, Y. Active Power Stratification Coordination Control Strategy for Large-Scale Cluster of Renewable Energy. *Power Grid Technol.* **2018**, *42*, 2160–2167.
20. Li, W.; Die, X.; Ma, Z. Two-Stage optimal dispatching of wind power-photovoltaic-solar thermal combined system considering economic optimality and fairness. *Energy Eng.* **2023**, *120*, 1001–1022. [CrossRef]
21. Sun, L.; Zhang, N.; Li, N.; Liu, X.L.; Li, W.D. An Intra-day Rolling Scheduling Considering Energy Equity Based on Gini Coefficient. *Power Grid Technol.* **2020**, *44*, 340–347.
22. Li, H.; Zhang, R.; Mahmud, M.A.; Hredzak, B. A novel coordinated optimization strategy for high utilization of renewable energy sources and reduction of coal costs and emissions in hybrid hydro-thermal-wind power systems. *Appl. Energy* **2022**, *320*, 119019. [CrossRef]
23. Zheng, X.; Khodayar, K.E.; Wang, J.; Yue, M.; Zhou, A. Distributionally Robust Multistage Dispatch with Discrete Recourse of Energy Storage Systems. *IEEE Trans. Power Syst.* **2024**, *39*, 6960–6973. [CrossRef]
24. Qiu, D.; Wang, Y.; Zhang, T.; Sun, M.; Strbac, G. Hierarchical multi-agent reinforcement learning for repair crews dispatch control towards multi-energy microgrid resilience. *Appl. Energy* **2023**, *336*, 120826. [CrossRef]
25. Jin, W.; Chen, L.; Chen, H.; Zheng, S. A Novel Evaluation Method of the Hosting Capacity of Multiple Renewable Energy Stations Accessed into a Power System. *Energy Rep.* **2023**, *9*, 56–65. [CrossRef]
26. Zhong, W.; Wang, L.; Liu, Z.; Hou, S. Reliability Evaluation and Improvement of Islanded Microgrid Considering Operation Failures of Power Electronic Equipment. *J. Mod. Syst. Clean Energy* **2020**, *8*, 111–123. [CrossRef]
27. Jiang, H.; Du, E.; He, B.; Zhang, N.; Wang, P.; Li, F.; Ji, J. Analysis and Modeling of Seasonal Characteristics of Renewable Energy Generation. *Renew. Energy* **2023**, *219*, 119414. [CrossRef]
28. Wang, W.; Li, H.; Hou, X. Multi-criteria evaluation of distributed energy system based on order relation-anti-entropy weight method. *Energies* **2021**, *14*, 246. [CrossRef]
29. Zeng, W.; Fan, J.; Ren, Z. Economic Evaluation Method of Modern Power Transmission System Based on Improved Technique for Order of Preference by Similarity to Ideal Solution (TOPSIS) and Best-Worst Method-Anti-Entropy Weight. *Energies* **2023**, *16*, 7242. [CrossRef]
30. Ding, Y. Ecological garden plant configuration method based on attribute analytic hierarchy process model. *Int. J. Environ. Technol. Manag.* **2021**, *24*, 168–183. [CrossRef]
31. Li, C.; Kang, Z.; Yu, H. Research on Energy Optimization Method of Multi-microgrid System Based on the Cooperative Game Theory. *J. Electr. Eng. Technol.* **2024**, *19*, 2953–2962. [CrossRef]
32. Wu, Y.; Wang, C.; Wang, Y. Cooperative game optimization scheduling of multi-region integrated energy system based on ADMM algorithm. *Energy* **2024**, *302*, 131728. [CrossRef]
33. Zhang, T.; Chen, C.; Li, Z. Shared energy storage-assisted and tolerance-based alliance strategy for wind power generators based on cooperative game and resource dependence theories. *Int. J. Electr. Power Energy Syst.* **2024**, *155*, 109605. [CrossRef]

34. Papavasiliou, A.; Mou, Y.; Cambier, L.; Scieur, D. Application of stochastic dual dynamic programming to the real-time dispatch of storage under renewable supply uncertainty. *IEEE Trans. Sustain. Energy* **2018**, *9*, 547–558. [CrossRef]
35. Chu, S.; Yang, D.; Ge, W.; Liu, C.; Cai, G.; Kou, L. Global Sensitivity Analysis of Voltage Stability in the Power System with Correlated Renewable Energy. *Electr. Power Syst. Res.* **2021**, *192*, 106916.

energies

MDPI

Article

Performance Analysis of a Parabolic Trough Collector with Photovoltaic—Thermal Generation: Case Study and Parametric Study

Benjamín Chavarría-Domínguez [1], Susana Estefany De León-Aldaco [1,*], Mario Ponce-Silva [1], Nicolás Velázquez-Limón [2], Jesús Armando Aguilar-Jiménez [2], Fernando Chavarría-Domínguez [3], Ernesto Raúl Rodríguez-García [3], Heriberto Adamas-Pérez [4], Ricardo Eliu Lozoya-Ponce [5] and Eligio Flores-Rodriguez [6]

1 Department of Electronic Engineering, National Technological Institute of Mexico—National Center for Research and Technological Development, Cuernavaca 62490, Mexico; d18ce092@cenidet.tecnm.mx (B.C.-D.); mario.ps@cenidet.tecnm.mx (M.P.-S.)

2 Center for the Study of Renewable Energies, Institute of Engineering, Autonomous University of Baja California, Mexicali 21280, Mexico; nicolas.velazquez@uabc.edu.mx (N.V.-L.); aguilarj30@uabc.edu.mx (J.A.A.-J.)

3 Faculty of Engineering, Veracruz University, Coatzacoalcos 96538, Mexico; fchavarria@uv.mx (F.C.-D.); ernesrodriguez@uv.mx (E.R.R.-G.)

4 Academic Division of Industrial Mechanics, Emiliano Zapata University of Technology, Emiliano Zapata 62765, Mexico; heribertoadamas@utez.edu.mx

5 Division of Graduate Studies and Research, National Technological Institute of Mexico Campus Chihuahua, Chihuahua 31310, Mexico; ricardo.lp@chihuahua.tecnm.mx

6 National Technological Institute of Mexico—Los Reyes Institute of Technology, Los Reyes 60300, Mexico; eligio.fr@losreyes.tecnm.mx

* Correspondence: susana.da@cenidet.tecnm.mx

Abstract: This study presents a mathematical model of a parabolic trough solar collector with photovoltaic cells integrated into its solar receiver. A case study is presented, utilizing meteorological data obtained from the localities of Cuernavaca and Mexicali in Mexico. The results demonstrate moderately variable electrical and thermal energy production for Cuernavaca (387.93 kWh to 239.38 kWh and 1036.11 kWh to 641.26 kWh, respectively). In contrast, the production of electrical and thermal energy in Mexicali exhibited considerably greater fluctuations (515.16 kWh to 177.69 kWh and 1424.39 kWh to 448.88 kWh, respectively). Furthermore, a parametric study is presented, which analyzes the impact of solar receiver geometry and mass flow on the model's behavior. The results demonstrate that the pipe length exerts the most significant influence on the electrical and thermal power output (1.21 kW to 2.22 kW and 3.7 kW to 6.9 kW, respectively). Additionally, the diameter has an impact on the thermal power output (5.23 kW to 7.1 kW) and the electrical and thermal efficiency (0.18 to 0.15 and 0.54 to 0.74, respectively). Modifying the mass flow facilitates the enhancement of electrical power and efficiency (1.54 kW to 1.72 kW and 0.16 to 0.18, respectively) while concurrently preventing a significant reduction in thermal power and efficiency (5.4 kW to 5.3 kW and 0.56 to 0.55, respectively). A script with the developed model is provided.

Keywords: concentrated photovoltaic/thermal; thermal modeling; solar energy

1. Introduction

Solar energy is obtained through the use of two distinct systems. The first type of system is a concentrated solar power (CSP) plant, which concentrates solar irradiance to obtain heat through the use of a heat transfer fluid (HTF). The second system is the photovoltaic (PV) cell, which can generate electrical energy through the photoelectric effect. CSP plants are primarily employed in industrial processes that require a considerable amount of heat, such as Rankine cycles, pasteurization, desalination, and others [1,2].

Photovoltaic (PV) systems provide the flexibility to meet a range of electricity supply needs. Such uses extend from microelectronics applications to supplying electricity to residential condominiums or similar buildings [3,4].

Researchers have concentrated their efforts on the development of new solutions that are complementary, versatile, and useful in a variety of scenarios. To this end, the feasibility of concentrating solar photovoltaic/thermal (CPV/T) systems has been studied in details. These technologies concentrate solar irradiance in a specific area, where the PV cells and a pipe through which the heat transfer fluid (HTF) circulates are placed [5,6].

The Parabolic Trough Collector (PTC) represents one of the most extensively developed CSP technologies. Consequently, it is of interest to analyze the performance of this collector when coupled to a solar photovoltaic thermal receiver (SRC-PVT). Table 1 provides a summary of the studies that have addressed this approach [7].

Table 1. Summary of studies focused on the performance of PTC collectors coupled to SRC-PVTs.

Year	Autor	Journal	Main Focus of the Study
2023	Renno [8]	Energies	Case study and Experimental testing
2023	Santana [9]	Energies	Parametric study
2022	Acosta-Pazmiño [10]	Renewable Energy	Case study
2022	Gorouh [11]	Renewable Energy	Parametric study
2021	Cabral [12]	Solar Energy	Experimental testing
2021	Herez [13]	Renewable Energy Focus	Case study and modeling
2021	Renno [14]	Energies	Experimental testing and optimization
2020	Acosta-Pazmiño [15]	Energies	Case study
2020	Felsberger [16]	Energies	Experimental testing
2020	Gakkhar [17]	Applied Thermal Engineering	Experimental testing and modeling
2020	Otanicar [18]	Applied Energy	Experimental testing
2020	Riahi [19]	Energy Conversion and Management	Experimental testing and modeling
2019	Karathanassis [20]	Renewable Energy	Optimization and modeling
2019	Maatallah [21]	Solar Energy	Performance analysis and modeling
2019	Valizadeh [22]	Renewable Energy	Parametric study and modeling
2018	Ben Youssef [23]	Solar Energy	Optimization and modeling
2017	Mohsenzadeh [24]	Renewable Energy	Experimental testing
2017	Yazdanifard [25]	Energy Conversion and Management	Parametric study
2016	Brekke [26]	Journal of Solar Energy Engineering	Parametric study and modeling
2015	Al-Nimr [27]	Solar Energy	Parametric study and modeling
2013	Calise [28]	Energy	Parametric study and modeling

Santana [9] evaluates collector performance through a case study of multiple coastal locations in Mexico and Singapore, covering a twelve-month period. The author also develops a parametric study to determine optimal collector sizing, focusing on the appropriate PTC surface area and inlet temperature.

Herez [13] conducted a case study of three towns: Baalbek, Angers, and Abu Dhabi. The temperature values given by the mathematical model, the efficiencies, and the energy production for all three cases were compared. The study identified solar radiation and fluid inlet temperature as the main factors affecting system performance. The highest electrical and thermal efficiencies were achieved in Lebanon during January (27.9%) and June (50.8%), respectively.

Acosta-Pazmiño [15] presents a case study under the meteorological conditions of Monterrey, Mexico. The results demonstrated that the CPV/T system could provide 72% of the required heating during the summer months and 24% during the winter.

Gorouh [11] carried out a parametric study that yielded results indicating that the sensitivity of thermal power to the HTF inlet temperature exceeds that of electrical power. Additionally, an increase in the mass flow rate of the HTF leads to an enhancement in the electrical and thermal power output of the collector, while a decrease in the HTF outlet temperature is observed. The increase in the difference between the HTF temperature and the ambient temperature has a significant impact on the thermal efficiency.

Ben Youssef [23] presented a parametric study that yielded the following conclusions. As wind speed decreases, the electrical efficiency of the PV system is reduced. A drastic decrease in the concentration ratio has a significant impact on heat exchange efficiency, which in turn affects the electrical and thermal efficiencies.

Yazdanifard [25] conducted a parametric investigation, the findings of which indicated that an increase in pipe length was associated with an increase in both HTF outlet temperature and PV temperature. An increase in pipe diameter was associated with a decrease in both electrical and thermal efficiencies. The results of this research underscored the importance of optimizing structural and geometrical parameters to achieve the desired performance.

Calise [28] developed a parametric study that provides the following findings. Increases in pipe length and diameter lead to a decrease in electrical and thermal efficiencies. Increasing the width of the prism faces increases the concentration ratio of the PTC collector, which is favorable for electrical efficiency. The model performs adequately even at temperatures over 100 °C. However, its efficiencies are very susceptible to changes in solar irradiance.

This study concerns the performance analysis of a PTC collector in conjunction with an SRC-PVT. A mathematical model of the collector is presented, based on a series of one-dimensional energy balance equations. The results of the study are summarized in the Conclusions Section. Considering the points made in previous studies and the results of the developed model, the contributions of this study are as follows:

1. Ben Youssef [23] has demonstrated that the wind speed at which photovoltaic (PV) systems operate has a significant impact on their efficiency. The model developed in this study addresses this issue by incorporating monthly variations in wind speed as one of the model's boundary conditions;
2. Calise [28] emphasizes the necessity of calibrating the convection coefficients incorporated within his zero-dimensional model), thereby facilitating the precise estimation of the yield. The model developed in this study (one-dimensional) addresses this issue by incorporating a recalculation function for the convection coefficients with each variation of the parameters, such as wind speed;
3. It is evident that meteorological conditions exert a considerable influence on the efficacy of solar collectors [9,13,15]. This paper presents a case study of the locations of Cuernavaca and Mexicali, which enables the performance of the model to be evaluated under the conditions of a relatively stable temperate climate and an extreme desert climate, respectively. This paper also provides criteria for the correct determination of climate data extracted from a TMY;
4. Parametric studies are essential for determining the behavior of the collector, and the information obtained from these studies can be used to size and optimize the collector [11,23,25]. This study presents a parameter study focused on the geometry of the SRC-PVT and the mass flow. The results obtained from this study are useful for the sizing of an experimental prototype.

2. Mathematical Model

A review of previous studies focusing on the analysis of PTC systems coupled to an SRC-PVT was conducted [7], and a specific mathematical model was selected from the studies consulted. The following presents a general description of the SRC-PVT and its mathematical modeling.

2.1. Structural Description of the SRC-PVT

Figure 1 illustrates the structural components of the SRC-PVT, which is distinguished by a triangular prism form. The two faces oriented towards the PTC comprise the PVs. The aforementioned faces receive the DNI concentrated by the PTC, as well as being in contact with the environment. The third face is situated in opposition to the PTC and incorporates a metallic absorber. This face is exposed to the GHI from the sky and is also in contact with the environment. Behind the prism faces is a metallic substrate that facilitates heat transfer from the PVs and absorber to a pipe located in the center of the prism. The pipe contains an HTF, which extracts the heat present in the SRC-PVT.

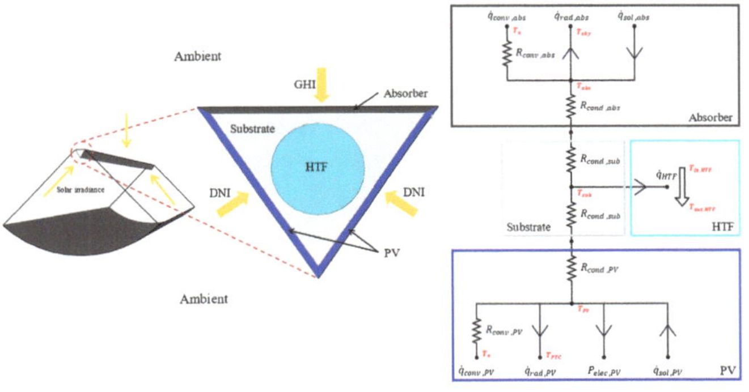

Figure 1. The SRC-PVT and its thermal resistance scheme.

The various modes of heat transfer (conduction, convection, and radiation) that occur between the different elements of the SRC-PVT, which is situated between the environment and the HTF, are represented as thermal resistances. The incident solar irradiance (GHI and DNI) is considered as incoming power. Based on the modes of heat transfer and the incidence of solar irradiance, the energy balance equations of the SRC-PVT are defined.

2.2. Description of the Mathematical Model

The thermal resistance model presented in Figure 1 and the series of energy balance equations that model the thermal behavior of the SRC-PVT were proposed by Herez [13]. It should be noted that this is a one-dimensional and stationary modeling approach. These energy balance equations are described below.

Equation (1) determines the surface temperature of the PTC, denoted by T_{PTC}:

$$\dot{q}_{sol,PTC} + \dot{q}_{rad,PV} = \dot{q}_{conv,front,PTC} + \dot{q}_{conv,back,PTC} + \dot{q}_{rad,PTC} \tag{1}$$

Equation (2) determines the temperature at the HTF outlet, denoted by $T_{out,HTF}$:

$$\dot{q}_{sol,PV} - P_{PV} - \dot{q}_{conv,PV} - \dot{q}_{rad,PV} + \dot{q}_{sol,abs} - \dot{q}_{conv,abs} - \dot{q}_{rad,abs} = \dot{q}_{HTF} \tag{2}$$

Equation (3) determines the temperature in the absorber, denoted by T_{abs}:

$$\dot{q}_{sol,abs} - \dot{q}_{conv,abs} - \dot{q}_{rad,abs} = \dot{q}_{cond,abs-sub} \tag{3}$$

Equation (4) determines the temperature in the substrate, denoted by T_{sub}:

$$\dot{m}_{HTF}C_{p,HTF}(T_{out,HTF} - T_{in,HTF}) = \in \dot{m}_{HTF} + C_{p,HTF}(T_{sub} - T_{in,HTF}) \tag{4}$$

Equation (5) determines the temperature at the PV, denoted by T_{PV}:

$$\dot{q}_{cond,PV-sub} + \dot{q}_{cond,abs-sub} = \dot{q}_{HTF} \tag{5}$$

In the case of the energy balance in Equations (1)–(3) and (5), the terms that comprise these equations represent the various modes of heat transfer and incident irradiance in the elements that constitute the SRC-PVT. The energy balance in Equation (4) corresponds to a fixed control volume (steady state analysis) for a heat exchanger, where the substrate and the pipe are the heat sources while the HTF is the heat receiver.

The terms that comprise the energy balance equations can be solved through the application of formulas, theorems, and correlations. This information is presented in greater detail in the studies of Herez [13] and thermodynamics textbooks such as Incropera [29,30] and Çengel [31].

The electrical and thermal power produced are estimated using Equation (6) and Equation (7), respectively. The electrical power is obtained from parameters affecting the PV, such as PTC geometry and DNI. The thermal power is derived from the energy balance in Equation (4), which allows the power absorbed by the HTF to be determined.

Equation (6) determines the electrical power of the PV system, denoted by P_{PV}:

$$P_{PV} = G_{DNI} * A_{PV} * CR_{PTC} * \eta_{opt} * IAM_{elec} * \eta_{PV} \tag{6}$$

Equation (7) determines the thermal power of the HTF, denoted by $\dot{q}_{HTF}$:

$$\dot{q}_{HTF} = \in *\dot{m}_{HTF} * C_{p,HTF} * (T_{sub} - T_{in,HTF}) \tag{7}$$

The electrical and thermal efficiencies are calculated using Equations (8) and (9), respectively. The efficiency is defined as the ratio between the power produced and the power input to the developed model. In this case, the power input is a system based on solar irradiation collection, which is obtained from the DNI and the collector area.

Equation (8) determines the electrical efficiency of the PV system, denoted by η_{elec}:

$$\eta_{elec} = \frac{P_{PV}}{G_{DNI} * A_{ap}} \tag{8}$$

Equation (9) determines the thermal efficiency of the PV system, denoted by η_{th}:

$$\eta_{th} = \frac{\dot{q}_{HTF}}{G_{DNI} * A_{ap}} \tag{9}$$

2.3. Strategy for the Solution of the Mathematical Model

The energy balance equations enable the estimation of temperatures in the different elements of the SRC-PVT (PV, absorber, and substrate). Additionally, they facilitate the calculation of the temperature of the HTF at the outlet of the SRC-PVT and the temperature at the surface of the concentrator. This is achieved through the following procedure:

1. An initial temperature value is assigned to T_{PV};
2. With the initial temperature of T_{PV} and Equation (1), the temperature of T_{PTC} is estimated;

3. With the temperatures of T_{PV} and T_{PTC}, a system of three equations and three unknowns is formed. The equation system consists of Equations (2)–(4), while the unknowns are the temperatures $T_{out,HTF}$, T_{abs}, and T_{sub};
4. Finally, with the temperatures $T_{out,HTF}$, T_{abs}, and T_{sub} and Equation (5), the appropriate value of T_{PV} is estimated.

The energy balance equations were solved with EES software [32]. It should be noted that the structure used within the script follows the four-step procedure previously proposed. The equations are organized into three subprograms following steps 2 to 4 of the aforementioned procedure. Outside the subprograms, the parameters, formulas, and correlations necessary to solve the equations are defined. The estimation of the unknown temperatures is calculated by the equation solver through an iterative process, which is applied automatically. The script with the developed model is provided in [33,34].

3. Fundamentals of PTCs and CPCs

The model validation process was designed following a strategy comparable to that utilized by Herez [13] and Calise [28]. This strategy consists of utilizing the boundary conditions and design parameters of a reference study as the input values for the mathematical model developed. The temperature values generated by the developed model are then compared with the temperature values presented in the reference study. If the divergence between both temperature values is less than 10%, the developed model is considered validated.

3.1. Boundary Conditions, Design, and General Model Parameters

The parameters that integrate the energy balance equations, the formulas, and the correlations necessary for their solution are organized into the following categories: boundary conditions, design parameters, and general parameters of the model.

The boundary conditions and design parameters utilized as a reference for the validation of the model developed in this study are those proposed by Herez [13] and Calise [28]. Table 2 presents the boundary conditions, while Table 3 presents the design parameters. It should be noted that the boundary conditions correspond to the meteorological conditions of the location where the model is analyzed.

Table 2. Boundary conditions.

Parameter	Value	Unit
GHI	1000	W/m^2
DNI	800	W/m^2
Sky temperature	25	°C
Ambient temperature	25	°C
HTF inlet temperature	70	°C
Air velocity	5	m/s

Table 3. Design parameters.

Parameter	Value	Unit
PTC aperture area	12	m^2
Absorber area	0.6	m^2
PV area	1.2	m^2
PTC concentration ratio	10	-
Receiver width (PV)	0.06	m
HTF pipe diameter	0.03	m
HTF-specific heat	4183	J/(kg × °K)
HTF mass flow rate	0.15	Kg/s

Table 3. *Cont.*

Parameter	Value	Unit
Thermal conductivity of the PV	50	W/m^2
Absorber thermal conductivity	205	W/m^2
Substrate thermal conductivity	250	W/m^2
Absorbance of the absorber	0.9	-
Absorbance of PTC	0.03	-
Absorbance of PV	0.97	-
Absorber emissivity	0.2	-
Concentrator emissivity	0.3	-
PV emissivity	0.2	-
Electrical coefficient IAM	0.28	-
Thermal coefficient IAM	0.14	-

The general parameters correspond to values that are not provided by Herez [13] but are necessary for solving the energy balance equations. These values are mainly related to the geometry of the PTC and SRC-PVT. Additionally, there are parameters such as the HTF (water) velocity, which has been cleared from the mass flow, and the PV and PTC efficiencies, which were determined by consulting technical data sheets and similar studies [25,35,36]. Table 4 presents the general parameters.

Table 4. General parameters.

Parameter	Value	Unit
Length of PTC and SRC-PVT	10	M
Width of PTC	1.2	M
Thickness of absorber and PV	0.003	M
Thickness of substrate	0.025	M
HTF pipe cross-sectional area	7×10^{-4}	m^2
HTF Velocity	2.17	m/s
Heat exchanger contact area	0.9425	m^2
PV efficiency	0.2998	-
PTC optical efficiency	0.83	-

3.2. Validation Strategy

The temperature values of the PV, absorber, substrate, and HTF outlet presented by Calise [28] were compared with the temperature values estimated by the model developed in this study. The results of this comparison are presented in Table 5, which also shows the percentage divergence between the two sets of values.

Table 5. Comparison of the values estimated by the developed model.

Element	Herez Reference (°C)	Model Development (°C)	Divergence (%)
PV	79.30	80.16	1.08
Absorber	82.30	79.56	3.33
Substrate	82.70	79.51	3.86
HTF outlet	80.50	78.44	2.56

The discrepancy between the reference values and those obtained by the developed model is less than 5%, indicating that the estimated temperatures are deemed satisfactory.

4. Case Study

One notable aspect of the study proposed by Herez [13] is its analysis of the performance of their model in various locations. However, when analyzing the methodology used in their study, it becomes evident that the following procedure was applied to calculate the energy production across each month:

- The mathematical model is fed with punctual values of solar irradiance (GHI and DNI). The data corresponds to the irradiance values characteristic of midday, which represents the moment of greatest solar irradiance throughout the day;
- The mathematical model generates punctual values for the electrical and thermal power produced from the solar irradiance values entered, along with the remaining meteorological conditions determined;
- To calculate the energy generated per month, the punctual values of power are multiplied by the number of daylight hours and the number of days in the specific month.

A direct analysis of a TMY file reveals a decline in solar irradiance between the first and last hours of the day. The methodology of multiplying the punctual power of midday by the number of daylight hours assumes that the maximum amount of solar irradiance is available throughout the day. This introduces a bias in the analysis, as it assumes conditions that are too favorable and unrealistic. To evade such biases, this model was provided with solar irradiance values (GHI and DNI) that were averaged over daylight hours and days within a given month. As demonstrated in Figure 2, the irradiance values employed by Herez are based on highly favorable assumptions to represent the irradiance per month at the Angers location, as opposed to the average values proposed in this study.

Meteorological values of Herez		
Month	GHI (W/m²)	DNI (W/m²)
January	356.45	225.15
February	472.85	299.33
March	580.6	361.21
April	679.57	432.82
May	644.11	356.06
June	670.98	383.1
July	695.56	408.63
August	697.95	417.09
September	676.25	436.14
October	525.64	331.23
November	448.42	299.01
December	434.85	297.97

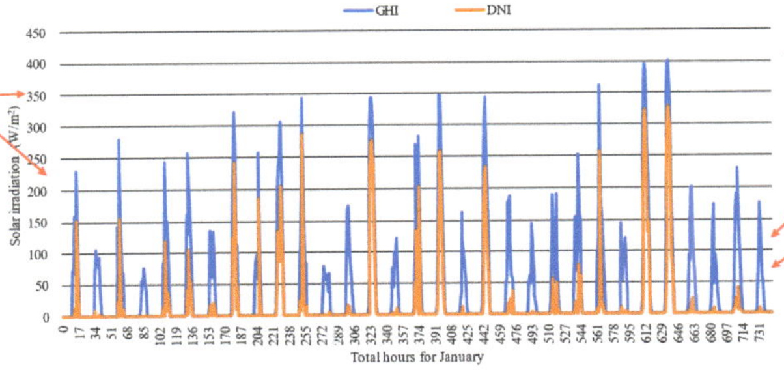

Meteorological values averaged		
Month	GHI (W/m²)	DNI (W/m²)
January	120.50	64.01
February	220.28	125.05
March	259.35	154.43
April	377.60	270.81
May	430.27	279.78
June	340.87	191.20
July	363.68	215.00
August	354.37	196.79
September	338.36	222.11
October	250.23	135.42
November	176.73	106.26
December	127.90	70.33

Figure 2. Comparison of the solar irradiation ranges presented by Herez and the averaged ones.

4.1. Meteorological Data for the Locations Studied

This study analyzes the behavior of the model developed for the specific meteorological conditions of the localities of Cuernavaca and Mexicali in Mexico. Cuernavaca has a temperate climate for the majority of the year, while Mexicali experiences extreme climatic conditions between summer and winter.

Meteorological conditions for Cuernavaca and Mexicali were obtained from the PVGIS database [37]. Using this information, meteorological data for each month of the year were determined. It should be noted that these data correspond to the boundary conditions of the model. The meteorological data for Cuernavaca are presented in Table 6, while those for Mexicali are presented in Table 7.

Table 6. Meteorological data for the locality of Cuernavaca.

Month	GHI (W/m^2)	DNI (W/m^2)	Amb./HTF Temp. (°C)	Sky Temp. (°C)	Air Vel. (m/s)
January	486.90	297.50	16.39	8.39	1.71
February	508.20	326.60	18.54	10.54	1.56
March	551.50	394.50	19.49	11.49	1.78
April	547.10	334.80	21.80	13.80	1.52
May	547.50	359.70	20.48	12.48	1.22
June	568.10	361.10	21.36	13.36	1.16
July	531.90	406.10	19.92	11.92	1.17
August	483.40	274.60	19.89	11.89	1.05
September	463.90	294.10	18.95	10.95	0.98
October	481.60	303.10	19.58	11.58	1.18
November	501.60	370.30	16.63	8.63	1.61
December	478.10	322.60	16.20	8.20	1.54

Table 7. Meteorological data for the locality of Mexicali.

Month	GHI (W/m^2)	DNI (W/m^2)	Amb./HTF Temp. (°C)	Sky Temp. (°C)	Air Vel. (m/s)
January	372.80	318.80	11.43	3.43	2.23
February	430.80	353.40	19.43	11.43	1.55
March	539.70	441.00	17.30	9.30	1.99
April	582.30	491.30	19.98	11.98	2.09
May	580.40	478.80	29.13	21.13	1.97
June	622.40	521.60	30.23	22.23	1.96
July	585.10	513.00	33.21	25.21	1.83
August	540.90	447.50	32.21	24.21	1.73
September	548.30	453.80	29.47	21.47	1.79
October	474.40	413.30	22.22	14.22	2.11
November	419.30	340.50	16.76	8.76	2.06
December	316.00	238.90	14.81	6.81	1.58

To determine the monthly values of the meteorological data, the following considerations were taken into account:

- The TMY file provides an hourly value for each of the meteorological data. The file comprises a total of 8738 h;
- As previously indicated, the solar radiation values are averaged directly from the TMY, with consideration given to the daylight hours covered by each of the months;
- The ambient temperature and HTF temperature correspond to the dry bulb temperature of the TMY file. These temperatures are averaged over the hours spanning each month;
- The temperature of the sky was determined to be 8 °C less than the ambient temperature, as indicated by Forristall [38].

4.2. Model Performance

- Figure 3 illustrates the monthly temperatures of the PV, absorber, and HTF at the outlet and inlet. In the case of Cuernavaca, the highest temperatures were recorded in April and the lowest in December. The temperature range throughout the year was 16 °C to 26 °C. The mean temperature gain of the HTF was 3.42 °C. In Mexicali, the highest temperatures were in July and the lowest in January. The temperature range throughout the year was 10 °C to 40 °C. The mean temperature gain of the HTF was 4.19 °C.

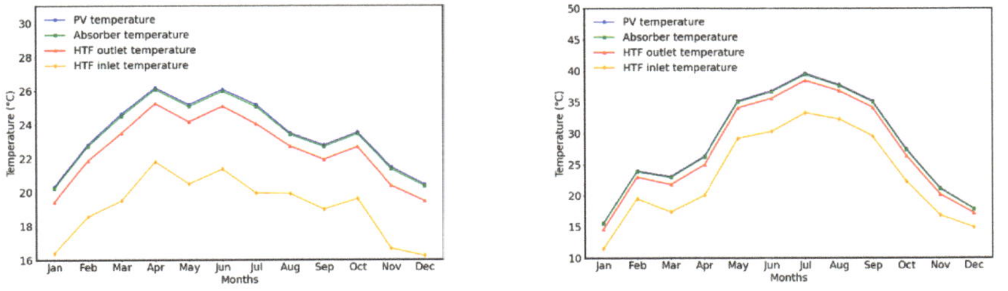

(**a**) (**b**)

Figure 3. Temperatures per month for the locations of (**a**) Cuernavaca and (**b**) Mexicali.

- Figure 4 illustrates the monthly production of electric and thermal power. In the case of Cuernavaca, the highest power produced was recorded in July, with a total of 0.96 kW of electric power and 2.57 kW of thermal power. The lowest power produced was recorded in August, with a value of 0.65 kW of electric power and 1.76 kW of thermal power. The total power produced was 9.61 kW of electric power and 25.77 kW of thermal power. The average power produced was 0.80 kW of electric power and 2.14 kW of thermal power. In the case of Mexicali, the highest power produced was observed in June, with a value of 1.21 kW of electric power and 3.34 kW of thermal power. The lowest power produced was observed in December, with a value of 0.57 kW for electric and 1.44 kW for thermal. The total power produced was 11.79 kW for electric and 31.62 kW for thermal. The average power produced was 0.98 kW for electric and 2.63 kW for thermal.

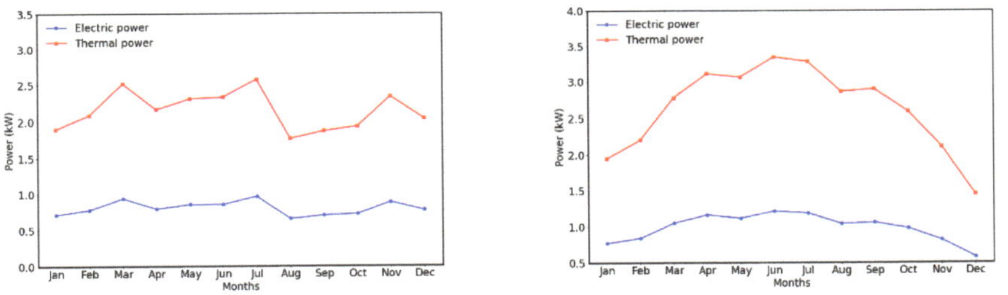

(**a**) (**b**)

Figure 4. Power per month for the locations of (**a**) Cuernavaca and (**b**) Mexicali.

- Figure 5 illustrates the electrical and thermal efficiency values on a monthly basis. In the case of Cuernavaca, the highest electrical efficiency was observed in January and December, with a value of 0.199, while the lowest electrical efficiency was recorded in April and June, with a value of 0.197. The highest thermal efficiency was observed in April, with a value of 0.537, while the lowest thermal efficiency was recorded in November, with a value of 0.525. The average electrical efficiency was 0.198, while the average thermal efficiency was 0.531. In the case of Mexicali, the highest electrical efficiency was observed in January, reaching 0.200, while the lowest electrical efficiency was recorded in July, at 0.192. The highest thermal efficiency was observed in June, at 0.534, while the lowest thermal efficiency was observed in December, at 0.505. The average electrical efficiency was 0.196, while the average thermal efficiency was 0.523.

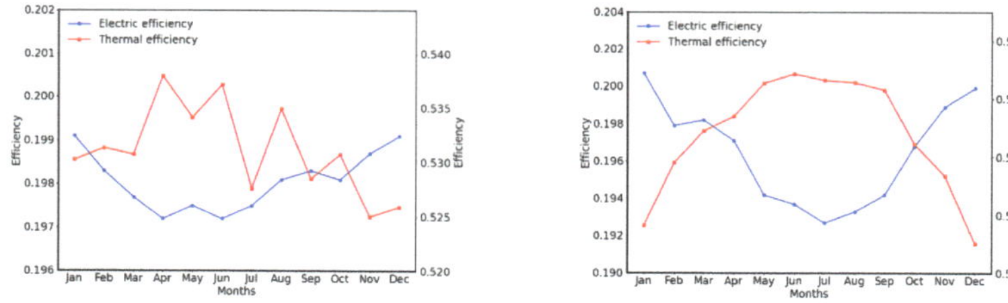

(**a**) (**b**)

Figure 5. Efficiencies per month for the locations of (**a**) Cuernavaca and (**b**) Mexicali.

- Figure 6 illustrates the values of electrical and thermal energy produced per month. The results were obtained by multiplying the point values of power, estimated by the model, by the number of daylight hours for each month and the number of days in each month [39,40]. In the case of Cuernavaca, the highest amount of energy produced was obtained in July, with a total of 387.93 kWh of electric energy and 1036.11 kWh of thermal energy generated. The lowest amount of energy produced was recorded in February, with a total of 239.38 kWh of electric energy and 641.26 kWh of thermal energy generated. The total energy produced was 3474.64 kWh of electricity and 9321.28 kWh of thermal energy. The average energy produced was 291.99 kWh of electric energy and 784.27 kWh of thermal energy. In the case of Mexicali, the highest amount of energy was produced in July, with a total of 515.16 kWh of electric energy and 1424.39 kWh of thermal energy. The lowest amount of energy produced was in December, at 177.69 kWh of electric energy and 448.88 kWh of thermal energy. The total energy produced was 4367.32 kWh of electric energy and 11,747.89 kWh of thermal energy. The average energy produced was 363.94 kWh of electric energy and 978.99 kWh of thermal energy.

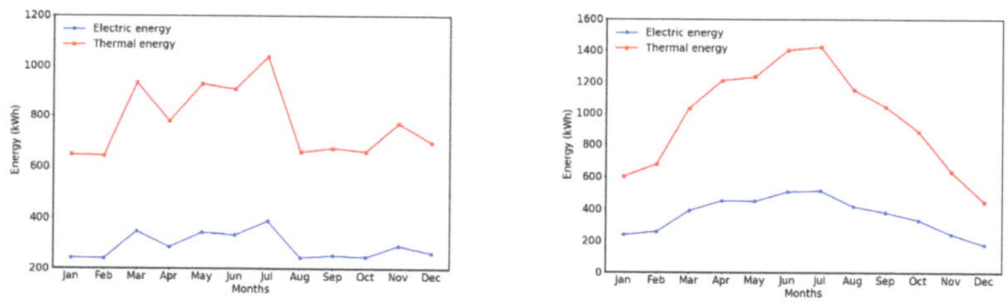

(**a**) (**b**)

Figure 6. Energy per month for the locations of (**a**) Cuernavaca and (**b**) Mexicali.

5. Parametric Study

The objective of the parametric study is to analyze the impact of geometric characteristics on the behavior of the model. The intention is to present information that will facilitate the choice of dimensions for an experimental prototype. In accordance with the studies presented by Calise [28] and Yazdanifard [25], it was considered appropriate to vary the values of the following parameters: the length of the pipe inside the SRC-PVT, the width of the faces of the SRC-PVT, the diameter of the pipe inside the SRC-PVT, and the mass flow of the HTF. The last parameter does not correspond to a geometrical characteristic of the SRC-PVT. Nevertheless, in practice, it is a widely used parameter as it allows for regulation from the system.

In order to develop the parametric study, the values indicated in Table 1 (boundary conditions), Table 2 (design parameters), and Table 3 (general parameters) were considered.

In establishing the ranges of variation of the geometric parameters analyzed and the mass flow, the typical dimensions of commercial receivers (solar panels and conventional solar collector pipes) were taken into account. This information is further detailed in the study by Tagle-Salazar [2].

5.1. Parameter Variations and Model Behavior

The objective of the parametric study is to analyze the impact of geometric characteristics on the behavior of the model. In light of the findings presented in the works of Calise [28] and Yazdanifard [25], it was deemed appropriate to vary the values of the following parameters: the pipe length inside the SRC-PVT, the width of the SRC-PVT faces, the pipe diameter inside the SRC-PVT, and the HTF mass flow. This last parameter does not correspond to a geometrical characteristic of the SRC-PVT. Nevertheless, in practice, it is a widely used parameter as it allows for the regulation of thermal energy extraction from the system. The variation ranges of the parameters were established by taking into account the dimensions of the model in its validation and the dimensions of commercial receivers (solar panels and conventional solar collector pipes), as presented by Tagle-Salazar.

- Figure 7 illustrates the temperature of the PV, absorber, and HTF outlet, as well as the electrical and thermal power and efficiency when varying the parameter of SRC-PVT pipe length. The value of this parameter in the validation is 10 m, which is comparable to the length of commercial receivers typically used in conventional collectors (approximately 8 m). To investigate the impact of varying pipe lengths, a range of +/−3 m was considered, resulting in a range of 7 m to 13 m.

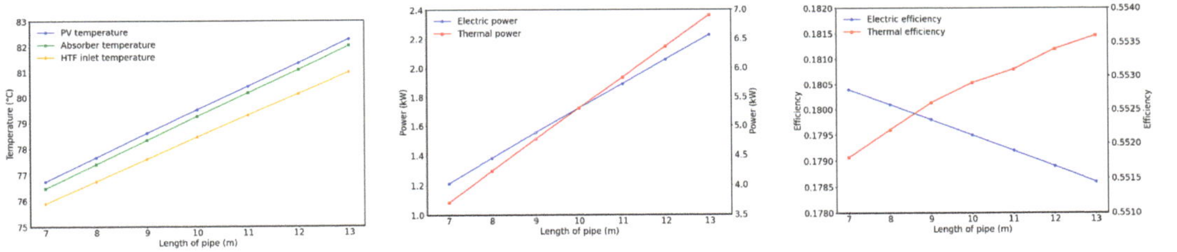

(a) (b) (c)

Figure 7. Results when varying the length of the pipe: (**a**) Temperature, (**b**) Power, and (**c**) Efficiency.

- Figure 8 illustrates the temperature of the PV, absorber, and HTF outlet, as well as the electrical and thermal power and efficiency when varying the SRC-PVT face width parameter. The value of this parameter in the validation is 6 cm. The solar panels have a width of 3 cm, while the diameter of a conventional SRC glass tube is 12 cm. Therefore, the operating range assigned to this parameter is from 3 cm to 12 cm.

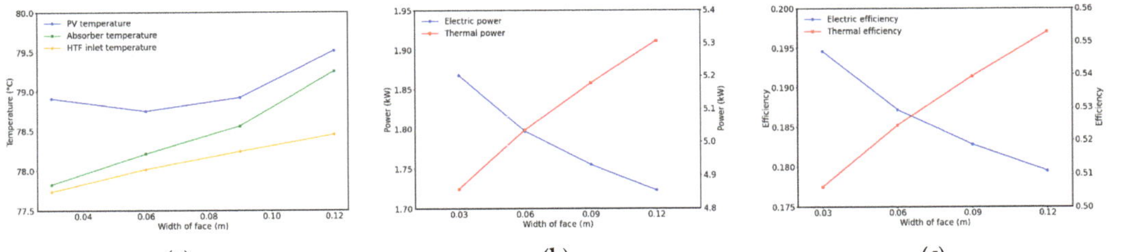

(a) (b) (c)

Figure 8. Results when varying the width of the face: (**a**) Temperature, (**b**) Power, and (**c**) Efficiency.

- Figure 9 shows the temperature of the PV, absorber, and HTF outlet, as well as the electrical and thermal power and efficiency when the pipe diameter parameter of the SRC-PVT is varied. The value of this parameter in the validation is 3 cm. The diameters of heat pipe collectors range between 8 mm and 14 mm while the diameter of a PTC inner tube can reach up to 9 cm. For this study, a range of 1.5 cm to 9 cm was deemed appropriate.

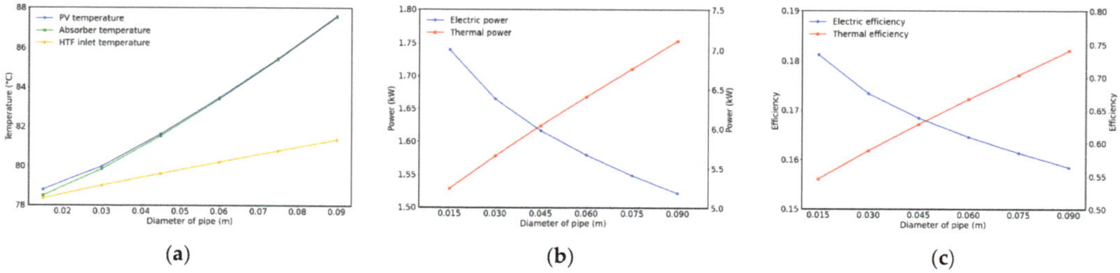

Figure 9. Results when varying the diameter of the pipe: (**a**) Temperature, (**b**) Power, and (**c**) Efficiency.

- Figure 10 illustrates the temperature of the PV, absorber, and HTF outlet, as well as the electrical and thermal power and efficiency when varying the HTF mass flow parameter. The value of this parameter in the validation is 0.15 kg/s. A brief consultation [41–43] revealed that the mass flow ranges for CPVT and PVT systems typically fall within the range of 0.02 kg/s to 0.18 kg/s. This range was adopted for the parametric study.

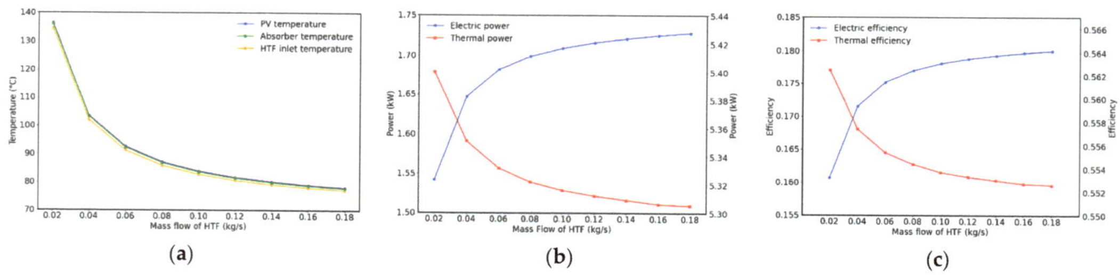

Figure 10. Results when varying the mass flow of HTF: (**a**) Temperature, (**b**) Power, and (**c**) Efficiency.

5.2. Analysis of the Behavior of the Model

- Figure 11 illustrates the comparative analysis of the initial and final values of power (electrical and thermal) when the analyzed parameters are increased. The percentage variation between the respective power values is also presented.
- It is evident that the parameter exhibiting the most significant influence on the electrical power is the length of the pipe, with a percentage variation of 83.92%. An increase in the length of the pipe and the mass flow of HTF has a favorable impact on thermal power, with values rising from 1.21 kW to 2.22 kW and 1.54 kW to 1.72 kW, respectively. Conversely, an increase in the width of the face and the diameter of the pipe has a detrimental effect on thermal power, with values declining from 1.86 kW to 1.72 kW and 1.74 kW to 1.52 kW, respectively.
- With respect to thermal power, the parameters that exhibit the greatest impact are the length of the pipe and the diameter of the pipe. These values represent percentage variations of 86.34% and 35.69%, respectively. An increase in the length of the pipe, the width of the face, and the diameter of the pipe results in an enhancement of thermal

power, with values rising from 3.7 kW to 6.9 kW, 4.85 kW to 5.3 kW, and 5.23 kW to 7.1 kW, respectively. Conversely, an increase in the mass flow of HTF has the opposite effect, reducing thermal power from 5.4 kW to 5.3 kW.

- Figure 12 illustrates the comparative analysis of the initial and final values of efficiency (electrical and thermal) when the analyzed parameters are increased. The percentage variation between the respective efficiency values is also presented.

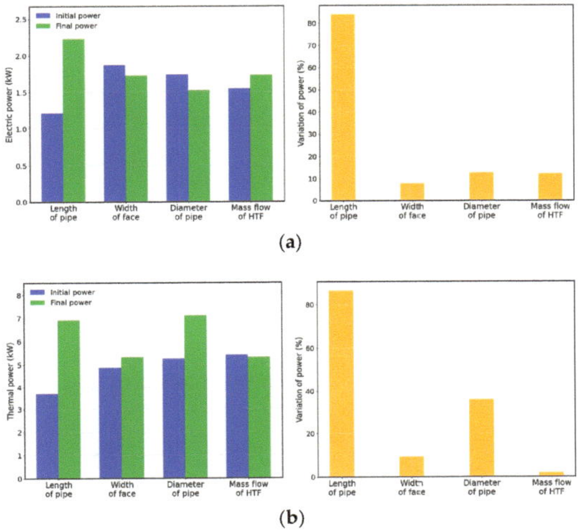

Figure 11. Initial and final power of the parameters evaluated, as well as their percentage variation: (**a**) Electrical power and (**b**) Thermal power.

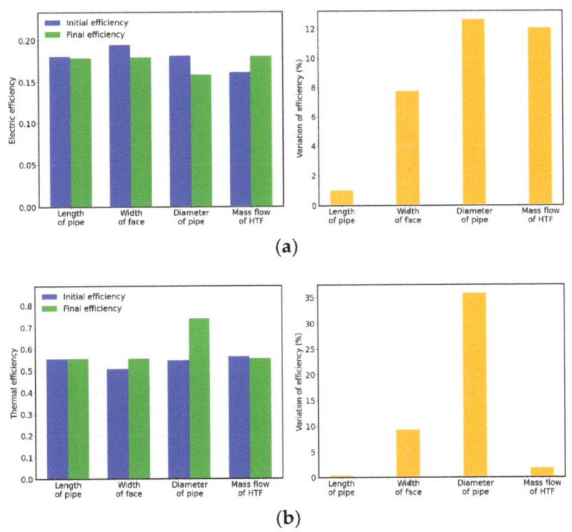

Figure 12. Initial and final efficiency of the parameters evaluated, as well as their percentage variation: (**a**) Electrical efficiency and (**b**) Thermal efficiency.

- It has been observed that the width of the faces, the diameter of the pipe, and the mass flow of HTF are parameters that have an impact on the electrical efficiency, with percentage variations of 7.75%, 12.57%, and 12.01%, respectively. An increase in the mass flow of HTF results in an improvement in electrical efficiency, with a corresponding rise from 0.16 to 0.18. Conversely, increases in the length of the pipe, the width of the face, and the diameter of the pipe have a detrimental impact on electrical efficiency, with a corresponding decline from 0.18 to 0.17, 0.19 to 0.17, and 0.18 to 0.15, respectively.

- In consideration of thermal efficiency, the parameter with the most significant impact is the diameter of the pipe, exhibiting a percentage variation of 35.69%. The length of the pipe is notable for its minimal impact, exhibiting a nearly imperceptible percentage variation of 0.32%. The enhancement in the width of the face and the diameter of the pipe contributes to an improvement in thermal efficiency, with values increasing from 0.50 to 0.55 and 0.54 to 0.74, respectively. Conversely, an increase in the mass flow of HTF has a detrimental effect on thermal efficiency, with an observed decrease from 0.56 to 0.55. However, this reduction is not particularly pronounced.

6. Conclusions

The performance of PTC solar collectors coupled to the SRC-PVT is particularly susceptible to variations in incident radiation and temperature. Such variations can be analyzed from the boundary conditions using the parameters of DNI, GHI, ambient temperature, and the HTF inlet temperature. In addition, a study of these variations can be conducted by taking into account the geometrical characteristics of the SRC-PVT and the mass flow, as these parameters are found to determine the use of incident radiation and heat transfer to the HTF, respectively.

The case study allowed the performance of the collector to be analyzed over the specified boundary conditions, which correspond to the particular meteorological conditions experienced in the locations of Cuernavaca and Mexicali. This study yielded the following results:

1. For Cuernavaca, the model demonstrated reduced variations per month in electrical and thermal power production (387.93 kWh to 239.38 kWh and 1036.11 kWh to 641.26 kWh);
2. For Mexicali, the model presented higher variations per month for thermal and electrical power production (515.16 kWh to 177.69 kWh and 1424.39 to 448.88 kWh);
3. It has been established that thermal energy production is higher than electrical energy production. However, it is more susceptible to variations caused by DNI, ambient temperatures, and the daylight hours of the month;
4. The average electrical efficiencies for both locations were found to be 0.196 and 0.198, respectively. This finding suggests that the PV stage, despite its comparatively lower production capacity, is less prone to fluctuations in environmental conditions.

The parametric study facilitated the identification of the design parameters exerting the most significant influence on collector performance. These parameters correspond to the geometric characteristics of the SRC-PVT and the mass flow, and their analysis yielded the following results:

- The pipe length exerts the most significant influence on the electrical and thermal power output (1.21 kW to 2.22 kW and 3.7 kW to 6.9 kW, respectively);
- The mass flow facilitates the enhancement of electrical power and efficiency (1.54 kW to 1.72 kW and 0.16 to 0.18, respectively), while concurrently preventing a significant reduction in thermal power and efficiency (5.4 kW to 5.3 kW and 0.56 to 0.55);
- It is evident that the accurate calibration of the HTF mass flow has the potential to enhance electrical power and efficiency without substantial compromise to thermal power and efficiency.
- The results of the case study provide a foundation for analyzing the viability and performance of the collector under study in locations in Mexico with conditions analogous to those presented. The parametric study provides a foundation for the sizing and construction of an experimental prototype of the studied collector. In

accordance with the aforementioned points, the script of the mathematical model developed is provided, so that interested researchers can carry out their own studies.

Author Contributions: Conceptualization, S.E.D.L.-A. and M.P.-S.; data curation, B.C.-D., E.F.-R., R.E.L.-P. and F.C.-D.; formal analysis, B.C.-D.; funding acquisition, H.A.-P., E.R.R.-G. and F.C.-D.; research, B.C.-D., F.C.-D. and E.R.R.-G. software J.A.A.-J. and N.V.-L.; methodology, B.C.-D., N.V.-L. and J.A.A.-J.; project management, S.E.D.L.-A.; resources, M.P.-S., H.A.-P. and J.A.A.-J.; supervision, S.E.D.L.-A.; validation, B.C.-D., J.A.A.-J. and N.V.-L.; visualization, R.E.L.-P., E.F.-R., H.A.-P. and B.C.-D.; writing: original draft, B.C.-D., M.P.-S. and S.E.D.L.-A.; writing: review and editing, B.C.-D.; writing: review and editing S.E.D.L.-A. and B.C.-D.; editing, S.E.D.L.-A. All authors have read and agreed to the published version of the manuscript.

Funding: This research received no external funding.

Data Availability Statement: The original data presented in the study are openly available in SRC-PVT Model (Fortran)—myCompiler. Available online: https://www.mycompiler.io/view/1NYQRM2YDkF (accessed on 13 January 2025).

Conflicts of Interest: The authors declare no conflict of interest.

Nomenclature and Subscripts

Acronyms	
CPV/T	Concentrated PhotoVoltaic/Thermal
CSP	Concentrated Solar Thermal Power
DNI	Direct Normal Irradiance
EES	Engineering Equation Solver
GHI	Global Horizontal Irradiance
MATLAB	MATrix LABoratory
HTF	Heat Fluid Transfer
PTC	Parabolic Trough Collector
PV	PhotoVoltaic cell
PVGIS	PhotoVoltaic Geographical Information System
SRC-PVT	Solar Receiver Collector—PhotoVoltaic Thermal
TRNSYS	Transient System Simulation Tool
Symbols	
A_{ap}	Aperture area, m^2
A_{PV}	Photovoltaic area, m^2
$C_{p,HTF}$	Specific heat of HTF, $J/kg \times K$
CR_{PTC}	Concentration ratio of PTC
G_{DNI}	Direct Normal Irradiance, W/m^2
IAM_{elec}	Electrical coefficient of incident angle modifier
P_{PV}	Photovoltaic power, W
$\dot{q}_{cond,abs-sub}$	Conductive heat transfer from absorber to substrate, W
$\dot{q}_{cond,PV-sub}$	Conductive heat transfer from photovoltaic cell to absorber, W
$\dot{q}_{conv,abs}$	Convective heat transfer from absorber, W
$\dot{q}_{conv,back,PTC}$	Convective heat transfer from the back face of PTC, W
$\dot{q}_{conv,front,PTC}$	Convective heat transfer from the front face of PTC, W
$\dot{q}_{conv,PV}$	Convective heat transfer from photovoltaic cell
$\dot{q}_{HTF}$	Heat fluid transfer power, W
$\dot{q}_{rad,abs}$	Radiative heat transfer from absorber, W
$\dot{q}_{rad,PTC}$	Radiative heat transfer from PTC, W
$\dot{q}_{rad,PV}$	Radiative heat transfer from photovoltaic cell, W
$\dot{q}_{sol,abs}$	Solar irradiance absorbed by absorber, W
$\dot{q}_{sol,PTC}$	Solar irradiance absorbed by PTC, W

$\dot{q}_{sol,PV}$	Solar irradiance absorbed by photovoltaic cell, W
T_{abs}	Temperature of absorber, °C
T_{PTC}	Temperature of concentrator, °C
$T_{in,HTF}$	Temperature of heat fluid transfer inlet, °C
$T_{out,HTF}$	Temperature of heat fluid transfer outlet, °C
T_{PV}	Temperature of photovoltaic cell, °C
T_{sub}	Temperature of substrate, °C
η_{elec}	Electrical efficiency
η_{opt}	Optical efficiency
η_{PV}	Photovoltaic efficiency
η_{th}	Thermal efficiency
$\in$	Heat transfer effectiveness
$\dot{m}_{HTF}$	Mass flow, Kg/s

References

1. Islam, M.T.; Huda, N.; Abdullah, A.B.; Saidur, R. A comprehensive review of state-of-the-art concentrating solar power (CSP) technologies: Current status and research trends. *Renew. Sustain. Energy Rev.* **2018**, *91*, 987–1018. [CrossRef]
2. Tagle-Salazar, P.D.; Nigam, K.D.P.; Rivera-Solorio, C.I. Parabolic trough solar collectors: A general overview of technology, industrial applications, energy market, modeling, and standards. *Green Process. Synth.* **2020**, *9*, 595–649. [CrossRef]
3. Pandey, A.K.; Tyagi, V.V.; Selvaraj, J.A.; Rahim, N.A.; Tyagi, S.K. Recent advances in solar photovoltaic systems for emerging trends and advanced applications. *Renew. Sustain. Energy Rev.* **2016**, *53*, 859–884. [CrossRef]
4. Gangwar, P.; Tripathi, R.P.; Singh, A.K. Solar photovoltaic tree: A review of designs, performance, applications, and challenges. *Energy Sources Part A Recovery Util. Environ. Eff.* **2021**, 1–28. [CrossRef]
5. Herez, A.; El Hage, H.; Lemenand, T.; Ramadan, M.; Khaled, M. Review on photovoltaic/thermal hybrid solar collectors: Classifications, applications and new systems. *Sol. Energy* **2020**, *207*, 1321–1347. [CrossRef]
6. Yazdanifard, F.; Ameri, M. Exergetic advancement of photovoltaic/thermal systems (PV/T): A review. *Renew. Sustain. Energy Rev.* **2018**, *97*, 529–553. [CrossRef]
7. Chavarría-Domínguez, B.; De León-Aldaco, S.E.; Velázquez-Limón, N.; Ponce-Silva, M.; Aguilar-Jiménez, J.A.; Chavarría-Domínguez, F. A Review of the Modeling of Parabolic Trough Solar Collectors Coupled to Solar Receivers with Photovoltaic/Thermal Generation. *Energies* **2024**, *17*, 1582. [CrossRef]
8. Renno, C.; Perone, A.; D'Agostino, D.; Minichiello, F. Performance Evaluation of a Linear CPV/T System in Different Working Conditions. *Energies* **2023**, *16*, 2115. [CrossRef]
9. Santana, J.P.; Rivera-Solorio, C.I.; Chew, J.W.; Tan, Y.Z.; Gijón-Rivera, M.; Acosta-Pazmiño, I. Performance Assessment of Coupled Concentrated Photovoltaic-Thermal and Vacuum Membrane Distillation (CPVT-VMD) System for Water Desalination. *Energies* **2023**, *16*, 1541. [CrossRef]
10. Acosta-Pazmiño, I.P.; Rivera-Solorio, C.I.; Gijón-Rivera, M. Hybridization of a parabolic trough-based thermal plant for industrial heat and power generation. *Renew. Energy* **2022**, *191*, 961–973. [CrossRef]
11. Gorouh, H.A.; Salmanzadeh, M.; Nasseriyan, P.; Hayati, A.; Cabral, D.; Gomes, J.; Karlsson, B. Thermal modelling and experimental evaluation of a novel concentrating photovoltaic thermal collector (CPVT) with parabolic concentrator. *Renew. Energy* **2022**, *181*, 535–553. [CrossRef]
12. Cabral, D.; Gomes, J.; Hayati, A.; Karlsson, B. Experimental investigation of a CPVT collector coupled with a wedge PVT receiver. *Sol. Energy* **2021**, *215*, 335–345. [CrossRef]
13. Herez, A.; Hage, H.E.; Lemenand, T.; Ramadan, M.; Khaled, M. Parabolic trough Photovoltaic/Thermal hybrid system: Thermal modeling, case studies, and economic and environmental analyzes. *Renew. Energy Focus* **2021**, *38*, 9–21. [CrossRef]
14. Renno, C.; Perone, A.; D'Agostino, D.; Minichiello, F. Experimental and Economic Analysis of a Concentrating Photovoltaic System Applied to Users of Increasing Size. *Energies* **2021**, *14*, 4968. [CrossRef]
15. Acosta-Pazmiño, I.; Rivera-Solorio, C.; Gijón-Rivera, M. Energetic and Economic Analyses of an LCPV/T Solar Hybrid Plant for a Sports Center Building in Mexico. *Energies* **2020**, *13*, 5681. [CrossRef]
16. Felsberger, R.; Buchroithner, A.; Gerl, B.; Wegleiter, H. Conversion and Testing of a Solar Thermal Parabolic Trough Collector for CPV-T Application. *Energies* **2020**, *13*, 6142. [CrossRef]
17. Gakkhar, N.; Soni, M.K.; Jakhar, S. Experimental and theoretical analysis of hybrid concentrated photovoltaic/thermal system using parabolic trough collector. *Appl. Therm. Eng.* **2020**, *171*, 115069. [CrossRef]
18. Otanicar, T.P.; Wingert, R.; Orosz, M.; McPheeters, C. Concentrating photovoltaic retrofit for existing parabolic trough solar collectors: Design, experiments, and levelized cost of electricity. *Appl. Energy* **2020**, *265*, 114751. [CrossRef]

19. Riahi, A.; Ali, A.B.H.; Fadhel, A.; Guizani, A.; Balghouthi, M. Performance investigation of a concentrating photovoltaic thermal hybrid solar system combined with thermoelectric generators. *Energy Convers. Manag.* **2020**, *205*, 112377. [CrossRef]
20. Karathanassis, I.K.; Papanicolaou, E.; Belessiotis, V.; Bergeles, G.C. Dynamic simulation and exergetic optimization of a Concentrating Photovoltaic/ Thermal (CPVT) system. *Renew. Energy* **2019**, *135*, 1035–1047. [CrossRef]
21. Maatallah, T.; Youssef, W.B. Simulation and performance analysis of concentrating photovoltaic/thermal collector (CPV/T) with three-sided thermal insulation based on coupled optothermal model. *Sol. Energy* **2019**, *181*, 308–324. [CrossRef]
22. Valizadeh, M.; Sarhaddi, F.; Adeli, M.M. Exergy performance assessment of a linear parabolic trough photovoltaic thermal collector. *Renew. Energy* **2019**, *138*, 1028–1041. [CrossRef]
23. Youssef, W.B.; Maatallah, T.; Menezo, C.; Nasrallah, S.B. Modeling and optimization of a solar system based on concentrating photovoltaic/thermal collector. *Sol. Energy* **2018**, *170*, 301–313. [CrossRef]
24. Mohsenzadeh, M.; Shafii, M.B.; Jafari mosleh, H. A novel concentrating photovoltaic/thermal solar system combined with thermoelectric module in an integrated design. *Renew. Energy* **2017**, *113*, 822–834. [CrossRef]
25. Yazdanifard, F.; Ebrahimnia-Bajestan, E.; Ameri, M. Performance of a parabolic trough concentrating photovoltaic/thermal system: Effects of flow regime, design parameters, and using nanofluids. *Energy Convers. Manag.* **2017**, *148*, 1265–1277. [CrossRef]
26. Brekke, N.; Otanicar, T.; DeJarnette, D.; Hari, P. A Parametric Investigation of a Concentrating Photovoltaic/Thermal System With Spectral Filtering Utilizing a Two-Dimensional Heat Transfer Model. *J. Sol. Energy Eng.* **2016**, *138*, 021007. [CrossRef]
27. Al-Nimr, M.A.; Dahdolan, M.-E. Modeling of a novel concentrated PV/T distillation system enhanced with a porous evaporator and an internal condenser. *Sol. Energy* **2015**, *120*, 593–602. [CrossRef]
28. Calise, F.; Vanoli, L. Parabolic Trough Photovoltaic/Thermal Collectors: Design and Simulation Model. *Energies* **2012**, *5*, 4186–4208. [CrossRef]
29. Incropera, F.P.; DeWitt, D.P.; Bergman, T.L.; Lavine, A.S. *Fundamentals of Heat and Mass Transfer*; Wiley: New York, NY, USA, 1996; Volume 6.
30. Bergman, T.L. *Fundamentals of Heat and Mass Transfer*; John Wiley & Sons: Hoboken, NJ, USA, 2011.
31. Cengel, Y. *Heat and Mass Transfer: Fundamentals and Applications*; McGraw-Hill Higher Education: New York, NY, USA, 2014.
32. EES: Engineering Equation Solver | F-Chart Software: Engineering Software. Available online: https://fchartsoftware.com/ees/ (accessed on 4 October 2023).
33. BenChavarria/SRC_PVT_Model. Available online: https://github.com/BenChavarria/SRC_PVT_Model/tree/master (accessed on 31 December 2024).
34. SRC-PVT Model (Fortran)—myCompiler. Available online: https://www.mycompiler.io/view/1NYQRM2YDkF (accessed on 12 November 2024).
35. Yingli Solar. *Ficha Técnica-Panel Solar Yingli*; Yingli Solar: Baoding, China, 2021.
36. Bernardo, L.R.; Perers, B.; Håkansson, H.; Karlsson, B. Performance evaluation of low concentrating photovoltaic/thermal systems: A case study from Sweden. *Sol. Energy* **2011**, *85*, 1499–1510. [CrossRef]
37. JRC Photovoltaic Geographical Information System (PVGIS)—European Commission. Available online: https://re.jrc.ec.europa.eu/pvg_tools/en/tools.html#MR (accessed on 7 October 2024).
38. Forristall, R. *Heat Transfer Analysis and Modeling of a Parabolic Trough Solar Receiver Implemented in Engineering Equation Solver*; NREL/TP-550-34169; National Renewable Energy Lab. (NREL): Golden, CO, USA, 2003. [CrossRef]
39. Weather Atlas. Anual y Mensual del Tiempo—Cuernavaca, México. Available online: https://www.weather-atlas.com/es/mexico/cuernavaca-clima (accessed on 17 October 2024).
40. Weather Atlas. Anual y Mensual del Tiempo—Mexicali, México. Available online: https://www.weather-atlas.com/es/mexico/mexicali-clima (accessed on 17 October 2024).
41. Kazemian, A.; Salari, A.; Hakkaki-Fard, A.; Ma, T. Numerical investigation and parametric analysis of a photovoltaic thermal system integrated with phase change material. *Appl. Energy* **2019**, *238*, 734–746. [CrossRef]
42. Hosseinzadeh, M.; Salari, A.; Sardarabadi, M.; Passandideh-Fard, M. Optimization and parametric analysis of a nanofluid based photovoltaic thermal system: 3D numerical model with experimental validation. *Energy Convers. Manag.* **2018**, *160*, 93–108. [CrossRef]
43. Gang, P.; Huide, F.; Huijuan, Z.; Jie, J. Performance study and parametric analysis of a novel heat pipe PV/T system. *Energy* **2012**, *37*, 384–395. [CrossRef]

energies

MDPI

Article

Evaluating the Impact of Dirt Accumulation on Photovoltaic Performance: Insights from an Experimental Plant in Brazil

Mylena Cruzinha da Silva [1], Dener A. de L. Brandao [1] and Igor A. Pires [1,2,*]

1 Graduate Program in Electrical Engineering (PPGEE), Federal University of Minas Gerais (UFMG), Belo Horizonte 31270-901, MG, Brazil; mi.lena.cruz@hotmail.com (M.C.d.S.); dalbrandao@ufmg.br (D.A.d.L.B.)

2 Department of Electronic Engineering (DELT), Federal University of Minas Gerais (UFMG), Belo Horizonte 31270-901, MG, Brazil

* Correspondence: iap@ufmg.br; Tel.: +55-31-3409-4874

Abstract: In recent decades, the use of photovoltaic (PV) modules as a source of electricity has grown significantly, driven largely by government incentives and regulatory advancements. However, merely installing these systems does not ensure optimal energy production, as maximizing solar energy capture requires additional measures. This study examines the impact of dirt accumulation on PV modules, focusing on a system installed at the School of Engineering of the Federal University of Minas Gerais (UFMG) in Belo Horizonte, Brazil. The research involved visual and thermographic analyses, as well as an evaluation of the I–V and P–V curve behavior for specific system arrays, which were cleaned during the 2024 dry season. Electrical parameters were compared between the dirty and cleaned states of the system, and the soiling ratio (SR) was calculated, ranging from 0.91 in the most affected case to 0.93 in the least affected.

Keywords: dirt accumulation; photovoltaic (PV) modules; qualitative analysis; quantitative analysis; I–V curve; P–V curve

1. Introduction

Photovoltaic (PV) solar energy is an increasingly prominent method of energy generation worldwide. According to the International Energy Agency (IEA), global renewable energy capacity increased by approximately 507 GW in 2023—nearly 50% higher than the previous year [1]. In Brazil, this growth is evident through the rising number of PV modules installed across various regions.

A significant portion of this growth in Brazilian territory can be attributed to policies and incentives established through regulations overseen by the National Electric Energy Agency (ANEEL). The agency began this process with Normative Resolution (RN) No. 482 [2], which has since undergone several updates, culminating in the current version, RN No. 1098, issued on 23 July 2024 [3].

When deciding to install a PV system, consumers should consider factors beyond regulations and initial operational conditions. Environmental factors, which contribute to the natural electrical wear of the equipment [4], are critical variables that affect the performance and longevity of any electrical component.

The module is a key component of a PV project, responsible for capturing solar radiation and converting it into electrical energy. Primarily made of semiconductor materials, the module features a multilayer structure [5], organized as follows:

https://doi.org/10.3390/en18051214

- Front cover: made of resistant and transparent glass, it has the function of protecting against the external action of the environment;
- Encapsulation: made of Ethylene-Vinyl Acetate (EVA), consisting of a double layer of insulation on the front and back, which coats and insulates the semiconductor material [5];
- PV cells: responsible for converting solar radiation into electricity through the photoelectric effect. Made of a semiconductor material, silicon, with monocrystalline, polycrystalline, or even a thin film structure, its terminals are interconnected by metallic conductors of copper, aluminum, or silver [6];
- Rear cover: lower layer of the modules with low thermal resistance for support purposes, made of polymer or acrylic;
- Junction box: electrical connection point for PV cells for the purpose of connection or association with other electrical points in the system.

The estimated useful life of a PV module is typically around 25 years. By the end of this period, its efficiency is expected to decrease by approximately 15%, depending on the manufacturer and the specific technology used [7,8]. However, this level of energy conversion efficiency is achieved only under conditions favorable to energy production. The efficiency testing of PV modules adheres to the standards set by IEC 61730-2 [9], which include evaluating electroluminescence (EL) to simulate conditions such as humidity and infrared (IR) exposure. These tests account for various temperature scenarios and assess factors like hot spots and light-induced degradation (LID).

The installation environment plays a crucial role in energy generation, and accurately assessing its impact on the PV module is essential for ensuring reliability, component integrity, safety during installation and use, as well as maximizing return on investment. Key environmental factors that contribute to losses and negatively affect productivity include variations in solar radiation, high operating temperatures of the modules, humidity, and the dust accumulation on the PV modules [4,10].

To illustrate the research data, the Tesla Experimental Photovoltaic Plant was used to demonstrate the effects of dirt, as the plant is strategically positioned for study due to its proximity to mining operations in neighboring cities and the urban nature of its own location. According to 2022 data from the Brazilian Institute of Geography and Statistics (IBGE) [11], Belo Horizonte is the sixth most populous municipality in the country.

It is estimated that there are regions where dirt, either directly or indirectly, causes losses of approximately 4–5% in just the year 2023 [12]. Therefore, given the initial context in which a photovoltaic module is tested and exposed to the environment throughout its useful life—often undergoing minimal maintenance—this study aims to investigate the impact of dirt and its effects on a system over the years. For this purpose, the Tesla Experimental Photovoltaic Plant, operational since 2016, was used as the object of study.

The main contributions of this paper are:

- To use the TESLA Experimental Photovoltaic Plant as a case study to elucidate the effects that the environment, particularly dirt, can have on a photovoltaic system exposed over the years without proper maintenance.
- To present a qualitative analysis of the photovoltaic modules studied to identify and validate the impact of dirt, thereby strengthening the reference literature for subsequent studies.
- To demonstrate, through quantitative analysis, the large-scale benefits of cleaning, as well as to determine the appropriate intervals for periodic maintenance, aiming to ensure the optimal performance of the system.

This paper is organized into the following sections: Section 2 provides a theoretical overview of loss factors, with a particular focus on dust accumulation, which is the main subject of this research and is further explored in the following section. Section 3 discusses the system under study and the methodology used for analysis. Section 4 presents the results obtained, followed by a discussion of these findings in Section 5. Finally, Section 6 offers the conclusion, summarizing the key takeaways from the study.

2. Analysis of Environmental Factors in PV Panels

In this paper, dirt is defined as the accumulation of particulate matter on PV modules, also referred to as dust accumulation. This buildup directly impacts electrical performance by continuously interfering with the absorption, scattering, and/or reflection of solar radiation. As a result, it reduces the amount of solar energy effectively absorbed by the PV module [13,14].

Research on this topic began in 1942 in the United States, when Hottel and Woertz linked the dust accumulation on solar collectors to a decrease in electrical performance. After three months of study, they estimated a performance drop of 4.7% due to soiling [15].

The production of electrical energy from PV modules is primarily influenced by temperature and solar irradiation. Ideal operating conditions are defined by standards and manufacturers, typically set at a PV cell temperature of 25 °C, an air mass of 1.5 AM, and a solar irradiation of 1000 W/m^2 [16]. However, due to environmental variability, maintaining these conditions consistently throughout the entire operational period is not possible.

Dirt accumulation on the modules is a variable that progressively reduces the efficiency of the system's electrical energy generation. This accumulation can also cause irreversible damage, such as accelerating the formation of hot spots, which may lead to the burning of internal module components [17,18]. An illustration of this effect is presented in Figure 1, where the lighter areas highlight regions with higher temperatures compared to the cooler, darker areas in blue. These variations occur within a module exposed to identical irradiance conditions [19]. Furthermore, the figure qualitatively demonstrates the distribution of dirt on the module. Prolonged exposure to rain and wind results in non-uniform dust accumulation, particularly concentrated at the right and lower edges.

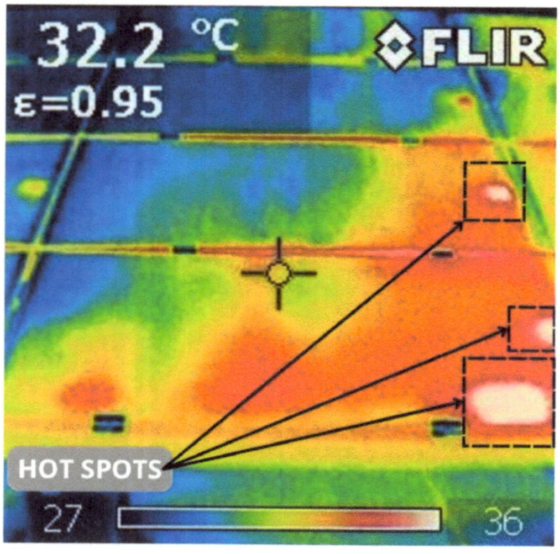

Figure 1. Example of a PV panel with hot spots.

The years of operation of a PV system, coupled with lack of regular maintenance such as periodic cleaning, can lead to the appearance of several visible defects. These include darkening of the EVA resin (browning), discoloration along the edges, and the formation of darkened lines (snail trails). Additionally, there are less visible issues, such as electrical mismatch, which is associated with high temperatures and accumulated dirt on the modules [20]. These defects and nonconformities are illustrated in Figure 2, captured during a visit to the Tesla PV Experimental Plant, which will be detailed in the next section.

Figure 2. Non-conformities present in PV cells.

The influence of individual cells on the performance of a PV module is significant, as the cell that absorbs the least amount of radiation determines the overall generation current of the module. When a system contains multiple damaged cells across various modules, the power output of the entire system can be severely impacted, leading to a substantial reduction in the overall efficiency of the project.

Dirt-related losses can vary significantly depending on the location of the study and the period of the analysis. As such, estimated loss values can differ widely. Desert environments or areas with high levels of air pollution generally experience more substantial losses and greater cleaning requirements. The literature reports losses ranging from 0.3% [21] to 1% [22] per day, with annual losses reaching up to 14% [14]. For a six-month period, a reduction of up to 80% is considered [21]. Therefore, it is critical to thoroughly assess the environment to ensure optimal performance of a PV system.

Understanding the particulate matter that typically accumulates on PV modules is crucial to determine the optimal cleaning frequency for the system. A related study conducted semi-quantitative tests to identify the primary chemical components present on a Tesla PV experimental plant. The analysis revealed that the most prominent elements in the collected samples included magnetite, hematite, clinochlore, alabandite, and quartz [23,24].

The compounds present are characteristic of the local urban environment and originated primarily from activities such as mining in the vicinity, civil construction, and natural soil erosion [23]. Magnetite and hematite, in particular, are materials known for their high emissivity, which means that they have a significant ability to emit energy through thermal radiation [25,26]. This characteristic suggests that these compounds contribute to heat retention in PV modules, potentially impacting their performance.

Urban environments and areas near mining zones have been previously studied and deserve attention for demonstrating a significant influence on the concentration of particulate matter suspended in the air, as well as requiring more frequent cleaning [18]. In these regions, suspended materials with approximate sizes of PM10 and PM2.5 have been described.

The composition of these particles is a separate study due to their interaction with the module surface. Deposition depends on factors such as installation geometry, sedimentation velocity, particle size, relative humidity, and precipitation. The effects of smaller particles are more harmful due to their strong adhesion to surfaces, making them harder to remove using natural agents, and their ability to cover the affected surface more uniformly. Conversely, larger particles tend to absorb and reflect more sunlight [18,27,28].

A study carried out on a PV plant located in a soccer stadium in Belo Horizonte, Brazil—just 1 km from the site of this research—examined not only the chemical compounds present in the modules but also the influence of rainfall on mitigating the impact of dirt. The study found that significant rainfall ($\geq$5 mm) played a significant role in reducing dust accumulation. During dry periods, the maximum power output was reduced by 13.7%, while after rainfall, this reduction was only 6.5% [24]. The inclination of the PV modules is another key factor that influences dust accumulation. Particulate materials tend to gather in specific areas of the modules, with the distribution largely determined by environmental factors such as wind, rain, and dew [21].

The influence of anthropogenic factors was exemplified through a study conducted in an urban area, where, during road repair work near the analyzed system, generation losses increased to 2.59% per day compared to 0.32% prior to the repairs [17].

A commonly used technique to identify issues caused by dust accumulation is the use of thermal or infrared imaging. Ref. [29] analyzed a PV plant located in Betim, a city close to the case study of this paper, using thermal imaging. It was found that 12.7% of the modules exhibited overheating points. The losses attributed to dirt were estimated to be 22.25%.

In Brazil, two standards are recommended for thermographic evaluations: NBR 15424 [30], which defines key terminology, and NBR 16292 [31], which outlines the procedures for safely conducting measurements using thermal images. The use of this evaluation method can indicate not only hot spots but also the distribution of dust accumulation on the PV modules, which may be uniform or non-uniform.

In addition to qualitative assessments, which include visual inspections and the use of thermal cameras, a quantitative analysis of the impact of dirt is essential to estimate system losses. It directly affects the module's I–V (current vs. voltage) and P–V (power vs. voltage) curves, which are essential indicators of the system's electrical performance.

These curves use PV module parameters such as maximum output power (P_{max}), maximum output current (I_{max}), maximum output voltage (V_{max}), open-circuit voltage (V_{oc}), and short-circuit current (I_{sc}). Figure 3 utilizes the experimental data used in this paper to illustrate, through simulation, the impact of dirt on the referenced curves of a module. The clean module's ideal operating condition is represented by the brown and green curves, where the maximum power point is achieved. In contrast, the curves in orange and blue represent the module's performance when the modules are dirty. Here, it is evident that the maximum operating point is not reached, and the curves exhibit deformations due to the reduced ability of the module to capture irradiance effectively [4].

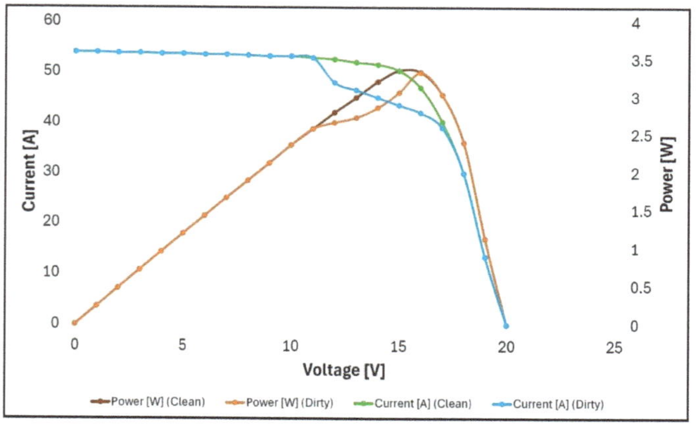

Figure 3. Comparison of I–V and P–V curves for clean and dirty PV module.

Performance Evaluation Coefficients of a PV System

The performance ratio (PR) is a key metric that represents the relationship between the energy generated under ideal conditions and the actual energy produced, providing insight into a system's efficiency. To measure the impact of dirt on a PV system, equations derived from international standards, such as IEC 61724, are employed [32]. In these calculations, the reference yield (Y_r) and final yield (Y_f) are essential parameters. This metric is crucial for estimating the actual energy generated, correlating it with the dust accumulation on PV modules, and determining whether system maintenance and cleaning are necessary to optimize performance [33].

$$PR = \frac{Y_r}{Y_f} \tag{1}$$

Measuring the short-circuit current of a PV system is crucial to evaluate the impact of energy efficiency losses in the system. This parameter, along with the maximum measured power, is one of the most effective methods for determining the soiling rate (SR). It is particularly advantageous because it does not require the direct removal of PV panels and does not physically interfere with the installation [14].

SR is a relationship obtained at the panel output and is based on the irradiance effectively received by the dirty panel (G_{dirt}) and the clean one (G_{clean}) [34], varying between 0 and 1, considering $SR = 1$ the clean system [32].

$$SR = \frac{G_{dirt}}{G_{clean}} \tag{2}$$

According to IEC 60891 [35], which defines the procedures to be followed to measure temperature and irradiance in the I–V curves in PV devices, the irradiance value G_x for dirty and clean panels is expressed as follows:

$$G_x = [I_{sc,M}(1 - \alpha(T_M - T_{ref}))]\left[\frac{G_{ref}}{I_{sc,cal}}\right], \tag{3}$$

where $I_{sc,M}$ and $I_{sc,cal}$ are the measured and calculated irradiance, respectively. The temperature coefficient of I_{sc} is α. T_M and T_{ref} correspond to the measured and reference temperature (25 °C). G_{ref} is the ideal irradiance of 1000 W/m^2.

Through the relationship between the irradiance effectively received by the panel and the SR for the dirty and clean panel, it is possible to obtain the soiling rate for short-circuit current $SR_{I_{sc}}$ and for maximum power $SR_{P_{max}}$ [34].

Although many studies report the influence of dirt on photovoltaic modules, there remains a gap regarding the impact of this factor over the years. This issue requires attention, as dirt that strongly adheres over time is not easily removed by rain or wind.

A study conducted by the European Solar Test Installation (ESTI) investigated the effects of dirt on 28 silicon modules exposed for approximately 30 years. According to the authors, the results were non-uniform; however, the I–V curves were significantly affected before cleaning and were almost fully restored after successive cleanings.

3. Tesla Photovoltaic Plant and Analysis Methods

The TESLA Power Engineering PV Experimental Plant, shown in Figure 4, was established as a key component of a research and development (R&D) project, a collaboration between the Federal University of Minas Gerais (UFMG) and the Bahia Electricity Company (COELBA). UFMG is located at Belo Horizonte, Brazil. This PV plant has been designed by the Tesla Power Engineering Laboratory, a research laboratory of the school of engineering of UFMG [36]. The plant began its operation in June 2016 and has an installed capacity of 37 kWp.

The plant was installed on the roof of the UFMG's School of Engineering, approximate location 19°52′10.81″ S 43°57′42.01″ W. The orientation of the building in relation to the geographic north, as well as that of the panels, is −7°, with a fixed inclination of the modules of 25° [37]. The city of Belo Horizonte, Brazil, has a highland tropical climate, characterized by distinct dry periods (autumn–winter) and rainy periods (spring–summer) [38]. In the southern hemisphere, summer occurs between December and March, while winter is concentrated between June and September.

Figure 4. TESLA power engineering PV experimental plant.

The plant consists of 154 polycrystalline silicon panels, a technology distinguished, among other factors, by the inclusion of a bypass diode in the junction box for protection. The PV modules used in the plant are Yingli Solar model 245P-32b, with a power rating of 245 W. Table 1 presents the electrical specifications based on the manufacturer's data [39].

Table 1. Yingli 245P-29b module parameters.

Peak power [Wp]	245
Max. power current [A]	7.6
Max. power voltage [V]	32
Short-circuit current [A]	8.22
Open circuit voltage [V]	40.8
Efficiency [%]	15.23

Since the plant was designed to meet academic scenarios, three different inverter models were chosen to investigate different performances. The panels are divided as shown in Figure 5 [40]. In this work, the impacts on the S1, F1, F2, and P2 arrangements are analyzed, whose connection diagrams are highlighted in green in Figure 5.

The first array corresponds to SMA inverter model STP12000TL-20 [41]. The second one is connected to Fronius inverter model IG Plus 150V-3 [42], while the third array uses PHB inverter model PHB20KNDT [43]. The respective electrical parameters, as provided by the manufacturers, are displayed in Table 2. The arrays are divided as follows:

- Array 1 (SMA inverter in purple): 45 panels, 3 parallel arrangements of 15 panels in series.
- Array 2 (Fronius inverter in orange): 50 panels, 5 parallel arrangements of 10 panels in series.
- Array 3 (PHP inverter in black): 57 panels, 3 parallel arrangements of 19 panels in series.
- Non-listed panels: two isolated modules, not connected to any of the inverters, intended for specific studies.

The TESLA PV Plant is equipped with a meteorological station and a small reference cell. These instruments measure irradiance and ambient temperature to which the system is exposed. The data collected from these measurements were used to support this research. To estimate the performance of the TESLA PV Experimental Plant in its current state of operation, qualitative and quantitative evaluations were used.

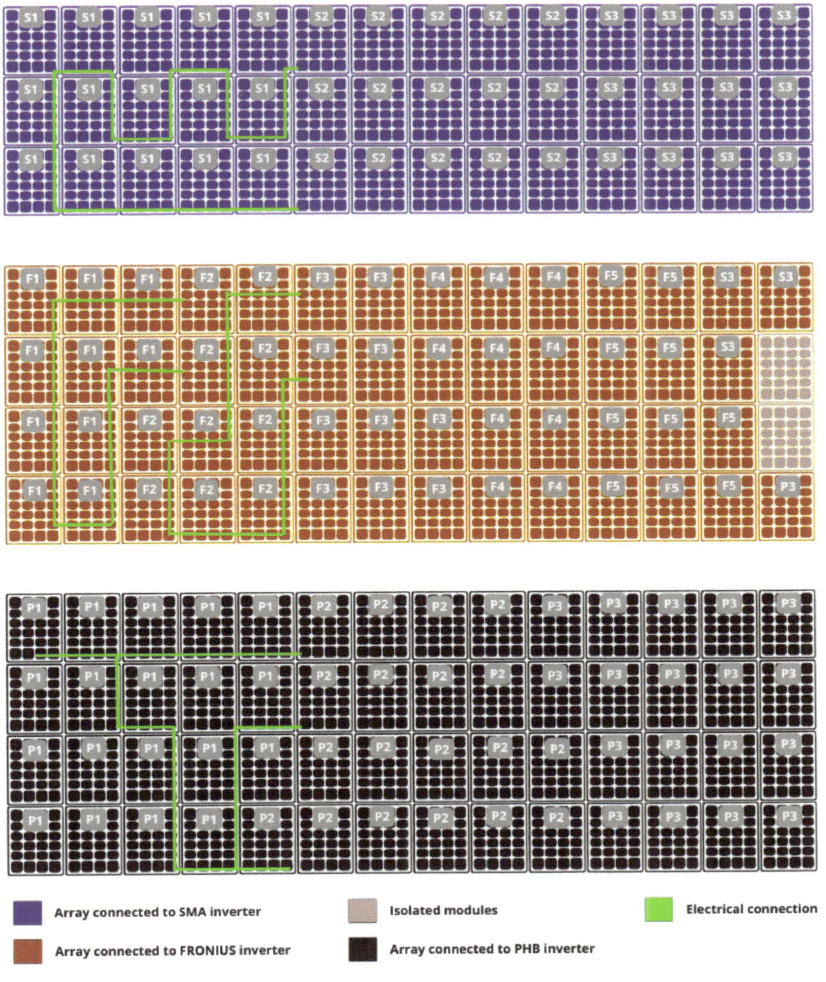

Figure 5. Panel distribution scheme by inverters.

Table 2. Inverters parameters.

Parameter	SMA	FRONIUS	PHB
Peak power [kW]	12	10	20
MPPT voltage range [V]	150–800	230–250	200–950
Max. input voltage [V]	1000	600	1000
Max. input current [A]	22	46.2	22
Number of DC inputs	2	6	4
AC output voltage [V]	400	400	380

3.1. Qualitative Analysis

The qualitative analysis of a PV plant relies on visual inspection to evaluate the impact of dirt on energy losses. This methodology uses thermal cameras to detect the presence of hot spots within the installation and to assess the distribution of dirt on the modules.

For this study, an infrared camera, model I60 from the manufacturer Flir [44], was used. Measurements were taken in accordance with the previously presented standards, both before and after cleaning. Table 3 presents the compiled camera data.

Table 3. I60 infrared camera parameters.

Infrared detector	25 micron pitch
Thermal sensitivity	0.1 °C to 25 °C
Measuring range	−20 °C to 350 °C
Thermal resolution	32,400 pixels
Spectral range	7.5 to 13 µm
Visual image resolution	2.3 Megapixels
Picture modes	Thermal, visual, fusion

3.2. Quantitative Analysis

The quantitative analysis aims to numerically quantify the approximate losses of the plant. The data were recorded using an I–V and P–V curve tracer, model PVA 600 PV Analyser, manufactured by Solmetric [45]. The parameters of the tracer are presented in Table 4.

Table 4. PVA 600 PV Analyser tracer parameters.

Current measurement range	0 to 20 A_{dc}
Voltage measurement range	0 to 600 V_{dc}
Minimum short circuit current	1 A_{dc}
Minimum open circuit voltage	20 V_{dc}
Measurement points per stroke	100
PV modules	Up to 420 modules
Operating temperature	0 °C a 50 °C

To use the equipment, a computer with the tracer software installed is required to enable wireless communication with the irradiance and temperature sensors. The PVA 600 PV Analyzer must be connected directly to the module, or in this case, to the array being analyzed, using MC4 connectors. The use of the device is exemplified in Figure 6. It is possible to observe the direct connection of the equipment being made in Figure 6b, the communication of the sensors in Figure 6c, the temperature sensor fixed to the bottom of the module in Figure 6d and the irradiance sensor in Figure 6e in the module next to the arrangement being tested.

Data acquisition began in May 2024, after the local rainy season, to account for the reduced influence of humidity on the modules and the increased concentration of particulate matter. The plant had not undergone maintenance for at least 6 years. According to data from the Brazilian National Institute of Meteorology (INMET) [46], the last significant rainfall in the region occurred on April 18, with 36 mm of rain. As a result, testing started on May 9 with visual inspections, and the first cleaning took place on May 13.

The data were collected both when the panels were dirty and after cleaning, always between 11 a.m. and 1 p.m. This time frame was chosen because it experiences less variation in temperature and irradiance, with levels concentrated between 900 and 1000 W/m^2.

For data collection, the methodology involved site visits for visual inspection of the modules, infrared camera recording, and data measurement using a curve tracer. Measurements were conducted both before and after each cleaning session following this approach: prior to cleaning, hot spots were identified through qualitative assessment and quantitative data collection. After cleaning, new qualitative and quantitative measurements were performed to facilitate a comparative analysis of the results.

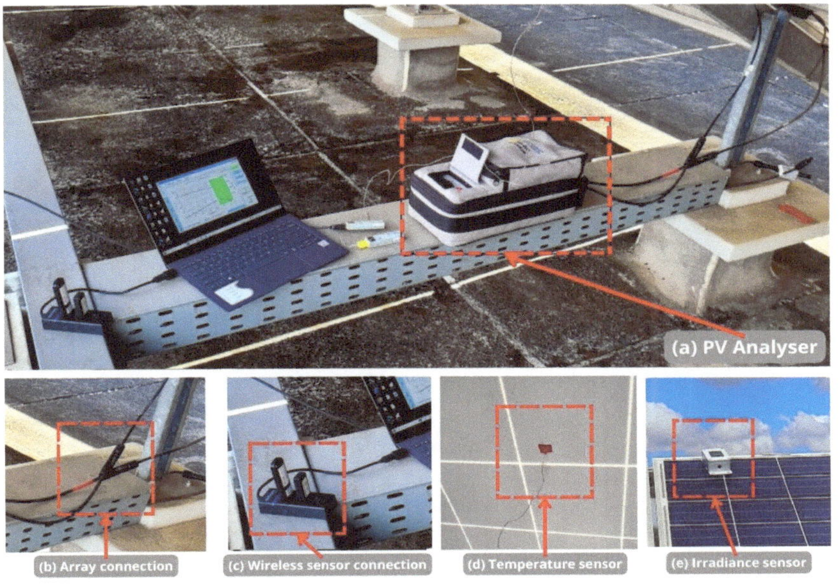

Figure 6. PVA 600 PV Analyser plotter in operation.

Among the various cleaning methods available in the literature, wet cleaning was selected, utilizing cloths and/or materials with soft bristles combined with water to remove and rinse the panels [29]. Three rounds of cleaning were carried out to achieve better results, considering the long period of exposure and the lack of maintenance of the system. The cleanings took place at intervals of 6 to 8 days.

4. Results

Figure 7 illustrates the various stages of dust accumulation on the panel belonging to the F1 arrangement. In the first stage, shown in Figure 7a, the dirt exhibits strong adhesion and was recorded in March 2024 during an initial visit to the TESLA PV Experimental Plant following a period of heavy rainfall. By May 13 (Figure 7b), the dirt had a dust-like appearance. Figure 7c,d depict the stages of dust accumulation between cleanings, where gradual deposition and formation of dirt are visible, including traces of water caused by dew formation.

Figure 7. Different stages of dirt on the PV Module.

Another critical aspect relates to the integrity of the modules. While some issues were observed during the initial visits, the defects became more evident after maintenance, revealing how dirt exacerbates these imperfections.

Among the observed nonconformities, including hot spots, browning, snail trails, and discoloration, it was found that 26.7% of the modules in the arrangement connected to the SMA inverter exhibited some form of imperfection. The Fronius inverter arrangement showed a higher percentage, with 45% of its modules affected. Similarly, the arrangement connected to the PHB inverter had 42% of its modules exhibiting these defects. The analysis of the arrangements individually will be presented below so that the results can be better visualized.

4.1. Arrangement S1

Using an infrared camera for qualitative analysis, the distribution of dirt on the PV modules in the S1 array, connected to the SMA inverter, was assessed. Figure 8 provides a comparison, showing the modules before cleaning, on the left, and after three rounds of cleaning, on the right.

According to data from the plant's meteorological station, the environmental conditions on May 9 and 28 at around 12:00 p.m. were relatively consistent, with temperatures averaging 27 °C. However, the qualitative assessment depicted in the Figure 8 highlights the effectiveness of dirt removal from the modules. Despite this, some hot spots persisted, such as the one observed on the right side of the lower module. This indicates that while cleaning improves overall module performance, underlying defects remain unaffected.

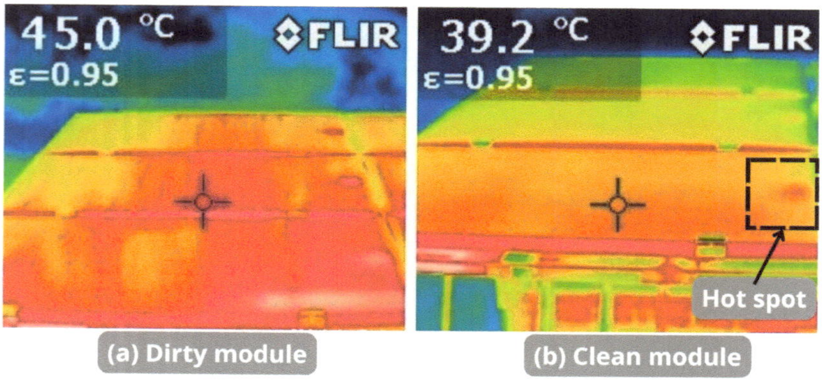

Figure 8. Comparison of thermal images of array S1.

Regarding the quantitative analysis, supported by the numerical data obtained through the curve tracer, the resulting curves are illustrated in Figure 9. This figure demonstrates the comparison of the S1 array's performance before maintenance, shown on the left, and after three rounds of cleaning, shown on the right. The I–V curves, depicted in red, and the P–V curves in blue were generated using the tracer's software, highlighting the impact of cleaning on the array's electrical performance.

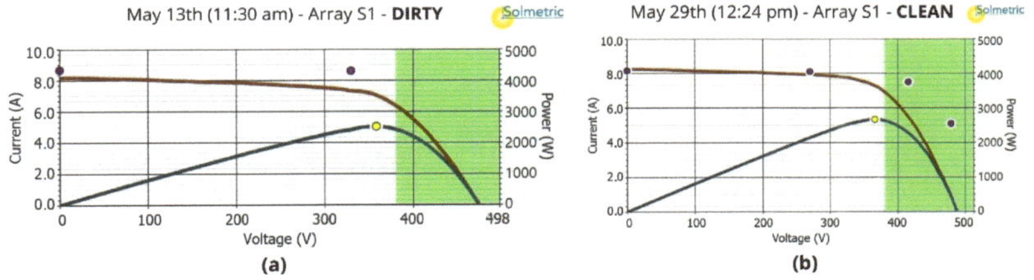

Figure 9. I–V and P–V curves for arrangement S1. Blue line: array P–V curve. Red line: array I–V curve. Yellow dot: array maximum power point. Purple dots: ideal I–V points.

The software estimates an ideal performance by considering previously input system parameters, including module specifications, quantity, inverter details, and location characteristics. This estimate is depicted with purple points representing ideal current tracking and a green zone indicating the optimal power operation range, highlighted by a yellow point. After proper cleaning, an improvement and adjustment of the curve can be observed, remaining closer to 8.22 for a longer duration—a value specified by the manufacturer as the standard operating short-circuit current. The numerical values obtained by the tracer indicate that the maximum power of the arrangement increased from 2389 W to 2708 W and the short-circuit current varied from 8.25 A to 8.36 A.

When analyzing the Soiling Ratio (SR), the key metric for determining the dust accumulation on the modules, the calculated value for the dirty state was $SR_{I_{sc}} = 0.93$, and after the complete cleaning of the arrangement, the SR became 1.02, being within the values estimated according to the literature.

However, $SR_{P_{max}}$, despite showing an increase, going from 0.77 to 0.85, a relative gain of 11.28%, still remained below the estimated value, indicating that some modules may have suffered permanent damage to their physical structure.

4.2. F1 and F2 Arrangement

Similar to the previous arrangement, the qualitative inspection yielded promising results for arrangements F1 and F2, as illustrated in Figure 10. A comparison was conducted between the modules that underwent the dirt removal process and those in arrangement F3, which remained consistently dirty. The results revealed a notable temperature difference of up to 7.5 °C at the panel's base, along with a visible change in the coloration of its structure.

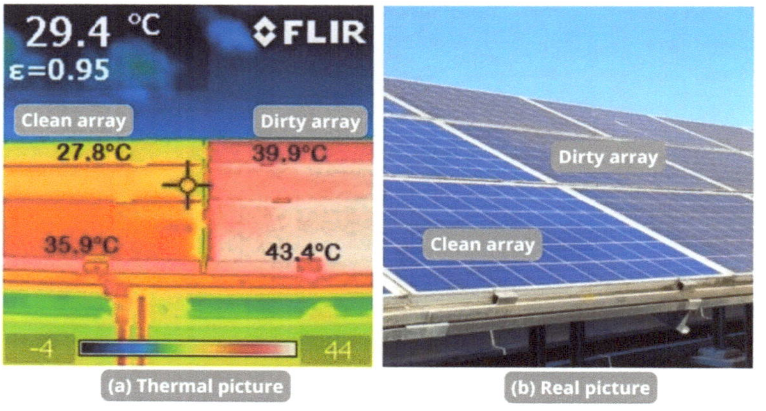

Figure 10. Comparison of arrangements F1 and F2: (**a**) thermal picture; (**b**) arrangements F1 (clean) and F2 (dirty).

The curve analysis also shows a positive improvement, with measurements taken at irradiances of 1029 W/m^2 and 944 W/m^2, respectively. Figure 11 displays the comparison between the clean and dirty systems. Initially, the current (shown in red) exhibited slight ripples, which improved after cleaning, along with an increase in both current and voltage at the maximum power point. Additionally, the module's operating temperature, measured using the sensor, decreased from 55.6 °C to 43.3 °C.

According to data from the local station on the measurement days, the ambient temperature was 26.2 °C and 23.8 °C, respectively. During the study period, which coincided with the autumn–winter season, the region experienced relatively stable climatic conditions, with no significant fluctuations in temperature, irradiance, humidity, or wind. As a result, the primary factor influencing the results was the maintenance performed on the system. After normalizing the irradiance values, it was concluded that there was a reduction of approximately 15% between the initial and final operating temperatures of the reference module.

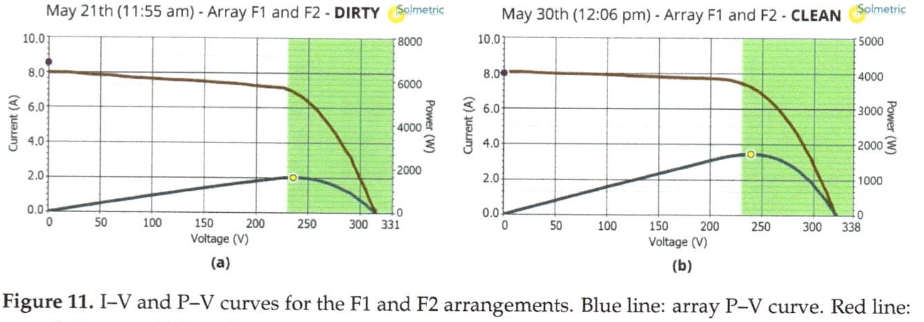

Figure 11. I–V and P–V curves for the F1 and F2 arrangements. Blue line: array P–V curve. Red line: array I–V curve. Yellow dot: array maximum power point. Purple dots: ideal I–V points.

The resulting analysis indicates that the maximum power increased from 1667 W to 1730 W, while the short-circuit current varied from 7.11 A to 8.07 A. The value of $SR_{I_{sc}}$ was positive, increasing from 0.92 to 1.02. As in the previous case, the analysis on the power soiling rate remained below expectations, with the initial $SR_{P_{max}}$ being 0.39 and the final $SR_{P_{max}}$ being 0.46.

4.3. Arrangement P1

During the visual inspection of the P1 arrangement, it was observed that the modules located at the ends had a greater accumulation of dirt compared to the others. This was likely due to the effects of wind over the years. As a result, the decision was made to perform the thermal analysis at this specific location. As shown in Figure 12, the results demonstrated an improvement similar to the other cases, with a significant reduction in dust accumulation.

In Figure 13a, deformities in the curves are evident, consistent with the previously identified non-uniform dirt distribution pattern. In Figure 13b, after the complete removal of dirt, there is a noticeable reduction in the deformation of the I–V curve, with values of I_{sc} remaining constant, close to 15 A, for a longer duration until reaching the maximum power point. This improvement is significantly beneficial for the inverter, as it indicates less difficulty in aligning the current and voltage, thereby ensuring more efficient power operation at the maximum power point.

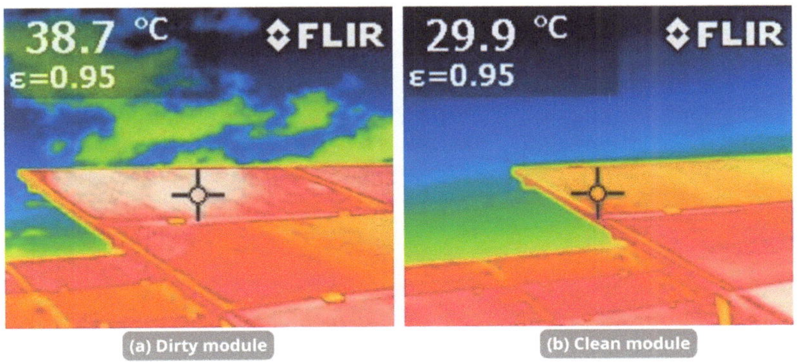

Figure 12. Comparison of thermal images of array P1.

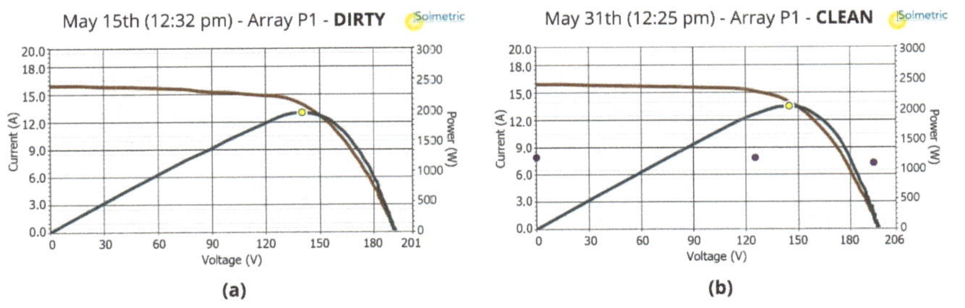

Figure 13. I–V and P–V curves for the P1 arrangement. Blue line: array P–V curve. Red line: array I–V curve. Yellow dot: array maximum power point. Purple dots: ideal I–V points.

Unlike the other systems analyzed, the P1 arrangement has a higher I_{sc}, which is a result of compliance with an internal project at the TESLA laboratory. Nevertheless, the analysis of the soiling rate can be conducted in the same manner as with the other systems, and the results obtained were equally consistent.

Based on the qualitative evaluation, $SR_{I_{sc}}$ improved from 0.91 to 1.02. However, as observed in the other arrangements, $SR_{P_{max}}$ did not reach the expected value of 1, remaining at a maximum of 0.72 for the clean system.

5. Discussion

Based on the qualitative analysis, the visual inspection revealed that all arrangements exhibited losses related to degradation over time, which were exacerbated by the prolonged period of exposure without maintenance. The removal of dirt resulted in a reduction in the operating temperature of the modules and a notable improvement in performance parameters. This improvement aligns with findings in the literature, where chemical compounds such as magnetite and hematite are known to cause high heat retention, further aggravating the degradation process.

In the quantitative analysis, which primarily focused on the SR factor that assesses the dirt present on the modules, all the configurations examined achieved a dirt rate of 1, considered the ideal value. The recovery of this factor, coupled with the export in the I–V curve, demonstrates the effectiveness of cleaning in regenerating $SR_{I_{sc}}$, even for systems exposed to environmental conditions over the years.

However, when analyzing the $SR_{P_{max}}$, none of the studied arrangements achieved values corresponding to 1, with results remaining at 0.85, 0.46, 0.46, and 0.72 for the S1, F1, F2, and P1 arrangements, respectively. This indicates that environmental factors are causing direct losses in the maximum power output of the module. The power reduction can be explained by the action of the bypass diode, which is present in silicon modules. This protective mechanism operates when shading and localized heating occur; however, as shown in the analysis, some areas experience permanent heating, leading to a decrease in the module's efficiency. It is also important to note that all arrangements exhibited non-conformities.

A comparison can also be made between the values obtained for $SR_{P_{max}}$ and the P–V curve of the arrangements. Arrangements F1 and F2 exhibit a lower soiling rate, which results in a lower peak in the P–V curve compared to the other arrangements.

Given the importance of properly cleaning PV modules to prevent premature damage and ensure optimal system performance, and based on the performance rate, it is recommended that the modules undergo annual maintenance in August, considering the location of the TESLA PV Experimental Plant.

The data analyzed indicate that in August, there is extended exposure to the local dry period, which requires attention, as it promotes the accumulation of particulate matter on the modules. Under these conditions, the recommended performance ratio (PR) is approximately 72.8%. The dry period typically ends with sporadic rains in November, which, if greater than 5 mm, should help maintain the effects of cleaning. The rainy season, concentrated from January to March, will likely lead to a gradual decrease in PR until the next scheduled maintenance in August.

The data clearly demonstrate the need for periodic maintenance to avoid long-term damage to the system, with the maintenance frequency being dependent on the environment the PV modules are exposed to. This necessity is further supported by the restoration of the short-circuit current, the improvement in the I–V curve for all studied arrangements, as well as the reduction in operating temperature and the improvement in parameters such as the irradiance effectively captured by the module and the performance rate.

Over the long term, regular annual maintenance should help restore the TESLA PV Experimental Plant to optimal operating conditions, compensating for losses in maximum power output.

6. Conclusions

In light of the increasing demand for PV solar energy for both small and large-scale generation, this study emphasizes the importance of regular maintenance to ensure optimal system performance and mitigate long-term losses, in addition to demonstrating the effectiveness of cleaning in restoring the parameters of systems exposed to prolonged environmental conditions, as evidenced by the reestablishment of the I–V curves, the analysis highlights the critical role of maintenance in maintaining optimal system performance.

The gradual accumulation of particulate matter on the modules directly impacts system performance over time. As dirt accumulates, it reduces the system's ability to capture irradiance, which is crucial for efficiently converting solar energy into electrical power.

Only installing a PV system does not ensure maximum efficiency throughout its lifespan. Natural wear and tear on electrical components can be significantly exacerbated by inadequate maintenance. This is evident through the evaluation of the soiling rate, which serves as a crucial indicator of system performance degradation.

The results indicate that the soiling rate $SR_{I_{sc}}$ for the analyzed arrangements ranged from 0.91 to 0.93 when measured with the modules in a dirty state. After successive cleanings, the SR consistently returned to its ideal value, as measurements were taken immediately following the cleaning process. Moreover, thermal imaging proved effective in visualizing dust accumulation on the modules. The quantitative analysis further demonstrated a reduction in operating temperature post-cleaning, a critical factor that directly impacts the energy generation efficiency of the system.

The soiling rate analyzed in terms of power ($SR_{P_{max}}$) for the three cases remained below the estimated value, indicating that prolonged exposure to environmental factors, without proper periodic maintenance, may have caused irreversible damage to the electrical components of the cells.

The analysis methods employed were based on established regulations and literature, aligning with expectations, particularly concerning the performance of $SR_{I_{sc}}$. For other aspects, the same theoretical framework was used to justify the system's performance in relation to $SR_{P_{max}}$.

Author Contributions: Conceptualization, M.C.d.S. and I.A.P.; methodology, M.C.d.S. and I.A.P.; software, M.C.d.S. and D.A.d.L.B.; validation, M.C.d.S. and I.A.P.; formal analysis, M.C.d.S., D.A.d.L.B., and I.A.P.; investigation, M.C.d.S.; resources, I.A.P.; data curation, M.C.d.S.; writing—original draft preparation, M.C.d.S. and D.A.d.L.B.; writing—review and editing, M.C.d.S., D.A.d.L.B., and I.A.P.; visualization, M.C.d.S., D.A.d.L.B., and I.A.P.; supervision, I.A.P.; project administration, I.A.P.; funding acquisition, I.A.P. All authors have read and agreed to the published version of the manuscript.

Funding: This study was financed in part by the Coordenação de Aperfeiçoamento de Pessoal de Nível Superior–Brasil (CAPES)–Finance Code 001.

Data Availability Statement: The data are contained within the article.

Acknowledgments: This work was carried out with the support of the Coordination for the Improvement of Higher Education Personnel—Brazil (CAPES) through the Academic Excellence Program (PROEX), in part by Conselho Nacional de Desenvolvimento Científico e Tecnológico (CNPq), and in part by the Minas Gerais state government agency Fundação de Amparo à Pesquisa do Estado de Minas Gerais (FAPEMIG).

Conflicts of Interest: The authors declare no conflicts of interest.

References

1. International Energy Agency (IEA). Renewables; Technical Report. 2023. Available online: https://www.iea.org/reports/renewables-2023 (accessed on 25 October 2024).
2. Agencia Nacional de Energia Eletrica (ANEEL). *Resolução Normativa N° 482*; Technical Report; ANEEL: Brasília, Brazil, 17 April 2012.
3. Agencia Nacional de Energia Eletrica (ANEEL). *Resolução Normativa N° 1098*; Technical Report; ANEEL: Brasília, Brazil, 23 July 2024.

4. Hickel, B.M.; Deschamps, E.M.; do Nascimento, L.R.; Rüther, R.; Simões, G.C. Análise da influência do acúmulo de sujeira sobre diferentes tecnologias de módulos fv: Revisão e medições de campo. In Proceedings of the Congresso Brasileiro de Energia Solar (CBENS), Belo Horizonte, Brazil, 4–7 April 2016; pp. 1–8. [CrossRef]
5. Kant, K.; Shukla, A.; Sharma, A.; Biwole, P.H. Thermal response of poly-crystalline silicon photovoltaic panels: Numerical simulation and experimental study. *Sol. Energy* **2016**, *134*, 147–155. [CrossRef]
6. Villalva, M.G. Chapter 3—Células e Módulos Fotovoltaicos. In *Energia Solar Fotovoltaica: Conceitos e Aplicações*; Gazolli, J.R., Ed.; Érica: Belo Horizonte, Brazil, 2012; pp. 65–98.
7. Klugmann-Radziemska, E.; Ostrowski, P. Chemical treatment of crystalline silicon solar cells as a method of recovering pure silicon from photovoltaic modules. *Renew. Energy* **2010**, *35*, 1751–1759. [CrossRef]
8. Cruz, F.T.; Isidoro, M.H.; Santos, I. Descarte, reciclagem e logística reversa: Análise do fim de vida útil dos painéis fotovoltaicos. *Braz. J. Dev.* **2020**, *6*, 73294–73309. [CrossRef]
9. *EN IEC 61730-2:2019*; Photovoltaic (PV) Module Safety Qualification—Part 2: Requirements for Testing. International Electrotechnical Commission (IEC): Geneva, Switzerland, 2016. Available online: https://iss.rs/sr_Cyrl/project/show/iss:proj:53597 (accessed on 25 October 2024).
10. Araújo, D.N.; de Carvalho, P.C.M.; Dupont, I.M. Efeitos da acumulação de sujeira sobre o desempenho de módulos fotovoltaicos. *Rev. Tecnol.* **2019**, *40*, 1–23. [CrossRef]
11. Brazilian Institute of Geography and Statistics (IBGE). Available online: https://cidades.ibge.gov.br/brasil/mg/belo-horizonte/panorama (accessed on 21 January 2024).
12. Kazem, H.A.; Chaichan, M.T.; Al-Waeli, A.H.; Sopian, K. A review of dust accumulation and cleaning methods for solar photovoltaic systems. *J. Clean. Prod.* **2020**, *276*, 123187. [CrossRef]
13. Bessa, J.G.; Micheli, L.; Almonacid, F.; Fernández, E.F. Monitoring photovoltaic soiling: Assessment, challenges, and perspectives of current and potential strategies. *Iscience* **2021**, *24*, 102165. [CrossRef]
14. Hickel, B.M. O Impacto do Desempenho de Sistemas Fotovoltaicos Causado pelo Acumulo de Sujeira Sobre os Módulos FV—Metodologia e Avaliação Através de Curvas IxV em Campo. Master's Thesis, Universidade Federal de Santa Catarina, Florianópolis, Brazil, 2017.
15. Hottel, H.C.; Woertz, B.B. The performance of flat-plate solar-heat collectors. *Trans. Am. Soc. Mech. Eng.* **1942**, *64*, 91–103. [CrossRef]
16. Silvestre, S. Chapter 7—Strategies for fault detection and diagnosis of PV systems. In *Advances in Renewable Energies and Power Technologies*; Yahyaoui, I., Ed.; Elsevier: Amsterdam, The Netherlands, 2018; pp. 231–255. [CrossRef]
17. da Fonseca, J.E.F.; de Oliveira, F.S.; Prieb, C.W.M.; Krenzinger, A. Degradation analysis of a photovoltaic generator after operating for 15 years in southern Brazil. *Sol. Energy* **2020**, *196*, 196–206. [CrossRef]
18. Bhattacharya, S.; Sadhu, P.K.; Singh, N.K. Effect of soiling loss in solar photovoltaic modules and relation with particulate matter concentration in nearby mining and industrial sites. *Microsyst. Technol.* **2024**. [CrossRef]
19. Tonolo, E.A.; Junior, J.U. Análise de perdas por degradação e descasamento de módulos em um sistema fotovoltaico da utfpr–campus curitiba. In Proceedings of the Anais Congresso Brasileiro de Energia Solar (CBENS), Fortaleza, Brazil, 1–5 June 2020. [CrossRef]
20. Carvalho, V.; Dantas, J.C.; Varella, F.K.D.O.M. Sistemas fotovoltaicos: Análise de perdas por Mismatch. *Rev. Eletrôn. Eng. Elétr. Eng. Mecân.* **2021**, *3*, 70–80. [CrossRef]
21. Klimm, E.; Lorenz, T.; Weiss, K.A. Can Anti-Soiling Coating on Solar Glass Influence the Degree of Performance Loss Over Time of PV Modules Drastically? In Proceedings of the 28th European Photovoltaic Solar Energi Conference and Exhibition, Amsterdam, The Netherlands, 22–26 September 2014. [CrossRef]
22. Khan, M.Z.; Ghaffar, A.; Bahattab, M.A.; Mirza, M.; Lange, K.; Abaalkheel, I.M.S.; Alqahtani, M.H.M.; Aldhuwaile, A.A.A.; Alqahtani, S.H.; Qasem, H.; et al. Outdoor performance of anti-soiling coatings in various climates of Saudi Arabia. *Sol. Energy Mater. Sol. Cells* **2022**, *235*, 111470. [CrossRef]
23. de Oliveira, F.A.F.; Pereira, R.G.S.; Porto, M.P.; Ferreira, R.A.M. Long-wavelength infrared optical properties of polycrystalline silicon PV modules under clean and soiled conditions for thermography analysis. *Sol. Energy* **2024**, *284*, 113028. . [CrossRef]
24. Fraga, M.M.; de Oliveira Campos, B.L.; de Almeida, T.B.; da Fonseca, J.M.F.; Lins, V.D.F.C. Analysis of the soiling effect on the performance of photovoltaic modules on a soccer stadium in Minas Gerais, Brazil. *Sol. Energy* **2018**, *163*, 387–397. [CrossRef]
25. Pabisiak, T.; Kiejna, A. Fe adsorption on hematite (α-Fe_2O_3)(0001) and magnetite (Fe_3O_4)(111) surfaces. *J. Chem. Phys.* **2014**, *141*, 134707. [CrossRef]
26. Oskay, C.; Meißner, T.M.; Dobler, C.; Grégoire, B.; Galetz, M.C. Scale formation and degradation of diffusion coatings deposited on 9% Cr steel in molten solar salt. *Coatings* **2019**, *9*, 687. [CrossRef]
27. Borah, P.; Micheli, L.; Sarmah, N. Analysis of Soiling Loss in Photovoltaic Modules: A Review of the Impact of Atmospheric Parameters, Soil Properties, and Mitigation Approaches. *Sustainability* **2023**, *15*, 16669. [CrossRef]

28. Ghosh, A. Soiling losses: A barrier for India's energy security dependency from photovoltaic power. *Challenges* **2020**, *11*, 9. [CrossRef]
29. Costa, A.L.C.; Hirashima, S.Q.d.S.; Ferreira, R.V. Operação e manutenção de sistemas fotovoltaicos conectados à rede: Inspeção termográfica e limpeza de módulos FV. *Ambiente Constr.* **2021**, *21*, 201–220. [CrossRef]
30. *NBR:15424*; Non-Destructive Testing—Infrared Thermography—Terminology. Technical report; ABNT—Brazilian Association of Norms Techniques: Rio de Janeiro, Brazil, 13 July 2022.
31. *NBR:16292*; Non-Destructive Testing—Thermography—Measurement and Compensation for Reflected Temperature Using Thermographic Cameras; Technical Report. ABNT—Brazilian Association of Norms Techniques: Rio de Janeiro, Brazil, 13 June 2014.
32. *IEC61724-1* ; Photovoltaic System Performance—Part 1: Monitoring. International Electrotechnical Commission: Geneva, Switzerland, 2021. Available online: https://webstore.iec.ch/en/publication/65561 (accessed on 25 October 2024).
33. Gaio, J.N.; Campos, K.M.A.D.; Tiepolo, G.M. Determination of the Ideal Time for Cleaning of Photovoltaic Panels Aiming at Better Productivity—Case Study of the Implanted Grid-Connected Photovoltaic System at the Federal University of Technology of Paraná. *Braz. Arch. Biol. Technol.* **2018**, *61*, e18000210. [CrossRef]
34. Costa, S.C.S.; Kazmerski, L.L.; Diniz, A.S.A. Impact of soiling on Si and CdTe PV modules: Case study in different Brazil climate zones. *Energy Convers. Manag. X* **2021**, *10*, 100084. [CrossRef]
35. *IEC60891*; Photovoltaic Devices—Procedures for Temperature and Irradiance Corrections to Measured I–V Characteristics; Technical Report; International Electrotechnical Commission (IEC): Geneva, Switzerland, 2021.
36. Antunes, H.M.A. Conversor Multifuncional Reconfigurável e Tolerante a Falhas para Microrredes de Energia Elétrica. Ph.D. Thesis, Universidade Federal de Minas Gerais, Belo Horizonte, Brazil, 2018.
37. Ferreira, R.V.; Silva, S.M.; Brandao, D.I.; Antunes, H.M.; Fernandes, N.T. Assessment of energy storage viability for a PV power plant injecting during peak load time. In Proceedings of the 2017 IEEE 8th International Symposium on Power Electronics for Distributed Generation Systems (PEDG), Florianópolis, Brazil, 17–20 April 2017; pp. 1–6. [CrossRef]
38. Assis, W.L.; Abreu, M.L.d. O clima urbano de Belo Horizonte: Análise têmporo-espacial do campo térmico e hígrico. *Rev. Ciências Humanas* **2010**, *10*, 47–63.
39. Datasheet Module Yingli. Available online: http://buildingenergy.be/wp-content/uploads/2015/06/Yingli_Poly_235W_240W_245W_250W_255W_EN.pdf (accessed on 14 November 2024).
40. Foureaux, N.C.; Antunes, H.M.A.; Silva, S.M.; de Jesus Cardoso Filho, B.; Júnior, F.M.R.; de Souza Brito, J.A. Decisoes de Projeto da Usina Experimental Fotovoltaica TESLA Engenharia de Potencia. In Proceedings of the Anais Congresso Brasileiro de Energia Solar (CBENS), Belo Horizonte, Brazil, 4–7 April 2016; pp. 1–8. [CrossRef]
41. Datasheet SMA Inverter. Available online: https://files.sma.de/downloads/STP12000TL-DEN1723-V10web.pdf (accessed on 14 November 2024).
42. Datasheet FRONIUS Inverter. Available online: https://www.enfsolar.com/pv/inverter-datasheet/68 (accessed on 14 November 2024).
43. Datasheet PHB Inverter. Available online: https://www.energiasolarphb.com.br/wp-content/uploads/2021/07/Inversor-PHB20KN-DT-datasheet-v2.3.pdf (accessed on 14 November 2024).
44. Datasheet InfraRed Camera. Available online: https://www.instrumart.com/assets/i60-datasheet.pdf (accessed on 14 November 2024).
45. Datasheet PV Analyze. Available online: https://www.solmetric.com/wp-content/uploads/SoftwareDownloads/Solmetric (accessed on 14 November 2024).
46. INMET (National Institute of Meteorology). Available online: https://portal.inmet.gov.br/dadoshistoricos (accessed on 4 June 2024).

energies

MDPI

Article

Stochastic Capacity Expansion Model Accounting for Uncertainties in Fuel Prices, Renewable Generation, and Demand

Naga Srujana Goteti [1], Eric Hittinger [2] and Eric Williams [2,*]

[1] Electric Power Research Institute, Washington, DC 20036, USA; ngoteti@epri.com
[2] Rochester Institute of Technology, Rochester, NY 14623, USA; eshgpt@rit.edu
* Correspondence: exwgis@rit.edu; Tel.: +1-(585)-475-7211

Abstract: Capacity expansion models for electricity grids typically use deterministic optimization, addressing uncertainty through ex-post analysis by varying input parameters. This paper presents a stochastic capacity expansion model that integrates uncertainty directly into optimization, enabling the selection of a single strategy robust across a defined range of uncertainties. Two cost-based risk objectives are explored: "risk-neutral" minimizes expected total system cost, and "risk-averse" minimizes the most expensive 5% of the cost distribution. The model is applied to the U.S. Midwest grid, accounting for uncertainties in electricity demand, natural gas prices, and wind generation patterns. While uncertain gas prices lead to wind additions, wind variability leads to reduced adoption when explicitly accounted for. The risk-averse objective produces a more diverse generation portfolio, including six GW more solar, three GW more biomass, along with lower current fleet retirements. Stochastic objectives reduce mean system costs by 4.5% (risk-neutral) and 4.3% (risk-averse) compared to the deterministic case. Carbon emissions decrease by 1.5% under the risk-neutral objective, but increase by 3.0% under the risk-averse objective due to portfolio differences.

Keywords: uncertainty; risk; demand variability; renewable adoption; carbon emissions; capacity expansion

1. Introduction

Electricity generation contributes significantly to climate change, accounting for 25% of global carbon emissions [1]. Electricity is also economically significant, e.g., generating USD 380 billion in revenue in 2016 in the U.S. and affecting the profitability of many industrial sectors [2]. Efforts are underway around the globe to decarbonize electricity systems. Federal, state, and provincial governments grant tax credits to select technologies, set targets for renewable energy adoption (renewable portfolio standards (RPS) [3]), and implement carbon taxes [4]. A variety of energy system models are used to inform future energy trends and decarbonization policies with differing scope and spatial scales. In the U.S., policy-linked modeling efforts are generally centered on government agencies, national laboratories, and electric utility companies, which often utilize these models for their integrated resource plans (IRPs). The National Energy Modeling System (NEMS), coordinated by the Energy Information Administration (EIA), is a complex multi-module model that simulates the entire U.S. energy system, including electricity infrastructure, used as the basis for the EIA Annual Energy Outlook [5]. Other examples include the Regional Energy Deployment System Model (ReEDS) by the National Renewable Energy Laboratory

(NREL) [6], Regional Economy, Greenhouse Gas, and Energy (REGEN) by the Electric Power Research Institute [7], and there are also commercial tools such as ProMod [8], EGEAS [9], Polaris [10], Plexos [11], and AURORA [12].

The capacity expansion of the power system in the models above is optimized such that the generators and infrastructure are chosen to minimize the total system cost to meet future demand. In most of the models listed above, the inputs to the capacity expansion model are deterministic: the output is a single set of generators and infrastructure that minimizes discounted system costs when faced with a set of deterministic inputs. Therefore, the models assume that decisions are made under perfect information, i.e., parameter values are fixed, and there is one generation plan that minimizes system costs.

Energy system futures are difficult to predict, and a real or hypothetical social planner should be concerned about lowering risk. Many parameters driving electricity system profitability are highly uncertain, including future fuel prices, technology prices, and policies. Uncertainty is typically treated by re-running the model with a set of alternative scenarios for future parameter values, followed by the comparison of model outcomes with different input assumptions [13]. There are two limitations to this approach. First is that the probabilities of alternative parameter values have not been well characterized, and thus the likelihood of different outcomes is unclear. Second, utility decision-makers are well aware that drivers are uncertain, and presumably account for this to some degree in choices for capacity expansion. It is thus well recognized that electricity system models ought to account for uncertainty and how uncertainty affects build decisions. This is increasingly important when resource planners are looking to rapidly transform the grid towards carbon neutrality.

Decision-making under uncertainty can be mathematically formulated using stochastic optimization on capacity expansion models [14,15]. In general, the idea is to characterize modeling inputs as distributions, then consider optimization objectives that account for the resulting distributions in outcomes. The goal is to find an expansion plan that optimizes an expected (but not specific) outcome. One approach to achieve this is stochastic linear optimization, which represents uncertainty as discrete scenarios with probabilities at various stages within the planning horizon, though still fundamentally constrained as a linear problem. For example, the Tools for Energy Model Optimization and Analysis (TEMOA) model [16] uses stochastic linear programming [17–20] for optimization with uncertainty. In this method, uncertain future outcomes are encoded as specific scenarios in an event tree with an assigned likelihood of occurrence. Other methods include robust optimization, probabilistic (chance-constrained) programming, fuzzy programming, and stochastic dynamic programming [21,22].

Metaheuristic search algorithms are often used in solving stochastic systems, especially when the derivative information is complex to formulate [23]. Some of the popular metaheuristic search optimization methods are particle swarm, genetic algorithm, simulated annealing, neural networks, and pattern search methods. To recount some examples of the application of metaheuristic search algorithms in energy systems modeling, [24] uses particle swarm optimization for identifying the placement of PV and wind turbines in a distributed grid network. Ref. [25] combines engineering principles and economics to maximize electric utility revenues in the European context, using metaheuristic search techniques; [26] uses a nonlinear programming approach for generating capacity expansion planning, with an objective to minimize the total cost consisting of investment cost plus generation cost for a multi-year planning horizon.

Stochastic optimization is growing in popularity, with research efforts focused on both improvement in the computational methods and applications to new research questions or topic areas. Prior work has applied different versions of stochastic optimization of electricity systems to understand how uncertainty interacts with optimal design choices. In ref. [27], Alvarez uses a two-stage mixed-integer linear programming approach with an IEEE 31-bus system to optimally dispatch generators, including storage. The author finds that the stochastic strategy saves 4% on overall costs and is better at retaining operational contingencies. In ref. [28], stochastic dual dynamic programming and Markov chains are used to study the optimal operation of microgrids containing renewables, thermal generation, and storage that can trade with the macrogrid. The authors find that cost reductions can be up to 70% when compared with a deterministic model of future prices, demand, and renewable generation. An integrated energy hub that includes electricity, heat, demand response, and electric vehicles is modeled in [29], and a slime mold algorithm is used to search for optimal operational strategies against uncertainty in both generation and demand. This work also uses a conditional value-at-risk (CVaR) approach to model risk aversion. A new computational method is developed in [30], which combines a distributionally robust optimization approach with a Stackelberg game for an energy hub that includes thermal and renewable generation, heating and cooling loads (including heat storage), and electric vehicle charging. Their computational approach improves on both computation time and convergence to optimal values, relative to traditional methods.

Stochastic optimization is inherently computationally intensive. As mentioned earlier, various mathematical and metaheuristic methods are available to solve complex stochastic problems in power systems, which are often multi-period and nonlinear. Our approach leverages nonlinear methods to achieve faster solve times, making it more practical for modelers to incorporate stochasticity into their analyses, considering uncertainties across multiple input parameters.

We demonstrate the method with a case study of capacity expansion of power generators for the Midcontinent Independent System Operator (MISO) region. The MISO grid portfolio snapshot considered in this study features a balanced mix of fossil fuel energy resources and wind energy, making it an ideal case to analyze the impacts of uncertainty in natural gas prices and the variability of renewables. The analysis also incorporates uncertainty in electricity demand. While the case study focuses on the MISO region, the methodology is flexible and can be adapted to other regions or countries with similar energy mixes, or facing comparable challenges in managing uncertainties in their long-term input assumptions.

The uncertainty in natural gas prices is treated with a mean reversion model, with historical data informing the mean and fluctuations. The uncertainty in wind generation is treated by applying historical hourly wind patterns in MISO to the model. Other driving variables, such as technology costs, learning rates, and discount rates, are deterministic. Technological progress is treated with an endogenous component via adoption in MISO, and this experience curve model leads to a nonlinear optimization problem. The result is a set of generator builds and retirements that minimizes the stochastic objective. The expansion of transmission and distribution infrastructure is not treated in the model, which effectively assumes that it is built to match the resulting generation capacity.

The method contribution of this work lies in developing a general and flexible approach to integrating uncertainty into capacity expansion decisions without the constraints or specialized expertise associated with linear programming techniques. While computationally intensive, this approach enables the incorporation of stochasticity into capacity expansion models and the evaluation of risk-based objective functions. Importantly, the method is agnostic to the complexity of the underlying operation model, whether linear or nonlinear. The proposed method can be applied as long as the electricity system model can accept inputs such as fuel prices, capacity build/retirement decisions, and similar data while producing operating costs as outputs. Additionally, the framework accommodates any set or form of uncertain inputs and objective functions, providing a versatile tool for decision-making. Alternative approaches, such as stochastic dual dynamic programming or Bender's decomposition [31,32], might offer greater computational efficiency but often require substantial restructuring of existing models. By contrast, this approach is particularly valuable to resource planners and practitioners for integrating uncertainty into their existing models without significant modifications. Furthermore, the robust technology choices identified through this method can be further analyzed within traditional deterministic models, yielding an overall robust plan with technology decisions tailored to varying levels of risk appetite.

For contributions in application, we investigate how accounting for three important uncertainties (demand, natural gas prices, and wind variability) influence capacity/retirement choices in a major real-world balancing region (MISO). There has been work addressing demand and gas price uncertainties using a stochastic linear approach [15] but, as will be seen, wind variability is a critical factor affecting renewable adoption. There is prior work investigating effects of wind variability on generator investment decisions, e.g., [33], but it is important to consider the effects of uncertainty in a main alternative to wind (natural gas). By including uncertainties that influence wind (variability), gas (price fluctuations), and both (demand), we can parse the combined effect on generator mix outcomes. One important caveat is that we only consider the building and retirement of generators in this case study, assuming that transmission is built to accommodate. In principle, the need for transmission affects build decisions, which is partially accounted for in some existing models [34,35]. However, the approach that we demonstrate can be expanded to consider transmission expansion, along with the generation decisions.

2. Materials and Methods

The model has several levels that interact with each other, as shown in Figure 1, which are explained below in the order of their operation.

Dispatch model (A in Figure 1): At the core of the framework is the dispatch model, which determines whether a set of generation technologies can meet demand and estimates the variable costs of doing so. This model operates on an hourly resolution, simulating the operation of the electricity system to meet demand.

Long-term assessment model (B in Figure 1): Building on the dispatch model, the long-term assessment model evaluates the discounted expected electricity system costs of an investment plan over a 30-year planning horizon. It incorporates fixed costs, capital costs of new power plants, and variable costs over time from the dispatch model. To address computational limitations and maintain practical resolution, the dispatch model runs for one representative year out of every five during the 30-year horizon.

Sampling model (C in Figure 1): The sampling model integrates the dispatch model and the long-term assessment model to handle uncertainty. It runs the combined models through a random sample of uncertain inputs, including fuel prices, electricity demand, and wind availability. This process produces an output distribution of the discounted total cost of electricity service for a given generation expansion plan, akin to a Monte Carlo analysis. The framework also incorporates endogenous changes to key variables: capital costs, based on technology-specific learning rates, and technology builds between periods. The endogenous treatment of variables can be extended to other input variables.

Optimization–objective evaluation model (D in Figure 1): The output distributions from the sampling model are aggregated into a single value using the risk preferences of the decision-maker or investor. This aggregation allows for decision-making based on a clearly defined objective. The optimization objective can also be adapted to other criteria depending on the specific goals of the analysis.

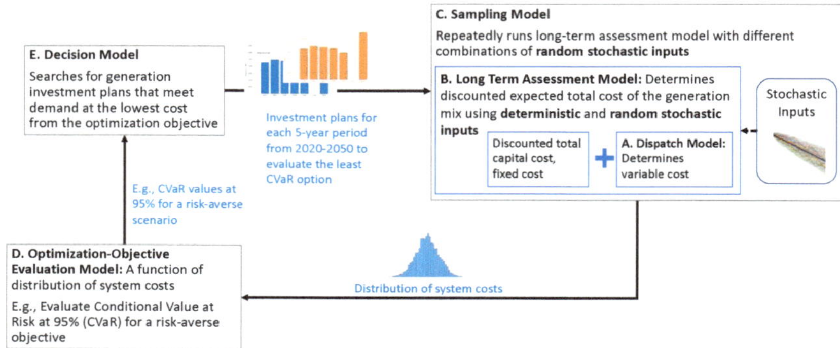

Figure 1. Framework for determining the optimized investment plan under uncertainty of inputs from 2020 to 2050 across the Midcontinent Independent System Operation (MISO) region. The sampling model (C) runs the long-term assessment model (B) over a random sample of input variables to calculate the distribution of the expected total cost of electricity for different stochastic inputs using fixed costs and the dispatch model (A) for variable costs. The optimization–objective evaluation model (D) produces a single cost value from the distribution of system costs for the optimization, based on the user-defined objective. This study uses conditional value-at-risk (CVaR) to assign a risk-adjusted value from the distribution of outputs from the cost model. The decision model (E) first uses genetic algorithm, then pattern search once the genetic algorithm finds an optimal neighborhood, to identify investment plans that minimize CVaR, given the uncertainty-driven distribution from C and the objective defined in A. An existing dispatch model can be used as is with this framework by calling on the relevant subfunctions in the long-term assessment model (B).

Decision model (E in Figure 1): Finally, the aggregated evaluation is passed to the decision model, which determines the optimal generation expansion plan based on the selected objective and risk preferences.

The decision model begins with a genetic algorithm search to identify the best set of generation technologies to build from 2020 to 2050, aligning with the decision-maker's risk preferences. The results from the genetic algorithm are then refined using a pattern search optimization to achieve an optimal solution. Additional numerical optimization techniques were also tested, as detailed in the Supplementary Information. Table S3 compares methods such as particle swarm, genetic algorithm, hybrid1 (genetic algorithm + pattern search, used in our main results), and hybrid2 (genetic algorithm + pattern search + simulated annealing), assessing costs and run times.

The hybrid approach proved effective, combining the global exploration capabilities of genetic algorithms with the local refinement of pattern search. While genetic algorithms excel at navigating complex fitness landscapes, they do not always guarantee convergence. To address this, the pattern search solver uses the genetic algorithm's results as a starting point, enabling faster and more precise local optimization. This combination balances efficiency and accuracy, though other techniques may be suitable depending on the problem and computational resources [36–42].

Heuristic methods like genetic algorithms efficiently manage computational time while accounting for uncertainties. During the evolution process, fitter solutions repeatedly emerge across iterations after being tested with random input samples, reducing computational demands. This approach avoids the linearization of uncertainty variables in the model. However, the parameters of the numerical techniques and the sample size of input distributions must be carefully tuned and tested to ensure consistent convergence. In this article, an "investment plan" refers to the quantities of each generation technology to build in each five-year period from 2020 to 2050, which defines the optimization space.

The dispatch model (A in Figure 1) can be replaced with any other black box model that can estimate the variable cost.

2.1. Inputs

Inputs to the model can be broadly categorized into stochastic and deterministic inputs. The stochastic inputs to the model are the distribution of expected natural gas prices, distribution of expected electricity demand as a function of season and hour-of-day, and random variations in wind profiles based on historical data. The model is capable of incorporating other uncertainties such as expected future subsidies, weather patterns, the distribution of capital costs, or RPS constraints, but we do not consider them in the current work. The deterministic inputs are expected capital costs, discount rate, and technology learning rates (but not future technology costs, which depend on adoption within each model run), as summarized in Table 1.

Table 1. Summary of the inputs and data sources used in the stochastic model (EIA = U.S. Energy Information Administration, NREL = National Renewable Energy Laboratory).

Deterministic Inputs	**Source**	**Note**
Capital cost of new power plants	EIA [43]	Shown in Table 2
Discount rate	EIA [43], NREL [44]	5%
Fixed and operating cost of new power plants	EIA [43]	Shown in Table 2
Fixed and operating cost of existing plants	eGrid [45]	Supplementary Information, Section S2
Heat rate of new technologies	EIA [43]	Shown in Table 2
Heat rate of existing plants	eGrid [45]	Supplementary Information, Section S1
Learning rates (single factor)	IEA [46], EIA [47], Rubin et al. [48]	Shown in Table 2
Existing power plant fleet	eGrid [45]	Supplementary Information, Section S1
Fuel prices	EIA [49]	Shown in Figure 2
Carbon emissions of existing plants	eGrid [45]	Supplementary Information, Section S1
Solar variability (hourly capacity factors)	NREL [50,51]	Supplementary Information, Section S3
Stochastic inputs	**Source**	**Note**
Natural gas prices	Simulated	Illustrated in Figure 3
Demand	Simulated	Illustrated in Figures 4 and 5
Wind profiles	Historical variations NREL [50,51]	Illustrated in Figure 6

2.1.1. Deterministic Inputs

This section covers the deterministic inputs used in the model, as summarized in Table 1. The model uses the existing portfolio of generation in the studied area, including the age, efficiencies, emissions, and generation capacities of each plant from the eGrid database [45] to define the starting point for future portfolios. Sample data from the eGrid database are shown in the Supplementary Information, Section S1.

New Power Plant Characteristics

New generation technologies and their corresponding capital cost, fixed cost, and efficiencies are considered based on the EIA's estimates used for modeling the NEMS electricity market module [52], shown in Table 2. Overnight capital costs are considered, excluding the financing/interest during the construction and development of the power plants. The discount rate in the model is 5% [44,53].

Table 2. Cost and efficiency characteristics of new generation technologies considered in the study, based on estimates from the U.S. Energy Information Administration [13]. All the costs are expressed in 2016 USD. Capital costs are overnight costs that exclude interest during the construction and development. The capital costs are for the year 2016 and the future capital costs are estimated accounting for technological progress through learning rates. (CCS = carbon capture and sequestration, CC = combined cycle, CT = combustion turbine).

Technology	Capital Cost (2016 USD/kW)	Fixed Cost (2016 USD/kW)	Operating Cost (2016 USD/MWh)	Heat Rate (Btu/kWh)	Fuel	Learning Rate (%)
Coal with CCS	4983	69	7.02	1070	Coal	8.3
Natural gas CC	962	10.80	3.47	6450	Gas	14
Natural gas CT	875	17.30	3.47	9900	Gas	15
Biomass	3757	110	5.46	13,500	Biomass	11
Wind	1622	47	0	0	Wind	12
Solar	1812	22	0	0	Solar	23
Nuclear	5822	99.17	2.27	232,930	Uranium	2

Capital Costs with Technological Progress

Rapid growth, competition, and technology improvements lead to cost reductions over time in a process described generally as "technological progress." Learning rates describe these cost reductions. The learning rates (*LR*) assumed in this study are based on the average learning rates observed from the literature review by Rubin et al. in their study [48], as given in Table 2.

The learning coefficient 'α' determines the capital cost of the technologies in any given year (CC_t) based on the cost during the previous year (CC_{t-1}), the global capacity of the technology during the previous year (P_{t-1}), and the current cumulative capacity after the new additions globally (P_t) (Equation (3)). The coefficient 'a' is determined from the learning rate of the technologies, which specifies the cost reduction rate as the technology's capacity doubles (*LR*) [54] (Equations (1) and (2)). The total installed capacity of the technology is determined based on the global level projections from the EIA data [47] and the local capacity additions in MISO, meaning that learning is endogenous within the model. The percentage change in capital cost over time for each tech is shown in Supplementary Information, Figure S3.

$$LR = 1 - 2^{\alpha} \tag{1}$$

$$\alpha = \frac{\ln(1 - LR)}{\ln(2)} \tag{2}$$

$$CC_t = CC_{t-1}\left(\frac{P_t}{P_{t-1}}\right)^{\alpha} \tag{3}$$

where

Subscript t—given year t
LR—learning rate
α—learning rate coefficient
P—cumulative global capacity
CC—capital cost

Fuel Prices

Fuel prices of coal, uranium for nuclear power, and oil are treated as deterministic and are taken from the EIA database [49], shown in Figure 2, with units expressed in USD/MMBtu.

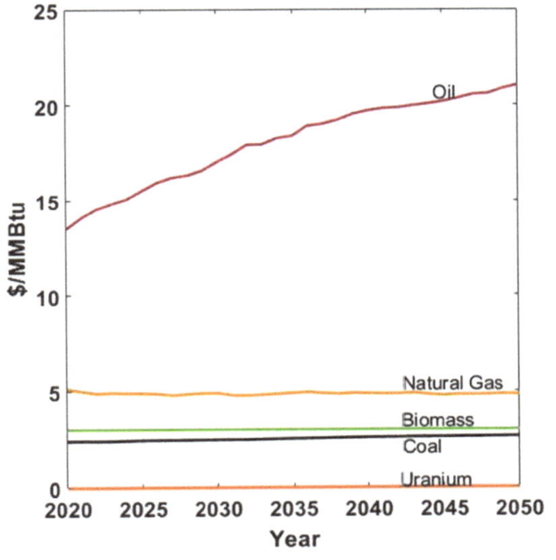

Figure 2. Fuel prices of coal, natural gas, uranium, and oil used in the deterministic scenario. The x-axis represents the year, and the y-axis represents the fuel price in USD per million British thermal units. All fuel prices except natural gas are based on U.S. Energy Information Administration data [49]. For natural gas, prices for the deterministic scenario are set to the mean of the distribution in stochastic scenario (see the Distribution of Natural Gas Prices section).

Solar Output

We estimate the hourly generation profiles of solar energy across various locations in MISO using the Eastern Solar Integration Data [51]. This dataset consists of meteorological data, 5 min resolution of solar power production, and capacity factors. We consider thirty potential locations in the midwest region and the corresponding hourly wind output per MW. The average of solar energy output (MWh/hour) per unit MW across these locations is used to generate the hourly variations in solar capacities considered in the study. Unlike wind, solar generation is not treated as a stochastic input in the model for two reasons: it is a less important input because the model does not choose much solar in any scenario (MISO has far more wind than solar), and focusing on three stochastic inputs allows us to more clearly explain the effects of each variable. Refer to Supplementary Information, Section S3 for more details.

2.1.2. Stochastic Inputs

In order to model the effects of uncertainty, the model requires some form of distribution of possible future values for each input whose uncertainty is considered. In this case study, we have chosen plausible methods for generating distributions of natural gas fuel prices, electricity demand, and wind variability that we use as inputs to the model alongside the deterministic inputs described above. To aid with interpretation and explanation, we only model uncertainty for natural gas price, electricity demand, and wind. It should also be noted that each of these three stochastic inputs takes a different form, designed to capture the realistic uncertainty/variability in that specific input. This poses no difficulty in our modeling approach: wind uncertainly comes from a set of prior year wind patterns, natural gas uncertainty follows a mean reversion time series with correlation between periods, and demand patterns are created by combining historical daily demand patterns with presumed growth rates over time.

Distribution of Natural Gas Prices

Volatility in natural gas prices demonstrate mean reversion and seasonality [55]. Mean reversion is the tendency of natural gas prices to revert to a long-term equilibrium value after fluctuations due to extreme weather, supply, or demand surges. Seasonality is the cyclic variation over the seasons because of the cyclic changes in demand [55]. In the current model, we consider only the annual variations using the Ornstein–Uhlenbeck mean reversion process [56]. Historical variations in the Henry Hub natural gas spot prices since 1986 are used to estimate future uncertainties [57].

The Ornstein–Uhlenbeck process is a variation of the Markov process, i.e., the future value is independent of the past but depends upon the present value [9]. Further information on the Ornstein–Uhlenbeck process and our application is provided in the Supplementary Information, Section S5. The simulated time series of natural gas prices is shown in Figure 3.

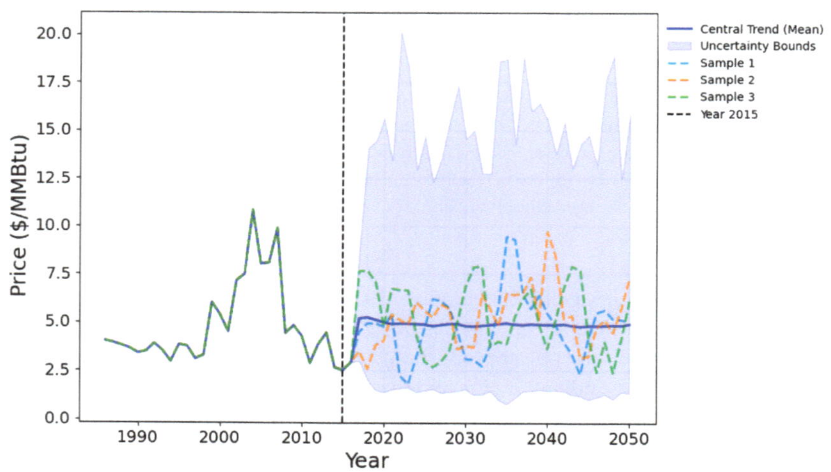

Figure 3. Natural gas Henry Hub spot prices from 1990 to 2018 and sample simulated price scenarios with uncertainty cone from 2018 to 2050. The x-axis represents the year, and the y-axis represents the natural gas price in USD per metric million British thermal units. Ornstein–Uhlenbeck mean-reversion process is used to create stochastic natural gas prices as an input to the long-term assessment model. Each colored dotted line indicates one possible sample price trajectory until 2050. Three out of a thousand samples are illustrated in the figure.

Stochastic Electricity Demand

We construct stochastic electricity demand over a year by multiplying normalized hourly demand values from historical data with a random annual average demand from a pre-defined distribution for the years 2020–2050 at five-year intervals. A simple Brownian motion is used to build a distribution of annual average demand. Further information on the modeling of stochastic demand is provided in the Supplementary Information, Section S6. The distribution of annual growth in demand is provided in Figure 4, and a sample of intra-annual random hourly demand patterns for two example days during the summer is provided in Figure 5. The model considers all 8760 h in the year, and demand profiles for two sample years are shown in Supplementary Information, Section S6, Figure S4.

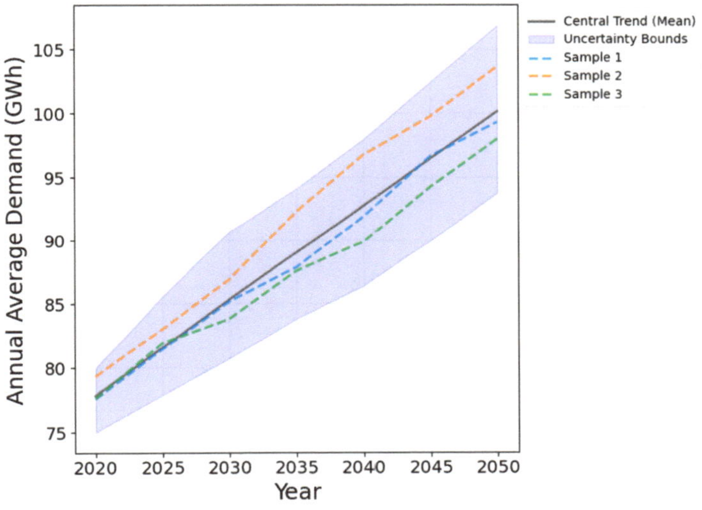

Figure 4. Uncertainty cone of average annual load growth from 2015 to 2050. The *x*-axis represents the year, and the *y*-axis represents the annual average hourly demand in gigawatt hours. Each colored dotted line indicates one possible average growth trajectory until 2050. Three out of a thousand samples are illustrated in the figure.

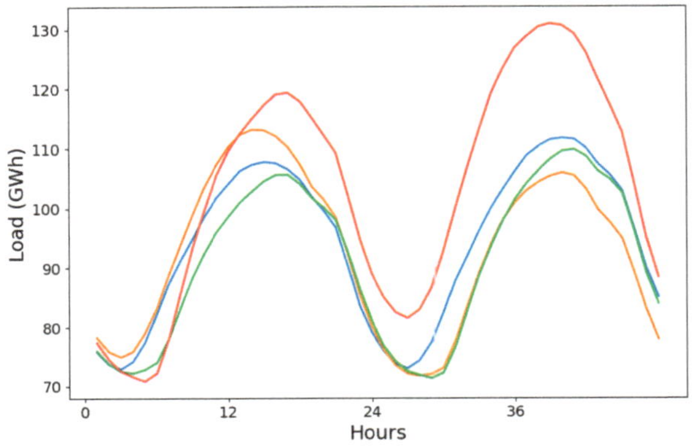

Figure 5. Four random samples of hourly load patterns for two summer days in the year 2035. The *x*-axis represents the hour, and the *y*-axis represents the hourly demand in gigawatt hours. The samples are created from multiplying a random point of annual demand growth distribution from the year 2035 with a random normalized hourly load pattern over a year from the historical data. Similar profiles are created at 5-year steps from 2020 to 2050 for 8760 h for every Monte Carlo simulation.

Wind Variability

We estimate the hourly generation profiles of wind energy across thirty locations in MISO using the Wind Integration National Database (WIND) toolkit [50] for the past 10 years. We average the wind output (MWh/hour) per unit MW across thirty potential locations for each year from the past ten years to generate different samples of hourly variations in wind as shown in Figure 6.

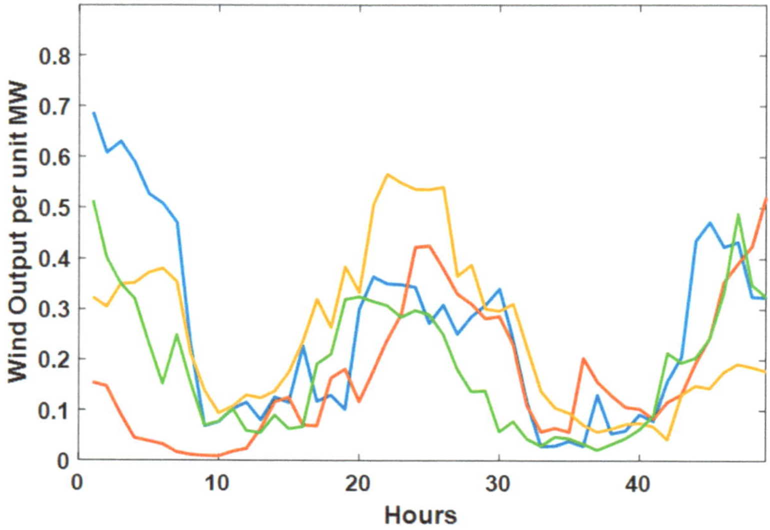

Figure 6. Four random samples of hourly wind output for two example days during the summer. The x-axis represents the sampled hours, and the y-axis represents the normalized wind output per unit megawatt of capacity. Each color represents one sample year of data describing the historical wind generation observed in the Midcontinent Independent System Operator (MISO) region.

2.2. Sampling Model

The sampling model (C in Figure 1) calculates a distribution of discounted expected total system costs for meeting load over the next 30-year horizon, given a specific investment plan over 30 years from 2020 to 2050. In this model, we consider a random sample of fifty different load profiles, and natural gas prices for each iteration of the genetic algorithm search and pattern search. For a scenario with wind uncertainty, we also consider a random sample of historical wind profiles along with the other stochastic variables.

2.2.1. Dispatch Model for Variable Costs

The lowest level of the cost model is a dispatch routine that uses simple rules to determine the variable cost of electricity generation over a year [58]. Due to the need to run many scenarios, both for the sampling model and the genetic algorithm search, there are limitations on computational time. Therefore, the dispatch model is limited to choosing the generation in each hour based on the marginal cost of operation for 8760 h in a year. We run the model for all the hours in a year to accurately capture the effects of variability of wind and solar energy.

The economic dispatch model is run with an objective of producing electricity at a minimum operating cost using linear optimization. The generators run with ramping constraints, shown in Equations (6) and (7). The total variable cost (VC) of the electricity generation is the marginal cost (MC) incurred by the power plants to produce electricity e summed over every hour in a year, as shown in Equation (9). We do not include more sophisticated electricity system elements, such as transmission constraints, startup time, and spinning reserves. However, the modular nature of the model permits upgrading to a sophisticated dispatch model without changes to the modeling framework.

The dispatch model uses the EPA's eGrid database for the current fleet of power plants for electricity generation [45], and the new generation fleet is added based on the inputs from the decision model and the plant characteristics from EIA data (Table 1) [43]. The marginal cost (assumed as bid price) of operation for each power plant is calculated based on the heat rate [45], and the subsequent fuel costs as given in Equation (4).

$$MC_{i,h,p}\left(\frac{\mathrm{USD}}{\mathrm{MWh}}\right) = \mathrm{HR_p} * \frac{\mathrm{Price_{i,t,f}}}{1000} + \mathrm{O\&M_p} \tag{4}$$

Objective function:

$$minimize\ C_{i,h,t} = \sum_p MC_{i,h,p,t} * e_{i,h,p,t} \tag{5}$$

Subject to:

$$e_{i,h,p,t} \geq e_{i,(h-1),p,t} - \frac{RD_p}{100} * P_p, \tag{6}$$

$$e_{i,h,p,t} \leq e_{i,(h-1),p,t} + \frac{RU_p}{100} * P_p, \tag{7}$$

$$e_{tp} > 0,\ t \leq 8760 \tag{8}$$

$$\mathrm{VC_{i,t}} = \sum_{h,t} C_{i,h,t} \tag{9}$$

where

Subscript h—hour
Subscript t—given year
Subscript i—ith Monte Carlo run
Subscript p—power plant
P—nominal capacity of power plant (MW)
e—energy output in hour h (MWh)
MC—marginal cost of operation of a power plant (USD/MWh),
HR—heat rate (Btu/kWh)
Price—average spot price of fuel (USD/MMBtu)
O&M—operations and maintenance cost of the power plant (USD/MWh)
RD_p—ramp down rate of power plant p (% of MW/h)
RU_p—ramp up rate of power plant p (% of MW/h)
VC—total variable cost

This study does not model imports of electricity from regions outside of MISO and penalizes the model with a cost of USD 10,000/MWh for any unserved energy.

2.2.2. Long-Term Assessment Model

The long-term assessment model calculates the discounted expected total system costs (Equation (11)) for meeting load over a 30-year horizon. The first step is to calculate the total system cost for each year (*TC*), which is a summation of capital cost of new power plants after taking into account technological progress (*CC*), fixed cost, and variable cost from the dispatch model (*VC*) (Equation (10)).

$$\begin{gathered}\text{Total system cost (USD)} = \text{capital cost} + \text{fixed cost} + \text{variable cost} \\ \mathrm{TC_{i,t}} = CC_t + FC_t + VC_{i,t}\end{gathered} \tag{10}$$

where

Subscript t—given year
Subscript i- ith Monte Carlo run
CC—capital cost (USD)
FC—fixed cost (USD)
VC—total variable cost (USD)

The model operates in 5-year intervals. The start year is 2015, and the costs of intermediate years are interpolated based on the costs estimated at 5-year intervals from 2020–2050.

Cashflow

We assume a discount rate (*r*) of 5% and calculate the total discounted cost of the electricity service (*C*) using the 2016 USD value, as shown in Equation (11) from the total cost of electricity service (*TC*) estimated for each year from 2020–2050.

$$C_i = \sum_{t=2015}^{2030} \frac{TC_{i,t}}{(1+r)^{(t-Y)}} \tag{11}$$

where

Subscript i—ith Monte Carlo run
C—discounted present value of the cost
TC—total cost
r—discount rate, 5%
t—given year
Y—reference year, 2016

2.2.3. Sampling Method

To include the effects of future uncertainties, we run the long-term assessment model iteratively with random combinations of natural gas prices, load, and wind and the output is a distribution of discounted total system costs. We consider a sample size of fifty, which takes about 16 h to solve on a system with Intel I9 10900K chip, with 10 cores and 20 threads. The run times vary depending upon the processor. Through a trial-and-error method using different sample sizes, a sample size of fifty, along with the tuning of other genetic algorithm parameters, gave optimal solutions. Also, evolutionary algorithms such as genetic algorithm run iteratively, and the fitter individuals/least cost solutions appear repeatedly many times across the iterations to determine the optimized results (i.e., the best solutions).

2.3. Deterministic Approach for Comparison

We compare the stochastic and deterministic approaches by running the model with deterministic inputs rather than the distribution for demand, natural gas price, and wind stochasticity. For this, the mean of the stochastic input distributions is used as a deterministic input to provide a fair comparison. The purely deterministic approach does not include any uncertain variables, and the output of the cost model is a single discounted total system cost of electricity service for an investment plan over 30 years from 2020 to 2050. The genetic algorithm and pattern search is applied in the same way as in the base case model, but is now searching for an investment plan that minimizes the single deterministic system cost figure rather than a distribution of costs.

2.4. Optimization–Objective Evaluation Model

The optimization–objective evaluation model in this study uses the conditional value-at-risk (CVaR) method to aggregate the distributions of net present costs from the cost model to a single value based on risk preference. For a risk-neutral scenario, the decision model minimizes the mean of the distribution, which is equivalent to a CVaR at 0%. For a risk-averse scenario, CVaR is set at 95%, meaning that the decision model is attempting to minimize the mean of the most expensive 5% of scenarios, which is a plausible goal for a risk-averse decision-maker (Equation (12)).

$$DC_{CVaR}\,(USD) = \frac{\sum_{(i=\frac{CVaR}{100}*n)}^{n} C_i}{\left(1-\frac{CVaR}{100}\right)*n} \text{ after arranging } C_i \text{ in ascending order} \tag{12}$$

where

Subscript i—ith Monte Carlo run
Subscript n—total Monte Carlo runs
C—discounted present value of the cost
DC—total discounted electricity cost based on CVaR

The choice of CVaR value directly affects the optimization outcomes. Lower CVaR values prioritize solutions with lower average costs across all scenarios, favoring investments that perform well under typical conditions. In contrast, higher CVaR values focus on minimizing potential losses in worst-case scenarios, leading to more conservative investment plans against extreme uncertainties. By adjusting the CVaR value, the decision-maker can explore a spectrum of trade-offs between cost efficiency and risk mitigation, tailoring the optimization to specific risk preferences and system requirements [59]. While CVaR is used in the current study, the optimization objective can be any other metric as a function of distribution of system costs or other outputs.

2.5. Decision Model Genetic Algorithm Search Plus Pattern Search

The decision model first uses genetic algorithm optimization to minimize the risk-adjusted output (average cost or CVaR at 95%) from the optimization–objective evaluation model. The genetic algorithm is good at rapidly identifying a set of reasonably fit solutions using a heuristic optimization algorithm derived from natural selection processes. The genetic algorithm iteratively modifies a population of individual solutions—in our case, investment plans. The default population size is 200 in the global optimization toolbox in the MATLAB R2017b version used in the model. The decision variables are the capacities (in MW) of each generation technology to be built from 2020 to 2050 at 5-year intervals, i.e., quantities of seven technologies to be built over seven periods (49 decision variables). Other parameters of the genetic algorithm, such as function tolerance (average relative change in the best least cost solution) and crossover fraction (to specify the fraction of the next generation produced by crossover), are tuned through cross-validation techniques. We use a heuristic initialization method, which seeds the initial random investment plan with a solution from the deterministic analysis and the rest with the random plans, as generated by the algorithm. After the genetic algorithm search, we use pattern search to ensure that a detailed search is completed around the most promising solutions. This algorithm iteratively finds a sequence of neighboring points around a least cost solution, until the objective function either decreases or remains the same from each point in the iteration to the next.

There are no environmental or renewable energy policies (e.g., renewable portfolio standard, emissions pricing) included as constraints, though these are perfectly compatible with the model structure. Supplementary Information, Section S5 provides a more detailed description of the objective function, and all the constraints used in the optimization. For more information on the genetic algorithm and its tuning, refer to the Supplementary Information, Sections S6–S8 (for the specific implementation in this research) and the MATLAB website (for general information about the method and application) [60].

2.6. Retirement

In reality, the retirement of power plants is case-specific and depends on several factors such as wholesale electricity prices, inefficiency, costs of operation, and environmental regulations [61]. In 2017, most of the retirement decisions in MISO were due to uneconomic power plants [62].

In the current study, the economics of power plants is based on the cost of operation every five years. The model endogenously retires the power plants by allowing the optimization algorithm to choose positive or negative capacity additions. Positive additions denote a new generation capacity of that technology, and negative additions denote the retirement of the power plants for a specific fuel type. For negative capacities, power plants with the highest annual cost of operation per unit nameplate capacity for a given fuel type are assumed to be the least economical to operate, and are retired until the retired capacities equal the negative capacities from the genetic algorithm. The total cost of electricity service is then calculated for the resultant investment plan.

2.7. Reporting the Investment Plan, Unserved Energy, and Distributions of Cost and Emissions

Distributions of discounted total system costs, unserved energy, and emissions of the resultant investment plans for different risk preferences are presented in Section 3 below. Probability distributions of system costs are plotted with two primary comparisons: deterministic versus uncertain stochastic scenarios, and risk-neutral versus risk-averse scenarios. The distributions are calculated by running the resultant investment plans for different risk preferences through the long-term assessment model and the Monte Carlo model. A fixed sample distribution of natural gas prices, demand, and wind energy is assumed for all the investment plans and run through 1000 Monte Carlo simulations, which provides a distribution of output discounted total system costs and emissions. This fixed sample ensures a fair comparison between scenarios.

3. Results

This work presents an alternative method for capacity expansion modeling in MISO under uncertainty of input load, natural gas prices, and wind. First, we present the results of the stochastic approach for the reference case. Second, a comparison is made between the outcomes of deterministic and stochastic approaches. Next, we compare the unserved energy of all the three cases. In the end, a summary of all three objective functions is presented.

3.1. Stochastic Scenario

The base case stochastic scenario is the risk-neutral scenario that minimizes the mean of the distribution of discounted, long-term electricity system cost. For this risk-neutral scenario, the capacity mix by 2050 is 43% wind (113 GW), 30% natural gas (79 GW), 19% coal (49 GW), 5% solar (13 GW), and 3% all other fuels (9 GW) (Figure 7, top).

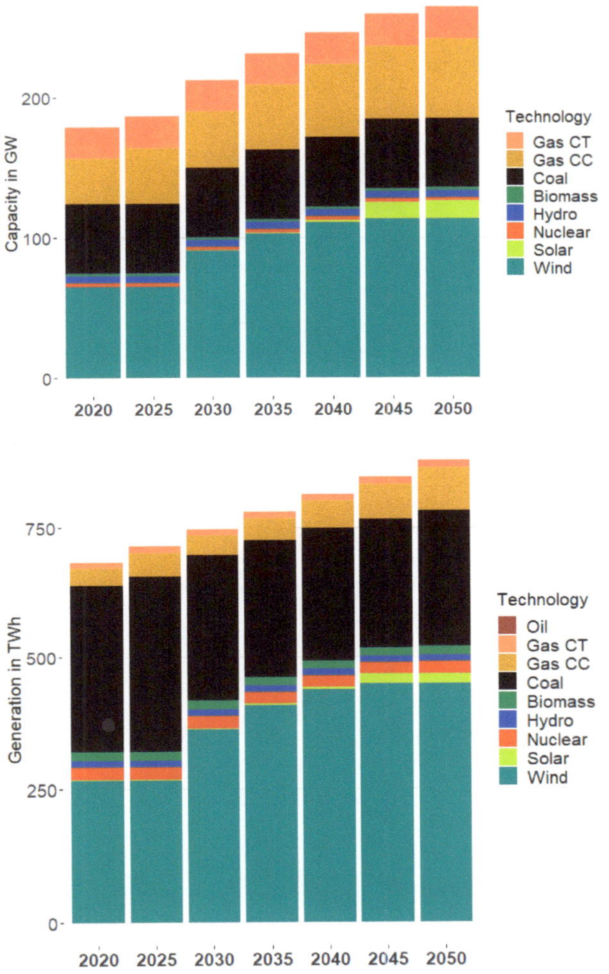

Figure 7. 2020–2050 generator capacity mix (gigawatts) (**top** figure) and generation (terawatt hours) (**bottom** figure) in the stochastic risk-neutral scenario (objective = minimize mean of cost distribution), using the mean natural gas prices and demand values from the input distributions. Colors of the bars indicate the generator type. (Gas CT = gas combustion turbine, Gas CC = gas combined cycle).

The energy mix that results from the capacity mix described above depends upon specific natural gas prices, demand, and wind generation profiles. While the capacity mix was chosen using stochastic inputs, calculating the energy mix requires specific fixed inputs. We use mean natural gas prices, demand, and wind variability values from the input distributions to find that wind dominates the energy generation mix in 2050 at 51% of total electricity, followed by coal (30%), natural gas (11%), and all other fuels at 8% (Figure 7, bottom).

3.2. *Comparing Deterministic and Stochastic Scenarios*

Relative to the deterministic optimization, the risk-neutral scenario adds 18 GW of additional natural gas combined cycle capacity, 8 GW of additional gas turbines, but 8 GW less wind capacity by 2050. We noticed that if wind is modeled as deterministic (with a fixed and known annual pattern) while demand and natural gas price are kept as stochastic, the risk-neutral optimizer chooses more wind generation than in the deterministic case. When wind generation, natural gas prices, and demand are all varied, the uncertainty regarding wind generation outweighs the importance of natural gas price uncertainty. Overall, wind capacity and generation are greater than natural gas-based capacity and generation in both deterministic and stochastic scenarios.

These results should be interpreted with caution, considering that they depend on factors such as resource availability in the area, the risk appetite of the modeler, and overall technology choices. The authors emphasize that these findings are intended to illustrate the results and sensitivity to various risk appetites, highlighting the importance of incorporating stochastic elements into models for grid planning purposes. Grid planners should consider these variations and uncertainties as part of their comprehensive modeling efforts.

Probability distributions of discounted total system costs of the investment plans illustrate the likelihood of different long-term system costs when subjected to input distributions of natural gas, demand, and wind variations. Figure 8 compares the probability distributions of these costs. The results show that the mean system cost of the deterministic scenario is higher than the risk-neutral scenario by USD 23 billion (5%). The mean system cost of the distribution for the risk-neutral scenario is USD 486 billion, and there is a lower likelihood of expensive system costs than for the deterministic scenario (Figure 8). The CVaR at 95% (the average of the 5% most expensive outcomes) for the risk-neutral scenario is USD 514 billion, versus USD 599 billion for the deterministic scenario. Overall, the risk-neutral scenario optimizes better for the expected distribution of load, natural gas prices, and wind variations than the deterministic approach, which does not consider this variability when it chooses an optimal system. This is not particularly surprising, considering that only the risk-neutral optimizer is designed to account for these uncertainties. But the magnitude of the differences is notable: a 5% difference in average long-term system cost and a 15% difference in the worst-case scenarios.

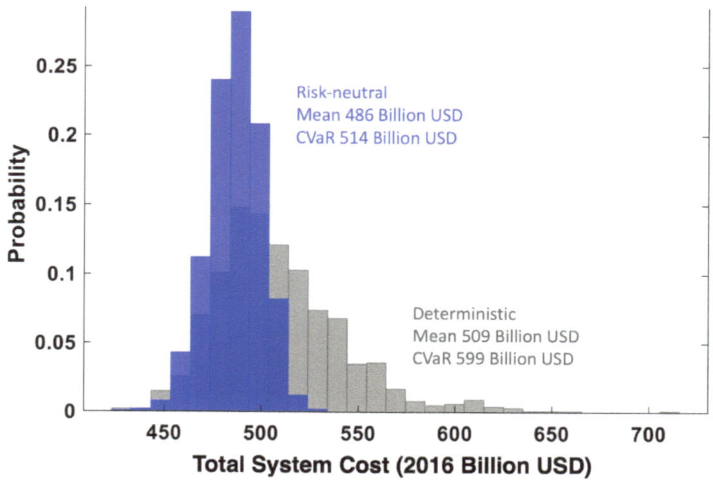

Figure 8. Probability distribution of discounted total system cost of electricity for risk-neutral and deterministic scenarios, generated when the resultant investment plans are run through a sample of 1000 random natural gas prices, demand, and historical wind variations. The x-axis represents the total discounted system cost of electricity in billions of USD, and the y-axis represents the probability. Colors represent the scenarios. The risk-neutral scenario is optimized for conditional value-at-risk (CVaR) at 0% (mean of the input distribution) and the deterministic scenario is optimized for average input values (not distributions).

3.3. Comparing Risk-Neutral and Risk-Averse Scenarios

In this section, results are compared for different risk preferences. The "risk-averse" scenario attempts to minimize the CVaR at 95% (the mean of the most expensive 5% of outcomes), and the "risk-neutral" scenario attempts to minimize the CVaR at 0% (mean of the entire system cost distribution). Probability distributions of discounted total system costs for the different risk preferences show that the mean cost of the risk-averse scenario is slightly higher than the risk-neutral scenario by USD 1 billion. The risk-averse scenario's CVaR value at 95% is USD 511 billion, lower than the risk-neutral scenario's CVaR value at 95% of USD 514 billion (Figure 9).

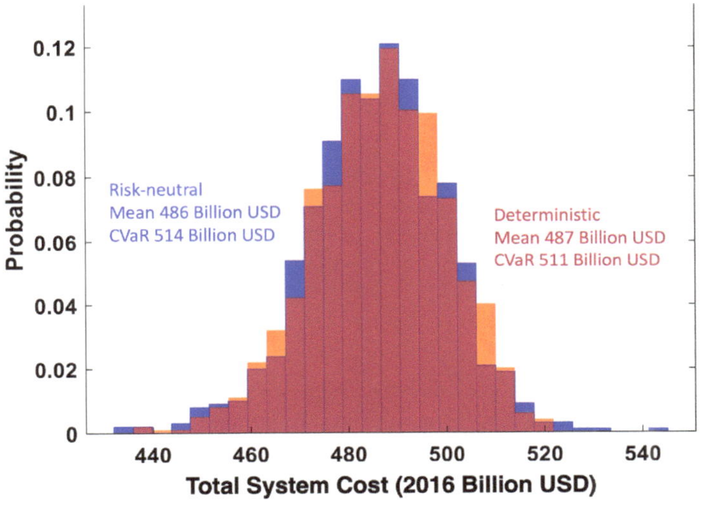

Figure 9. Probability distribution of the discounted total cost of the electricity system for risk-neutral and risk-averse scenarios. The x-axis represents the total discounted system cost in billions of USD, and the y-axis represents the probability. The risk neutral scenario is optimized for conditional value-at-risk (CVaR) at 0%, which is the mean of the distribution. The risk averse scenario is optimized for CVaR at 95%.

Relative to the risk-neutral optimizer, the risk-averse scenario proposes higher capacity additions of solar capacity by 6 GW, biomass by 3 GW, and lower additions of wind by 12 GW, fewer gas turbines by 8 GW, and a lower retirement of coal, by 2050. Minimizing the probability of high system costs generally shifts the preferred system towards diversification: greater use of lightly used resources and slightly lower deployment of wind and natural gas generation, which tend to dominate in any scenario.

3.4. Unserved Energy

In this section, results of unserved energy are compared between all the scenarios. The boxplots in Figure 10 shows a median unserved energy of less than 0.5 TWh for all scenarios, which is less than 0.5% of the average annual demand. For all the modeled time periods, the mean unserved energy of the deterministic scenario is higher than the stochastic scenarios. We also observed similar trends for unserved energy costs.

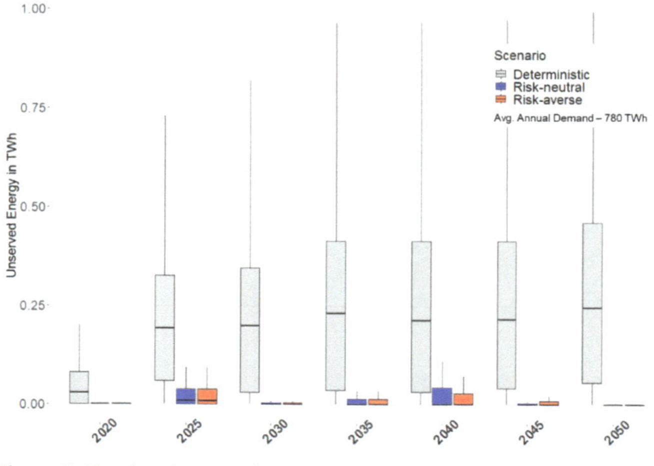

Figure 10. Boxplot of unserved energy (not produced by MISO generators, purchased at cost of USD 10,000/megawatt hours) comparing the deterministic, risk-neutral, and risk-averse scenarios. Deterministic scenarios have higher unserved energy values due to not accounting for high-demand/low wind generation scenarios. The x-axis represents the year, and the y-axis represents the unserved energy in terawatt hours. Colors represent different scenarios. (MISO = Midcontinent Independent System Operator).

3.5. Summary of Results

Between the three optimization goals described in this work, the deterministic scenario has the lowest capacity additions, a higher mean system cost, and highest high-risk values because it neither optimizes for the expected distribution of load, gas prices, and wind variability, nor considers high-risk scenarios. Since the penalty of unserved energy at USD 10,000/MWh is much higher than energy costs under uncertain natural gas prices, the risk-neutral scenario adds more dispatchable natural gas (of both types) than the deterministic scenario to address the uncertainty in demand and wind variability. Therefore, the amount of unserved energy is much lower for the stochastic scenarios than the deterministic scenario. The risk-averse scenario adds the highest solar capacity of all the scenarios. It also adds the most natural gas combined cycle capacity of all the scenarios. The cumulative capacity additions in all three scenarios are plotted in Figure 11.

Figure 12 shows the cumulative distribution function of emissions for all three scenarios by the year 2050, given uncertainty in the three examined stochastic variables. The results show that the average emissions are close for all three scenarios, but lowest for the risk-neutral scenario at an average of 260 kg/MWh, second for the deterministic scenario at an average of 264 kg/MWh, and highest for the risk-averse scenario at 272 kg/MWh (Figure 12). However, the distributions in each value are more interesting: each of the three vary by more than a factor of two, driven mainly by uncertainty about total electricity demand in the future. The risk-neutral and risk-averse scenarios have more variable emissions levels than the deterministic one because they both build greater amounts of gas generation. This means that they have a greater ability to trade-off between coal and gas generation depending on the prevailing natural gas price, which is uncertain and variable.

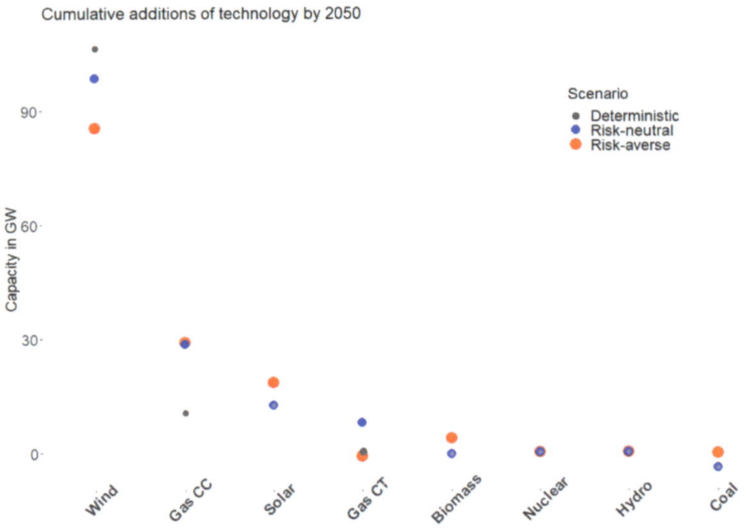

Figure 11. Scatter plot of cumulative additions of different generation technologies by 2050, comparing the deterministic, risk-neutral, and risk-averse scenarios. The risk-averse scenario is optimized for conditional value-at-risk (CVaR) at 95% and risk-neutral scenario is optimized for CVaR at 0% of the system cost distribution. The *x*-axis represents the generation technologies, and the *y*-axis represents the cumulative capacity additions (or retirements as negative values) in gigawatts. Colors represent different scenarios.

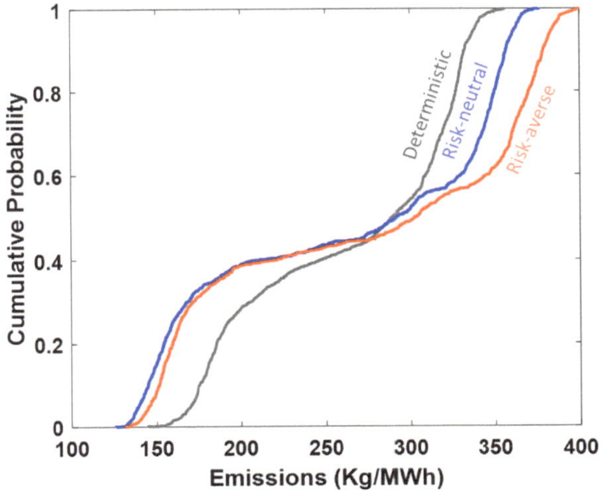

Figure 12. Cumulative probability distributions of the output emissions for deterministic, risk-neutral and risk-averse scenarios by the year 2050. The *x*-axis represents the emissions in kg/megawatt hours and the *y*-axis represents the cumulative probability. Colors represent the scenarios. The risk-neutral scenario is optimized for conditional value-at-risk (CVaR) at 0% and risk-averse scenario is optimized for CVaR at 95%.

A summary of the results comparing the deterministic, risk-neutral, and risk-averse scenarios is shown in Table 3.

Table 3. Comparison of summary results for three model cases. Results are obtained with heuristic optimization, meaning that there is small variability in results depending on run. Therefore, all results show the mean of ten runs. (CVaR = conditional value-at-risk).

Model Case	Discounted Total System Cost (2016 Billion USD)	CVaR@95% (2016 Billion USD)	Average 2050 Emissions Factor (kg/MWh)	Total Capacity of Wind and Solar by 2050 (GW)
Deterministic	509	599	264	134
Risk-neutral	486	514	260	126
Risk-averse	487	511	272	120

4. Discussion

This work develops a flexible approach for integrating stochastic factors and alternative objective functions into capacity expansion modeling, which allows use of "black box" operation/production cost models that are time-intensive or unavailable for modification. The flexibility does come at a computational cost: the requirement that the operation model underneath the capacity expansion model must be run iteratively, depending upon the size of the input distributions.

For the case study of MISO, results show that accounting for uncertainty in decision-making reduced the adoption of wind power in both risk-neutral and risk-averse cases. In contrast, running the model without wind variability led to increased wind power adoption, suggesting that the impact of wind variability in reducing adoption outweighs the influence of natural gas price variability, which favors wind. Despite lower wind capacity, the risk-neutral case results in lower carbon emissions than the deterministic scenario, driven by the operation of new, efficient natural gas plants replacing older units. However, if emission constraints were strictly enforced, the model would likely favor dispatchable, low-emitting resources to account for wind variability. This highlights a key area for future research, exploring how emission limits interact with the need for firm, low-carbon generation in the presence of renewable variability.

There are unresolved questions on the applicability of the approach in more complex situations. Stochastic models are more computationally intensive than deterministic ones. In our application, the optimization, run on a personal computer, was manageable with three stochastic inputs for capacity expansion of generators in MISO. The case study model did not include expansion of transmission and distribution, storage, or demand-response, all of which increase computational complexity. It is not clear yet what combination of optimization technique and increased computing power will enable which additions to be computationally feasible. As with other meta-heuristic optimization, run-times depend on the specific approach applied and there is risk of being caught in a local minimum (see Supplemental Information S12 for related results on this topic). Due to the flexibility they introduce, storage and/or demand response could change the current result that stochastic decision-making reduces wind adoption.

This said, the additional functionality to analyze cost risk represents, we argue, an important feature of real-life capacity expansion decisions. The usual deterministic approach assumes that the grid is built with perfect foresight of fuel prices, technology prices, and demand. Decision-makers who plan grids do have some knowledge of future uncertainties that can inform the way that these uncertainties are modeled. For example, natural gas prices and demand are certain to be more volatile than the marginal costs of operating wind and solar power plants, though the generation patterns of the latter result in uncertainty of their own. Grid buildout decisions ought to account for this knowledge of the future by considering cost risk in addition to minimizing average expected cost.

5. Conclusions

This work demonstrates the importance of incorporating stochastic factors and alternative objective functions into capacity expansion modeling, especially in the context of uncertain future energy scenarios. By applying this approach to the MISO region, the study highlights significant trade-offs between deterministic, risk-neutral, and risk-averse optimization strategies. The deterministic scenario, while simpler and computationally less intensive, fails to account for uncertainties in natural gas prices, wind variability, and electricity demand, resulting in higher system costs and unserved energy.

The stochastic scenarios, particularly the risk-neutral case, achieve lower system costs and emissions on average by incorporating uncertainty into decision-making. This comes at the expense of reduced wind power adoption due to wind variability but is offset by the deployment of efficient natural gas plants. The risk-averse scenario prioritizes reducing extreme cost outcomes, leading to higher capacity additions of natural gas and solar, but results in slightly higher average emissions compared to the other scenarios.

The results show that deterministic approaches oversimplify real-world decision-making by neglecting uncertainties that significantly impact cost, emissions, and resource mix. Stochastic modeling adds complexity but enables a more realistic assessment of cost risk, better aligning with the uncertainties faced by decision-makers. However, the computational intensity of stochastic models and the omission of key factors such as transmission, storage, and demand response suggest that further research is needed to enhance the practicality and applicability of this approach to broader, more complex systems.

In conclusion, this study emphasizes that future capacity expansion decisions should account for both average costs and cost risks, leveraging stochastic approaches to address the inherent uncertainties in fuel prices, renewable variability, and demand. This flexibility not only improves the robustness of energy planning, but also highlights opportunities for further exploration, particularly in integrating emission constraints and low-carbon, firm generation resources in the presence of renewable variability.

Supplementary Materials: The following supporting information can be downloaded at: https://www.mdpi.com/article/10.3390/en18051283/s1, References [43,45,50,51,57,60,62–66] are cited in Supplementary Materials.

Author Contributions: Conceptualization, N.S.G., E.H. and E.W.; Methodology, N.S.G., E.H. and E.W.; Software, N.S.G.; Formal analysis, N.S.G.; Data curation, N.S.G.; Writing—original draft, N.S.G.; Writing—review & editing, N.S.G., E.H. and E.W.; Supervision, E.H. and E.W.; Funding acquisition, E.H. and E.W. All authors have read and agreed to the published version of the manuscript.

Funding: This work was supported by the Civil Infrastructure Systems program of the National Science Foundation (grant# CMMI 1436469) and Rochester Institute of Technology.

Data Availability Statement: All the data are downloaded from the publicly available resources and cited in the article.

Conflicts of Interest: The authors declare no conflict of interest.

References

1. Edenhofer, O.; IPCC (Eds.) Climate Change 2014: Mitigation of Climate Change. In *Working Group III Contribution to the Fifth Assessment Report of the Intergovernmental Panel on Climate Change*; Cambridge University Press: New York, NY, USA, 2014.
2. U.S. Energy Information Administration. Electric Power Monthly. November 2016. EIA. Available online: http://www.eia.gov/electricity/monthly/pdf/epm.pdf (accessed on 11 February 2017).

3. National Conference of State Legislation. State Renewable Portfolio Standards and Goals. Available online: https://www.ncsl.org/energy/state-renewable-portfolio-standards-and-goals (accessed on 28 April 2019).
4. Carbon Pricing Dashboard | Up-to-Date Overview of Carbon Pricing Initiatives. Available online: https://carbonpricingdashboard.worldbank.org/ (accessed on 24 May 2019).
5. EIA. Annual Energy Outlook 2023—U.S. Energy Information Administration (EIA). Available online: https://www.eia.gov/outlooks/aeo/index.php (accessed on 13 February 2024).
6. Brown, M. *Regional Energy Deployment System (ReEDS) Model Documentation: Version 2019*; National Renewable Energy Laboratory: Golden, CO, USA, 2020.
7. US-REGEN Model Documentation. Available online: https://www.epri.com/research/products/000000003002016601 (accessed on 14 March 2021).
8. PROMOD | Hitachi Energy. Available online: https://www.hitachienergy.com/us/en/products-and-solutions/energy-portfolio-management/enterprise/promod (accessed on 18 February 2025).
9. EPRI. EGEAS. Available online: https://www.epri.com/#/pages/product/3002001929/?lang=en-US (accessed on 25 April 2020).
10. Polaris—Power Systems Optimization. Available online: http://psopt.com/ (accessed on 25 April 2020).
11. PLEXOS Market Simulation Software, Energy Exemplar. Available online: https://energyexemplar.com/solutions/plexos/ (accessed on 25 April 2020).
12. AURORA Electric Modeling Forecasting and Analysis Software, Energy Exemplar. Available online: https://energyexemplar.com/products/aurora-electric-modeling-forecasting-software/ (accessed on 25 April 2020).
13. Yue, X.; Pye, S.; DeCarolis, J.; Li, F.G.; Rogan, F.; Gallachóir, B.Ó. A review of approaches to uncertainty assessment in energy system optimization models. *Energy Strat. Rev.* **2018**, *21*, 204–217. [CrossRef]
14. DeCarolis, J.; Daly, H.; Dodds, P.; Keppo, I.; Li, F.; McDowall, W.; Pye, S.; Strachan, N.; Trutnevyte, E.; Usher, W.; et al. Formalizing best practice for energy system optimization modelling. *Appl. Energy* **2017**, *194*, 184–198. [CrossRef]
15. Jin, S.; Ryan, S.M.; Watson, J.-P.; Woodruff, D.L. Modeling and solving a large-scale generation expansion planning problem under uncertainty. *Energy Syst.* **2011**, *2*, 209–242. [CrossRef]
16. Hunter, K.; Sreepathi, S.; DeCarolis, J.F. Modeling for insight using Tools for Energy Model Optimization and Analysis (Temoa). *Energy Econ.* **2013**, *40*, 339–349. [CrossRef]
17. Schröder, A. An electricity market model with generation capacity expansion under uncertainty. *Energy Syst.* **2014**, *5*, 253–267. [CrossRef]
18. Ehrenmann, A.; Smeers, Y. Generation Capacity Expansion in a Risky Environment: A Stochastic Equilibrium Analysis. *Oper. Res.* **2011**, *59*, 1332–1346. [CrossRef]
19. Patankar, N.; de Queiroz, A.R.; DeCarolis, J.F.; Bazilian, M.D.; Chattopadhyay, D. Building conflict uncertainty into electricity planning: A South Sudan case study. *Energy Sustain. Dev.* **2019**, *49*, 53–64. [CrossRef]
20. Bertocchi, M. (Ed.) Stochastic optimization methods in finance and energy: New financial products and energy market strategies. In *International Series in Operations Research & Management Science*; Springer: New York, NY, USA, 2011; Volume 163.
21. Steeger, G.; Lohmann, T.; Rebennack, S. Strategic bidding for a price-maker hydroelectric producer: Stochastic dual dynamic programming and Lagrangian relaxation. *IISE Trans.* **2018**, *50*, 929–942. [CrossRef]
22. Hu, Q.; Huang, G.; Cai, Y.; Xu, Y. Energy and Environmental Systems Planning with Recourse: Inexact Stochastic Programming Model Containing Fuzzy Boundary Intervals in Objectives and Constraints. *J. Energy Eng.* **2013**, *139*, 169–189. [CrossRef]
23. Wildman, R.; Gaynor, A. 11—Topology optimization for robotics applications. In *Robotic Systems and Autonomous Platforms*; Walsh, S.M., Strano, M.S., Eds.; Woodhead Publishing in Materials; Woodhead Publishing: Cambridge, UK, 2019; pp. 251–292. [CrossRef]
24. Kayal, P.; Chanda, C. Optimal mix of solar and wind distributed generations considering performance improvement of electrical distribution network. *Renew. Energy* **2015**, *75*, 173–186. [CrossRef]
25. Leuthold, F.U.; Weigt, H.; von Hirschhausen, C. A Large-scale spatial optimization model of the European electricity market. *Netw. Spat. Econ.* **2012**, *12*, 75–107. [CrossRef]
26. Yakin, M.Z.; McFarland, J.W. Electric generating capacity planning: A nonlinear programming approach. *Electr. Power Syst. Res.* **1987**, *12*, 1–9. [CrossRef]
27. Alvarez, G.E. Stochastic optimization considering the uncertainties in the electricity demand, natural gas infrastructures, photovoltaic units, and wind generation. *Comput. Chem. Eng.* **2022**, *160*, 107712. [CrossRef]
28. Aaslid, P.; Korpas, M.; Belsnes, M.M.; Fosso, O.B. Stochastic Optimization of Microgrid Operation with Renewable Generation and Energy Storages. *IEEE Trans. Sustain. Energy* **2022**, *13*, 1481–1491. [CrossRef]
29. Yang, C.; Wu, Z.; Li, X.; Fars, A. Risk-constrained stochastic scheduling for energy hub: Integrating renewables, demand response, and electric vehicles. *Energy* **2023**, *288*, 129680. [CrossRef]

30. Zhong, J.; Li, Y.; Wu, Y.; Cao, Y.; Li, Z.; Peng, Y.; Qiao, X.; Xu, Y.; Yu, Q.; Yang, X.; et al. Optimal Operation of Energy Hub: An Integrated Model Combined Distributionally Robust Optimization Method with Stackelberg Game. *IEEE Trans. Sustain. Energy* **2023**, *14*, 1835–1848. [CrossRef]
31. Rebennack, S. Combining sampling-based and scenario-based nested Benders decomposition methods: Application to stochastic dual dynamic programming. *Math. Program.* **2016**, *156*, 343–389. [CrossRef]
32. Lohmann, T.; Rebennack, S. Tailored Benders Decomposition for a Long-Term Power Expansion Model with Short-Term Demand Response. *Manag. Sci.* **2017**, *63*, 2027–2048. [CrossRef]
33. Park, H.; Baldick, R. Optimal capacity planning of generation system integrating uncertain solar and wind energy with seasonal variability. *Electr. Power Syst. Res.* **2020**, *180*, 106072. [CrossRef]
34. Hemmati, R.; Hooshmand, R.-A.; Khodabakhshian, A. Coordinated generation and transmission expansion planning in deregulated electricity market considering wind farms. *Renew. Energy* **2016**, *85*, 620–630. [CrossRef]
35. Gonzalez-Romero, I.; Wogrin, S.; Gómez, T. Review on generation and transmission expansion co-planning models under a market environment. *IET Gener. Transm. Distrib.* **2020**, *14*, 931–944. [CrossRef]
36. Cantoni, M.; Marseguerra, M.; Zio, E. Genetic algorithms and Monte Carlo simulation for optimal plant design. *Reliab. Eng. Syst. Saf.* **2000**, *68*, 29–38. [CrossRef]
37. Marseguerra, M.; Zio, E.; Podofillini, L. A multiobjective genetic algorithm approach to the optimization of the technical specifications of a nuclear safety system. *Reliab. Eng. Syst. Saf.* **2004**, *84*, 87–99. [CrossRef]
38. Zhou, Z.; Zhang, J.; Liu, P.; Li, Z.; Georgiadis, M.C.; Pistikopoulos, E.N. A two-stage stochastic programming model for the optimal design of distributed energy systems. *Appl. Energy* **2013**, *103*, 135–144. [CrossRef]
39. Marseguerra, M.; Zio, E. Optimizing maintenance and repair policies via a combination of genetic algorithms and Monte Carlo simulation. *Reliab. Eng. Syst. Saf.* **2000**, *68*, 69–83. [CrossRef]
40. Jellouli, O.; Chatelet, E. Monte Carlo Simulation and Genetic Algorithm for Optimising Supply Chain Management in a Stochastic Environment. In Proceedings of the 2001 IEEE International Conference on Systems, Man and Cybernetics, Tucson, AZ, USA, 7–10 October 2001; Volume 3, pp. 1835–1839. [CrossRef]
41. Park, J.-B.; Kim, J.-H.; Lee, K. Generation expansion planning in a competitive environment using a genetic algorithm. In Proceedings of the IEEE Power Engineering Society Summer Meeting, Chicago, IL, USA, 21–25 July 2002; Volume 3, pp. 1169–1172. [CrossRef]
42. Banzhaf, W. Genetic Programming for Pedestrians. In Proceedings of the 5th International Conference on Genetic Algorithms, San Francisco, CA, USA, 1 June 1993; Morgan Kaufmann Publishers Inc.: San Francisco, CA, USA, 1993; p. 628. Available online: http://dl.acm.org/citation.cfm?id=645513.657432 (accessed on 27 April 2019).
43. U.S. Energy Information Administration (EIA). Updated Capital Cost Estimates for Utility Scale Electricity Generating Plants. 2013. U.S. Department of Energy. Available online: http://www.eia.gov/outlooks/capitalcost/pdf/updated_capcost.pdf (accessed on 1 August 2017).
44. NREL. Cost-Benefit Analysis of Flexibility Retrofits for Coal and Gas-Fueled Power Plants. Available online: http://www.nrel.gov/docs/fy14osti/60862.pdf (accessed on 23 June 2020).
45. US Environmental Protection Agency. eGRID2018. Available online: https://www.epa.gov/energy/egrid (accessed on 1 December 2019).
46. IEA. WEO Model. Available online: https://iea.blob.core.windows.net/assets/77ecf96c-5f4b-4d0d-9d93-d81b938217cb/World_Energy_Outlook_2018.pdf (accessed on 25 May 2019).
47. EIA. *World Energy Projection System Plus: Overview*; EIA: Washington, DC, USA, 2017; p. 35.
48. Rubin, E.S.; Azevedo, I.M.; Jaramillo, P.; Yeh, S. A review of learning rates for electricity supply technologies. *Energy Policy* **2015**, *86*, 198–218. [CrossRef]
49. EIA. Annual Energy Outlook. 2017. Available online: https://www.eia.gov/pressroom/presentations/sieminski_01052017.pdf (accessed on 13 May 2018).
50. Draxl, C.; Clifton, A.; Hodge, B.-M.; McCaa, J. The Wind Integration National Dataset (WIND) Toolkit. *Appl. Energy* **2015**, *151*, 355–366. [CrossRef]
51. National Renewable Energy Laboratory (NREL). Solar Power Data for Integration Studies. 2006. Available online: http://www.nrel.gov/grid/solar-power-data.html (accessed on 10 November 2016).
52. EIA. *Assumptions to the Annual Energy Outlook 2017*; EIA: Washington, DC, USA, 2017; p. 29.
53. Availability of the National Energy Modeling System (NEMS) Archive. Available online: https://www.eia.gov/outlooks/aeo/info_nems_archive.php (accessed on 23 March 2019).
54. Williams, E.; Hittinger, E.; Carvalho, R.; Williams, R. Wind power costs expected to decrease due to technological progress. *Energy Policy* **2017**, *106*, 427–435. [CrossRef]
55. Ranjbari, L.; Bahar, A.; Aziz, Z.A. Stochastic Models of Natural Gas Prices. *J. Karya Asli Lorekan Ahli Mat.* **2013**, *6*, 50–65.

56. Keles, D.; Genoese, M.; Möst, D.; Fichtner, W. Comparison of extended mean-reversion and time series models for electricity spot price simulation considering negative prices. *Energy Econ.* **2012**, *34*, 1012–1032. [CrossRef]
57. Henry Hub Natural Gas Spot Price (Dollars per Million Btu). Available online: https://www.eia.gov/dnav/ng/hist/rngwhhdm.htm (accessed on 29 April 2019).
58. Blair, N.; Zhou, E.; Getman, D.; Arent, D.J. *Electricity Capacity Expansion Modeling, Analysis, and Visualization: A Summary of Selected High-Renewable Modeling Experiences*; National Renewable Energy Laboratory (NREL): Golden, CO, USA, 2015; p. 39.
59. Rockafellar, R.T.; Uryasev, S. Optimization of conditional value-at-risk. *J. Risk* **2000**, *2*, 21–41. [CrossRef]
60. Mathworks. Genetic Algorithm. Available online: https://www.mathworks.com/discovery/genetic-algorithm.html (accessed on 28 April 2019).
61. Mills, A.D.; Wiser, R.H.; Seel, J. *Power Plant Retirements: Trends and Possible Drivers*; Lawrence Berkeley National Laboratory: Berkeley, CA, USA, 2017; p. 1411667. [CrossRef]
62. MISO. 2017 MISO State of the Market Report.pdf. Available online: https://www.potomaceconomics.com/wp-content/uploads/2018/07/2017-MISO-SOM_Report_6-26_Final.pdf (accessed on 8 July 2019).
63. US Energy Information Administration. Cost and Performance Characteristics of New Generating Technologies, Annual Energy Outlook 2017. 2017. Available online: https://www.eia.gov/outlooks/aeo/assumptions/pdf/table_8.2.pdf (accessed on 12 February 2018).
64. Midcontinent Independent System Operator (MISO). Daily Regional Forecast and Actual Load. Available online: https://www.misoenergy.org/markets-and-operations/#t=10&p=0&s=&sd= (accessed on 9 June 2018).
65. Adler, D. Genetic algorithms and simulated annealing: A marriage proposal. In Proceedings of the IEEE International Conference on Neural Networks, San Francisco, CA, USA, 28 March–1 April 1993; Volume 2, pp. 1104–1109.
66. Genetic Algorithm Options-MATLAB & Simulink. Available online: https://www.mathworks.com/help/gads/genetic-algorithm-options.html#f6593 (accessed on 28 August 2017).

energies

MDPI

Article

Reduced-Order Modeling and Stability Analysis of Grid-Following and Grid-Forming Hybrid Renewable Energy Plants

Yue Ma *, Ning Chen and Luming Ge

China Electric Power Research Institute, Nanjing 210003, China; chen_ning@epri.sgcc.com.cn (N.C.); geluming@epri.sgcc.com.cn (L.G.)
* Correspondence: mayue_421@outlook.com

Abstract: The control methods of energy systems can be categorized into grid-following and grid-forming types. The grid-following control method relies on grid synchronization and is prone to stability issues in weak grid conditions. By contrast, the grid-forming control method exhibits synchronous machine characteristics, providing voltage support to the system, but potentially introducing stability risks under strong grid conditions. Constructing a grid-following and grid-forming hybrid renewable energy plant can effectively enhance the system's support capability and ensure reliable operation. However, the interactions among multiple inverters are complex, and traditional modeling methods are inadequate to meet the modeling requirements for such systems. To effectively address this problem, this paper presents a reduced-order modeling method that simplifies the complex system into a simple system consisting of an equivalent grid-following, an equivalent grid-forming, and grid impedance through frequency decoupling and the aggregation of similar inverters. Furthermore, this study employs both the Nyquist stability criterion and the harmonic characteristic analysis method to elucidate how the capacity ratio between grid-following and grid-forming affects system stability.

Keywords: grid-following; grid-forming; renewable energy plants; reduced-order modeling method; stability analysis

1. Introduction

In response to the increasing strain on conventional energy supplies and the worsening environmental crisis, the transition to low-carbon power systems is accelerating, with renewable energy playing an ever-growing role in the grid [1]. While the integration of wind, solar, and other renewable sources helps mitigate environmental issues, it also poses new challenges to the safe and stable operation of power systems [2].

Renewable energy control methods are primarily classified into grid-following (GFL) and grid-forming (GFM) types. The GFL method synchronizes with the grid using the phase locked loop (PLL) [3] and depends on system voltage sources to maintain synchronization at the point of interconnection. However, it only passively adapts to grid conditions, offering limited support in weak grid environments, ultimately reducing system stability [4,5]. By contrast, the grid-forming (GFM) method exhibits damping and inertia characteristics similar to synchronous machines, allowing it to actively provide voltage support to the grid and maintain a higher stability margin under weak grid conditions [6,7]. However, it is prone to stability issues in strong grid environments [8]. Grid-following and grid-forming systems are suited to different grid conditions. By combining both types within

https://doi.org/10.3390/en18071752

a single system, the application range of inverters can be expanded, ensuring stability even when impedance fluctuates significantly due to changes in operating modes or load conditions. The hybrid application of grid-following and grid-forming can be realized through switching, integration, or hybrid control methods. The mode-switching approach automatically switches the control mode based on the detected grid impedance, though this transition may introduce unknown stability risks [9]. The control-loop integration method merges the control mechanisms of both types of inverters within the controller, which integrates their characteristics; however, this increases the difficulty of system characteristic analysis [10]. The hybrid control method, which parallels grid-following and grid-forming systems, not only improves the reliability of inverter operation but enhances flexibility in renewable energy plants [11]. The inclusion of grid-forming can resolve the issue of insufficient stability margin in grid-following renewable energy plants under weak grid conditions. In [12], the impedance characteristics of grid-following and grid-forming systems are compared using impedance modeling by Xu et al. Under weak grid conditions, the negative damping frequency range of the grid-forming system is narrower, and its amplitude characteristic is lower, compared to the grid-following system. When the two systems are hybridized, the negative damping frequency range is reduced, effectively enhancing system stability. This issue has been qualitatively explained from the perspective of a short-circuit ratio, suggesting that the grid-forming system, when modeled as a voltage source, enhances system strength and leads to an equivalent increase in the short-circuit ratio, as discussed in [13,14].

The aforementioned studies analyze the stability of grid-following and grid-forming two-machine systems through modeling. However, in renewable energy plants, multiple inverters of different types coexist. Traditional modeling methods can lead to overly complex models, making it challenging to fully capture the characteristics and interactions of each inverter. Meanwhile, neglecting certain parameters may introduce errors, reducing modeling accuracy and compromising the reliability of system analysis. In [15], Sun et al. developed a sixth-order transient model for photovoltaic-storage systems, and its stability was analyzed using the region of attraction. Nonetheless, the model's high complexity and computational burden make it difficult to apply to renewable energy plant modeling. Some studies reduce the system order at the control level, such as Akwasi et al., who adopted new control methods for model simplification. In [16], the reduced-order Luenberger observer (ROLO) is applied to a grid-forming inverter control. The ROLO simplifies the model while preserving key dynamics and reduces computational complexity. However, as a method that removes the PI controller, it requires further experiments and validation. Meanwhile, in [17], Du et al. proposed a model reduction method for inverter-dominated microgrids by simplifying the external area model, which reduces computational complexity while preserving key dynamic responses under disturbances. However, the model includes only grid-forming inverters, without considering interactions between different types. In addition, although the dynamic performance was verified, no effective stability assessment method for multi-machine systems was provided. In [18], Rosso et al. derived the transfer function of a multi-machine system, and the stability of a grid-following and grid-forming hybrid system was evaluated using the μ analysis method. This approach, based on multivariable control theory, facilitates the assessment of robust stability under various conditions. The lack of a clear physical interpretation, however, limits the ability to conduct further mechanism analysis of the system. In [19], a hybrid system support index was introduced by Wang et al. to address the qualitative stability analysis of renewable energy plants, characterizing the stability of hybrid systems. The inclusion of a grid-forming inverter adds a polynomial with positive coefficients to both the hybrid short-circuit ratio and the support index. Nevertheless, the calculation of these indicators depends on accurate system parameters, their general applicability and adaptability still need further verification. In [20], an impedance-based composite grid modeling method was proposed to construct the admittance matrix of a renewable energy plant with multiple inverters. The system's stability was determined using eigenvalue analysis. This approach, though, ignores the interactions between inverters, which compromises the accuracy of the model. Additionally, solving the eigenvalues of high-order matrices is a challenging task, adding complexity to the method. In [21], a model reduction method for grid-forming inverters based on frequency-domain perturbation operators is presented. It simplifies stability analysis and effectively identifies factors affecting single-machine impedance. However, the method is not extended to renewable energy plants. Overall, the existing modeling methods are relatively complex and fail to account for the interactions between inverters, making it difficult to accurately reflect the impedance characteristics of grid-following and grid-forming hybrid renewable energy plants.

In a nutshell, this paper addresses the issues of high complexity and low versatility in existing modeling methods for renewable energy plants by constructing reduced-order models for grid-following and grid-forming hybrid renewable energy plants. It also explores the capacity ratio problem of two types of inverters in these plants from the perspectives of the harmonic suppression using the Nyquist stability criterion and harmonic analysis methods. Additionally, the conclusions drawn from the analysis are validated through simulation. Finally, the research findings of this paper are summarized.

2. Topology and Control Block Diagram of Grid-Following and Grid-Forming Hybrid Renewable Energy Plants

This section presents the hybrid system topology and control block diagrams for both grid-following and grid-forming inverters. The grid-following inverter uses a phase-locked loop to measure the point of common coupling, employing closed-loop control to synchronize with the grid's voltage and frequency. The grid-forming inverter, on the other hand, actively controls system frequency and voltage. In weak grid environments, it mimics synchronous machine behavior, providing inertia and damping support to maintain system stability.

2.1. Topology of Renewable Energy Plant

The capacity of renewable energy plants typically ranges from 100 MW to 1000 MW, significantly larger than single-machine systems, and the greater number of inverters further increases modeling complexity. As technology advances, grid-following and grid-forming hybrid renewable energy plants have emerged, enabling stable operation under varying short-circuit ratios. However, the complex interactions among inverters pose additional challenges for modeling and stability analysis. Figure 1 illustrates the topology of a grid-following and grid-forming hybrid renewable energy plant, comprising converter groups, transformers, transmission lines, and an AC power grid.

In renewable energy plants, inverters convert direct current power into alternating current power, which is more suitable for transmission. Grid-following and grid-forming hybrid renewable energy plants incorporate both grid-following inverters and grid-forming inverters. Grid-following inverters perform well and maintain good stability in strong power grids, while grid-forming inverters provide better stability under weak power grid conditions. By combining the advantages of both types of inverters, such hybrid renewable energy plants can achieve stable operation even when the short-circuit ratio of the power grid changes. The following study focuses on the modeling and stability analysis of this type of inverter.

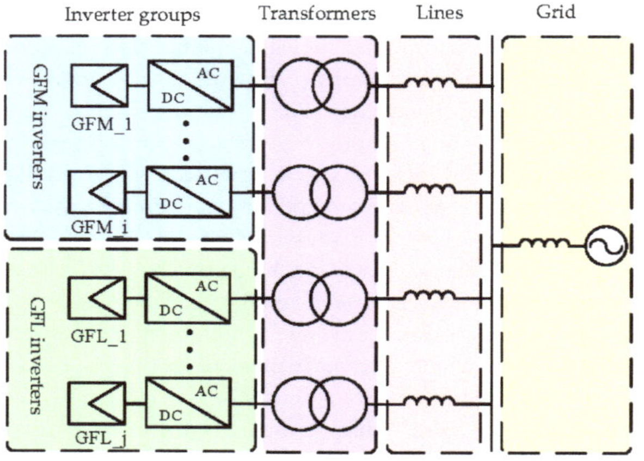

Figure 1. Topology diagram of the grid-following and grid-forming hybrid renewable energy plant.

2.2. Control Structure of Grid-Following/Grid-Forming Inverter

The control block diagrams for grid-following and grid-forming systems are shown in Figures 2 and 3. The grid-following inverter utilizes a phase-locked loop (PLL) to achieve synchronization by detecting the grid phase. In Figure 2, L_1, L_2, and C_1 represent the filter inductors and capacitors, L_g is the grid inductance, u_{abc} is the port voltage, i_{abc} is the port current, i_{dref} and i_{qref} are the reference values for the *d*-axis and *q*-axis currents, u_{dref} and u_{qref} are the reference values for the *d*-axis and *q*-axis voltages.

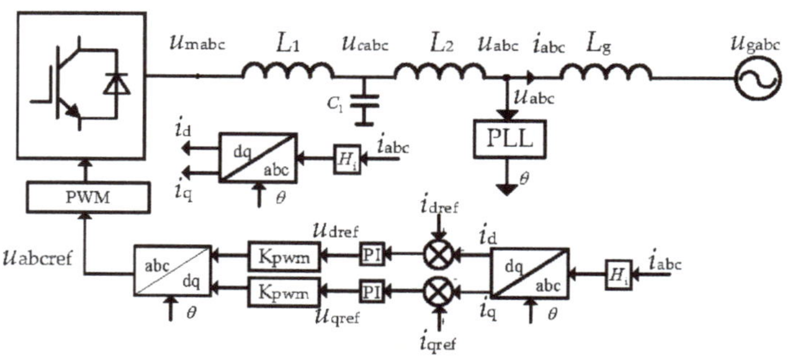

Figure 2. Control diagram of the grid-following inverter.

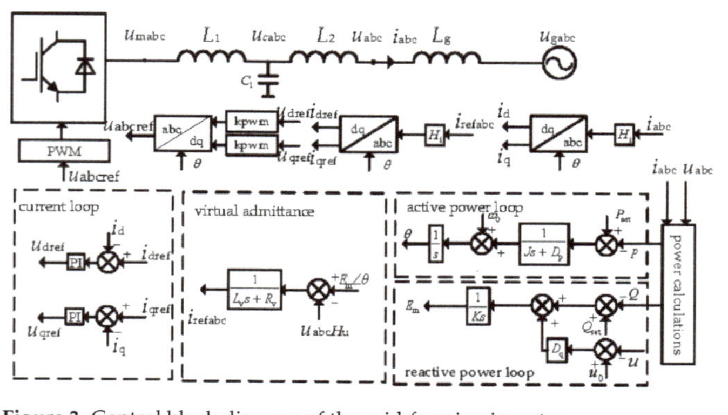

Figure 3. Control block diagram of the grid-forming inverter.

The grid-forming inverter uses virtual synchronous generator control (VSG), as shown in Figure 3. In this diagram, L_1, L_2, and C_1 represent the filter inductors and capacitors; u_{mabc} is the bridge arm voltage; u_{abc} is the grid connection point voltage; i_{abc} is the grid connection point current; P_{set} and Q_{set} are the rated active and reactive powers; P and Q represent the active and reactive power generated by GFM; J represents inertia; D_p and D_q are the active and reactive damping coefficients; u and u_0 are the rated voltage RMS value and the output voltage RMS value; K is the reactive inertia coefficient; θ and E_m are the output power angle and output voltage RMS value of the VSG; L_v and R_v are the virtual inductance and virtual resistance, which together form the virtual admittance. The control loop consists of active power, reactive power, virtual admittance, and current loops.

3. Reduced-Order Modeling Method for Grid-Following and Grid-Forming Hybrid Renewable Energy Plants

The grid-following and grid-forming hybrid renewable energy plant represents a promising solution that aligns with the current trends in energy development. However, conventional modeling methods are often complicated, lack general applicability, and are difficult to adapt to the stability analysis requirements of such hybrid systems. To overcome these limitations, this paper proposes a model order reduction method tailored for grid-following and grid-forming hybrid renewable energy plants. The equivalent admittance matrices of the grid-following and grid-forming inverters are derived, and the accuracy of the proposed model is verified through comparison between calculated and measured results.

3.1. Modeling of Grid-Following/Grid-Forming Inverters

The impedance method is based on harmonic linearization theory [22]. By injecting a harmonic signal into the terminal voltage and measuring the corresponding harmonic current at the grid connection point, the system's impedance model can be derived. In the grid-following and grid-forming hybrid renewable energy plant, transformers and transmission lines result in phase angle discrepancies between the terminal voltages of different types of inverters. Therefore, when calculating the impedance models for grid-following and grid-forming inverters, it is essential to account for the phase of their respective terminal voltages. Frequency coupling effects exist in both grid-following and grid-forming inverters. When a positive-sequence voltage with frequency f_p is injected at the grid connection point, it generates a positive-sequence current component at f_p and a negative-sequence current component at f_p-$2f_1$ [23]. Taking the grid-forming inverter as an example, when a positive-sequence disturbance signal is injected into phase-a of the terminal voltage, the time-domain expressions for the voltage and current are as follows:

$$\begin{aligned} u_a(t) &= U_1\cos(2\pi f_1 t+\varphi_u)+U_p\cos(2\pi f_p t+\varphi_{up})+U_n\cos(2\pi f_n t+\varphi_{un}) \\ i_a(t) &= I_1\cos(2\pi f_1 t+\varphi_i)+I_p\cos(2\pi f_p t+\varphi_{ip})+I_n\cos(2\pi f_n t+\varphi_{in}) \end{aligned} \tag{1}$$

The above expressions are transformed into the frequency domain, and the expressions are as follows:

$$\begin{aligned} u_a &= \begin{cases} \boldsymbol{U}_1, & f=\pm f_1 \\ \boldsymbol{U}_p, & f=\pm f_p \end{cases} \\ i_a &= \begin{cases} \boldsymbol{I}_1, & f=\pm f_1 \\ \boldsymbol{I}_p, & f=\pm f_p \\ \boldsymbol{I}_n, & f=\pm (f_p-2f_1) \end{cases} \end{aligned} \tag{2}$$

In Equation (2), $\boldsymbol{U}_1 = (U_1/2)e^{\pm j\varphi_u}$, $\boldsymbol{U}_p = (U_p/2)e^{\pm j\varphi_{up}}$, $\boldsymbol{I}_1 = (I_1/2)e^{\pm j\varphi_i}$, $\boldsymbol{I}_p = (I_p/2)e^{\pm j\varphi_{ip}}$ and $\boldsymbol{I}_n = (I_n/2)e^{\pm j\varphi_{in}}$; U_1, U_p represent the fundamental voltage and the amplitude of the positive-sequence disturbance voltage; I_1, I_p, I_n represent the amplitudes of the fundamental current, positive-sequence disturbance current, and negative-sequence disturbance current; φ_u represents the initial phase angle of the fundamental voltage; φ_{up} represents the initial phase angle of the disturbance voltage; φ_i represents the initial phase angle of fundamental current; φ_{ip} and φ_{in} represent the initial phase angles of the positive-sequence disturbance current and the negative-sequence disturbance current. Based on the control block diagrams of grid-following and grid-forming inverters, the admittance models incorporating the terminal voltage phase angle can be derived as follows:

$$\begin{bmatrix} U_{p\_GFL}(s) \\ U_{n\_GFL}(s) \end{bmatrix} = \begin{bmatrix} Y_{11\_GFL}(s) & Y_{21\_GFL}(s) \\ Y_{12\_GFL}(s) & Y_{22\_GFL}(s) \end{bmatrix} \begin{bmatrix} I_{p\_GFL}(s) \\ I_{n\_GFL}(s) \end{bmatrix} \tag{3}$$

$$\begin{bmatrix} U_{p\_GFM}(s) \\ U_{n\_GFM}(s) \end{bmatrix} = \begin{bmatrix} Y_{11\_GFM}(s) & Y_{21\_GFM}(s) \\ Y_{12\_GFM}(s) & Y_{22\_GFM}(s) \end{bmatrix} \begin{bmatrix} I_{p\_GFM}(s) \\ I_{n\_GFM}(s) \end{bmatrix} \tag{4}$$

In Equation (3), $Y_{11\_GFL}(s)$, $Y_{21\_GFL}(s)$, $Y_{22\_GFL}(s)$, and $Y_{12\_GFL}(s)$ represent the impedance expressions for the grid-following inverter, as detailed in Appendix A, Equations (A2)–(A5). Similarly, in Equation (4), $Y_{11\_GFM}(s)$, $Y_{21\_GFM}(s)$, $Y_{22\_GFM}(s)$, and $Y_{12\_GFM}(s)$ correspond to the impedance expressions for the grid-forming inverter, provided in Appendix A, Equations (A14)–(A17).

3.2. Overview of Reduced-Order Modeling Methods

The grid-following and grid-forming hybrid renewable energy plant is a highly complex system. If the entire plant is modeled directly, the model complexity becomes extremely high, as shown in [24], and you would need to spend a significant amount of time handling this complex matrix, which could even lead to a dimensional explosion. In the topology shown in Figure 1, directly modeling a grid-following and grid-forming hybrid renewable energy plant with i grid-forming inverters and j grid-following inverters results in an impedance model of order (2i + 2j) × (2i + 2j), as given in Equation (5). As i and j increase, the difficulty of analyzing the impedance model rises significantly. Therefore, reducing the model order is essential to simplify the analysis of impedance characteristics.

$$\begin{bmatrix} U_{p\_GFM1}(s) \\ U_{n\_GFM1}(s) \\ \cdots \\ U_{p\_GFLj}(s) \\ U_{n\_GFLj}(s) \end{bmatrix} = \begin{bmatrix} Y_{11\_GFM1}(s) & Y_{21\_GFM1}(s) & & & \\ Y_{12\_GFM1}(s) & Y_{22\_GFM1}(s) & & & \\ & & \cdots & & \\ & & & Y_{11\_GFLj}(s) & Y_{11\_GFLj}(s) \\ & & & Y_{11\_GFLj}(s) & Y_{11\_GFLj}(s) \end{bmatrix} \begin{bmatrix} I_{p\_GFM1}(s) \\ I_{n\_GFM1}(s) \\ \cdots \\ I_{p\_GFLj}(s) \\ I_{n\_GFLj}(s) \end{bmatrix} \tag{5}$$

During the plant modeling process, if the impedance of inverters and transformers is known, their series and parallel connections can be determined based on their electrical interconnections, ultimately establishing the impedance model of the grid-following and grid-forming hybrid renewable energy plant. As shown in Equation (5), the admittance matrix of inverters contains coupling terms. When inverters are connected in series with transformers and transmission lines, coupling effects alter the impedance of both the inverters and the lines. These changes also occur between inverters and the grid. Neglecting such effects will introduce errors in the impedance matrix, affecting the accuracy of stability analysis.

Based on the above analysis, implementing an appropriate model order reduction in the modeling process of grid-following and grid-forming hybrid renewable energy plants can significantly simplify the impedance model. At the same time, considering the coupling effects between inverters and other components helps maintain the accuracy of the modeling results.

The flowchart of the reduced-order modeling method is shown in Figure 4. First, based on the impedance models of grid-following/grid-forming inverters, transformers, and transmission lines, the admittance model of inverters with collector lines is derived, while also taking into account the coupling effects between the inverters and the transmission lines. Next, inverters of the same type are categorized and aggregated to obtain the equivalent admittance matrices for grid-following and grid-forming inverters. Finally, the system is connected to the grid, and the equivalent admittance matrices for grid-following and grid-forming inverters are derived, while considering frequency coupling effects both among inverters and between inverters and the grid. This model reduction approach maintains accuracy, while simplifying the renewable energy plant into a simple circuit, where the equivalent grid-following and grid-forming inverters operate in parallel with the power grid. The following sections provide a detailed analysis of this model reduction approach based on specific system models.

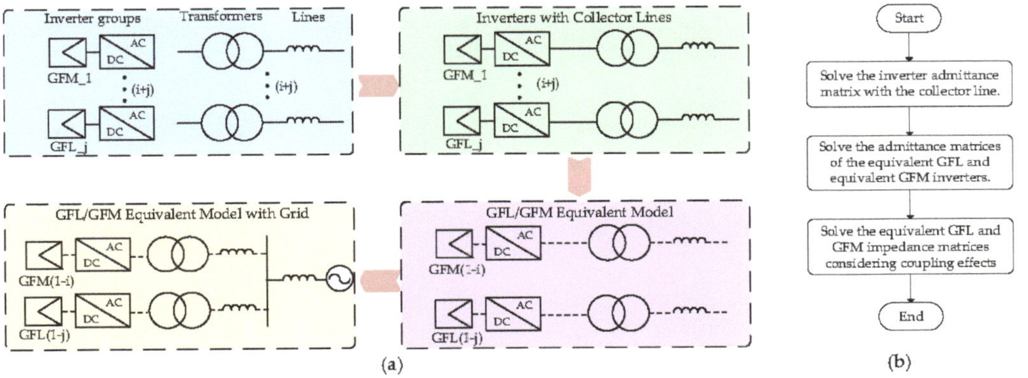

Figure 4. Schematic diagram of model order reduction: (**a**) topology evolution diagram of renewable energy plants; (**b**) flowchart of the model order reduction method.

3.3. Detailed Description of Reduced-Order Modeling Methods

Referring to the flowchart in Figure 4, the first step is to calculate the admittance matrix of inverters with the collection line. As shown in Figure 5, when the inverter is in series with the line, the coupled current flows through the collection line, creating a current loop. The collection line impedance affects the inverter's admittance, so it is essential to account for these coupling effects during the derivation to ensure the accuracy of the results.

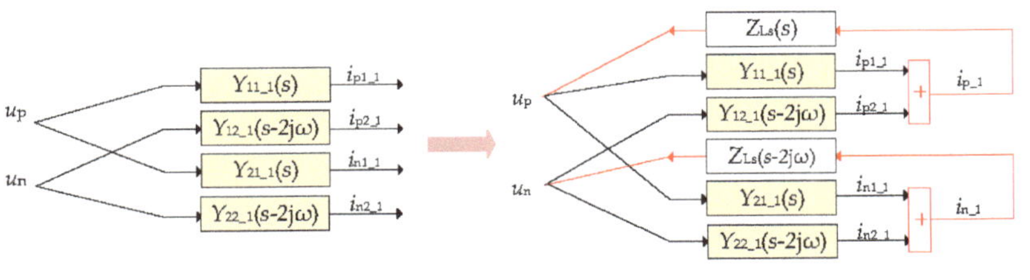

Figure 5. Frequency coupling between the inverter and the collection line.

In Figure 5, $Y_{11\_1}(s)$, $Y_{12\_1}(s)$, $Y_{22\_1}(s)$, and $Y_{21\_1}(s)$ represent the admittance matrix of the inverter in an ideal grid system, while $Z_{Ls}(s)$ is the sum of the line impedance and transformer impedance. According to Ohm's law, $Y = U/I$, the admittance matrix of the inverter operating in stand-alone mode is given by the following:

$$
\begin{aligned}
Y_{11\_1}(s) &= -\frac{i_{\mathrm{p1\_1}}(s)}{u_{\mathrm{p}}(s)} \\
Y_{21\_1}(s) &= -\frac{i_{\mathrm{n2\_1}}(s)}{u_{\mathrm{p}}(s)} \\
Y_{22\_1}(s) &= -\frac{i_{\mathrm{n1\_1}}(s)}{u_{\mathrm{n}}(s)} \\
Y_{12\_1}(s) &= -\frac{i_{\mathrm{p2\_1}}(s)}{u_{\mathrm{n}}(s)}
\end{aligned} \tag{6}
$$

After the collector lines are integrated, the coupling paths within the system are modified. The admittance matrix of the inverter with collection line is as follows:

$$
\begin{aligned}
Y_{11\_\mathrm{coupling}}(s) &= -\frac{i_{\mathrm{p\_1}}(s)}{u_{\mathrm{p}}(s)} \\
Y_{21\_\mathrm{coupling}}(s) &= -\frac{i_{\mathrm{n\_1}}(s)}{u_{\mathrm{p}}(s)} \\
Y_{22\_\mathrm{coupling}}(s) &= -\frac{i_{\mathrm{n\_1}}(s)}{u_{\mathrm{n}}(s)} \\
Y_{12\_\mathrm{coupling}}(s) &= -\frac{i_{\mathrm{p\_1}}(s)}{u_{\mathrm{n}}(s)}
\end{aligned} \tag{7}
$$

Based on the relationships shown in Figure 5 and Ohm's law, $I = U \times Y$, the relationship between current and voltage can be expressed as follows:

$$
\begin{aligned}
& i_{\mathrm{p1\_1}}(s) = -u_{\mathrm{p}}(s)Y_{11\_1}(s) \\
& i_{\mathrm{p2\_1}}(s) = -u_{\mathrm{n}}(s)Y_{12\_1}(s-2j\omega) \\
& i_{\mathrm{n1\_1}}(s) = -u_{\mathrm{p}}(s)Y_{21\_1}(s) \\
& i_{\mathrm{n2\_1}}(s) = -u_{\mathrm{n}}(s)Y_{22\_1}(s-2j\omega) \\
& u_{\mathrm{p}}(s) = Z_{\mathrm{Ls}}(s)[i_{\mathrm{p1\_1}}(s)+i_{\mathrm{p2\_1}}(s)] \\
& u_{\mathrm{n}}(s) = Z_{\mathrm{Ls}}(s-2j\omega)[i_{\mathrm{p1\_1}}(s)+i_{\mathrm{p2\_1}}(s)]
\end{aligned} \tag{8}
$$

Based on Equation (8), the inverter admittance matrix considering frequency coupling can be obtained through the following calculation:

$$
\begin{aligned}
Y_{11\_\mathrm{coupling}}(s) &= Y_{11}(s) - \frac{Y_{12\_1}(s-2j\omega)Y_{21\_1}(s)}{Y_{\mathrm{Ls}}(s-2j\omega)+Y_{22\_1}(s-2j\omega)} \\
Y_{21\_\mathrm{coupling}}(s) &= Y_{21}(s) - \frac{Y_{22\_1}(s-2j\omega)Y_{21\_1}(s)}{Y_{\mathrm{Ls}}(s-2j\omega)+Y_{22\_1}(s-2j\omega)} \\
Y_{22\_\mathrm{coupling}}(s) &= Y_{22}(s) - \frac{Y_{21\_1}(s+2j\omega)Y_{12\_1}(s)}{Y_{\mathrm{Ls}}(s+2j\omega)+Y_{11\_1}(s+2j\omega)} \\
Y_{12\_\mathrm{coupling}}(s) &= Y_{12}(s) - \frac{Y_{11\_1}(s+2j\omega)Y_{12\_1}(s)}{Y_{\mathrm{Ls}}(s+2j\omega)+Y_{11\_1}(s+2j\omega)}
\end{aligned} \tag{9}
$$

In Equation (9), $Y_{11\_\mathrm{coupling}}(s)$, $Y_{21\_\mathrm{coupling}}(s)$, $Y_{22\_\mathrm{coupling}}(s)$, and $Y_{12\_\mathrm{coupling}}(s)$ denote the inverter admittance matrix considering frequency coupling. The admittance matrix of the collection line incorporating these coupling effects can be expressed as follows:

$$
\begin{aligned}
Y_{11\_\mathrm{Lscoupling}}(s) &= Y_{\mathrm{Ls}}(s) \\
Y_{21\_\mathrm{Lscoupling}}(s) &= \frac{Y_{21\_\mathrm{coupling}}(s)}{Y_{11\_\mathrm{coupling}}(s)}Y_{\mathrm{Ls}}(s) \\
Y_{22\_\mathrm{Lscoupling}}(s) &= Y_{\mathrm{Ls}}(s) \\
Y_{12\_\mathrm{Lscoupling}}(s) &= \frac{Y_{12\_\mathrm{coupling}}(s)}{Y_{22\_\mathrm{coupling}}(s)}Y_{\mathrm{Ls}}(s)
\end{aligned} \tag{10}
$$

In Equation (10), $Y_{11\_\mathrm{Lscoupling}}(s)$, $Y_{21\_\mathrm{Lscoupling}}(s)$, $Y_{22\_\mathrm{Lscoupling}}(s)$, and $Y_{12\_\mathrm{Lscoupling}}(s)$ denote the collection line admittance matrix considering frequency coupling. Since the inverter and collection line impedance are in series, the admittance is given by $Y_{\mathrm{equ}}(s) = Y_{\mathrm{Lscoupling}}(s) \times Y_{\mathrm{coupling}}(s)/[Y_{\mathrm{Lscoupling}}(s) + Y_{\mathrm{coupling}}(s)]$. The admittance matrix for the series combination of the collection line and inverter can be derived from Equations (9) and (10), as follows:

$$
\begin{aligned}
Y_{11\text{equ}}(s) &= \frac{[Z_{\text{Ls}}(s-2\text{j}\omega)Y_{11\_1}(s)Y_{22\_1}(s-2\text{j}\omega)+Y_{11\_1}(s)-Z_{\text{Ls}}(s)(s-2\text{j}\omega)Y_{12\_1}(s-2\text{j}\omega)Y_{21\_1}(s)]}{[1+Y_{22\_1}(s-2\text{j}\omega)Z_{\text{Ls}}(s-2\text{j}\omega)+Z_{\text{Ls}}(s)Y_{11\_1}(s)+Z_{\text{Ls}}(s-2\text{j}\omega)Z_{\text{Ls}}(s)Y_{11\_1}(s)Y_{22\_1}(s-2\text{j}\omega)-Y_{12\_1}(s-2\text{j}\omega)Y_{21\_1}(s)Z_{\text{Ls}}(s-2\text{j}\omega)Z_{\text{Ls}}(s)]} \\
Y_{21\text{equ}}(s) &= \frac{Y_{21\_1}(s)}{[1+Y_{22\_1}(s-2\text{j}\omega)Z_{\text{Ls}}(s-2\text{j}\omega)+Z_{\text{Ls}}(s)Y_{11\_1}(s)+Z_{\text{equ}}(s-2\text{j}\omega)Z_{\text{Ls}}(s)Y_{11\_1}(s)Y_{22\_1}(s-2\text{j}\omega)-Y_{12\_1}(s-2\text{j}\omega)Y_{21\_1}(s)Z_{\text{Ls}}(s-2\text{j}\omega)Z_{\text{Ls}}(s)]} \\
Y_{12\text{equ}}(s) &= \frac{Y_{12\_1}(s)}{[1+Y_{11\_1}(s+2\text{j}\omega)Z_{\text{Ls}}(s+2\text{j}\omega)+Z_{\text{Ls}}(s)Y_{22\_1}(s)+Y_{22\_1}(s)Z_{\text{Ls}}(s+2\text{j}\omega)Z_{\text{Ls}}(s)Y_{11\_1}(s+2\text{j}\omega)-Y_{21\_1}(s+2\text{j}\omega)Z_{\text{Ls}}(s+2\text{j}\omega)Y_{12\_1}(s)Z_{\text{Ls}}(s)]} \\
Y_{22\text{equ}}(s) &= \frac{[Z_{\text{Ls}}(s+2\text{j}\omega)Y_{22\_1}(s)Y_{11\_1}(s+2\text{j}\omega)+Y_{22\_1}(s)-Z_{\text{Ls}}(s+2\text{j}\omega)Y_{21\_1}(s+2\text{j}\omega)Y_{12\_1}(s)]}{[1+Y_{11\_1}(s+2\text{j}\omega)Z_{\text{Ls}}(s+2\text{j}\omega)+Z_{\text{Ls}}(s)Y_{22\_1}(s)+Y_{22\_1}(s)Z_{\text{Ls}}(s+2\text{j}\omega)Z_{\text{Ls}}(s)Y_{11\_1}(s+2\text{j}\omega)-Y_{21\_1}(s+2\text{j}\omega)Z_{\text{Ls}}(s+2\text{j}\omega)Y_{12\_1}(s)Z_{\text{Ls}}(s)]}
\end{aligned}
\tag{11}
$$

In the Equation (11), $Y_{11\text{equ}}(s)$, $Y_{12\text{equ}}(s)$, $Y_{21\text{equ}}(s)$, and $Y_{22\text{equ}}(s)$ represent the admittance matrix of the inverter with the collection line.

According to the flowchart in Figure 4, after deriving the inverter admittance matrix with the collection line, the system can be simplified to multiple parallel inverters connected in series with the grid. In the plant, inverters are categorized into grid-following and grid-forming based on their control methods. Aggregating inverters with the same control method helps reduce the order of the admittance matrix. However, coupling effects exist both among inverters and between inverters and the power grid. Therefore, these coupling effects should be considered when deriving the equivalent grid-following and grid-forming admittances. Figure 6 illustrates the coupling relationships between positive and negative sequence voltages, as well as their corresponding currents in the renewable energy plant.

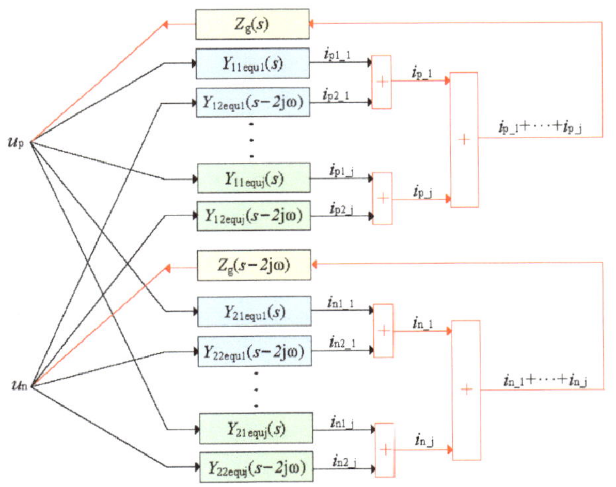

Figure 6. Frequency coupling relationship in the renewable energy plant.

In Figure 6, $Z_g(s)$ represents the power grid impedance matrix, while $Y_{11\text{equk}}(s)$, $Y_{21\text{equk}}(s)$, $Y_{22\text{equk}}(s)$, and $Y_{12\text{equk}}(s)$ denote the admittance matrices of inverters with the collection lines. The voltage and current of each inverter in the coupling loop are related as follows:

$$
\begin{aligned}
i_{\text{p1\_equGFM1}}(s) &= -u_{\text{p}}(s)Y_{11\_\text{equGFM1}}(s) \\
i_{\text{p2\_equGFM1}}(s) &= -u_{\text{n}}(s)Y_{12\_\text{equGFM1}}(s-2\text{j}\omega) \\
i_{\text{n1\_equGFM1}}(s) &= -u_{\text{p}}(s)Y_{21\_\text{equGFM1}}(s) \\
i_{\text{n2\_equGFM1}}(s) &= -u_{\text{n}}(s)Y_{22\_\text{equGFM1}}(s-2\text{j}\omega) \\
\\
i_{\text{p1\_equGFLj}}(s) &= -u_{\text{p}}(s)Y_{11\_\text{equGFLj}}(s) \\
i_{\text{p2\_equGFLj}}(s) &= -u_{\text{n}}(s)Y_{12\_\text{equGFLj}}(s-2\text{j}\omega) \\
i_{\text{n1\_equGFLj}}(s) &= -u_{\text{p}}(s)Y_{21\_\text{equGFLj}}(s) \\
i_{\text{n2\_equGFLj}}(s) &= -u_{\text{n}}(s)Y_{22\_\text{equGFLj}}(s-2\text{j}\omega)
\end{aligned}
\tag{12}
$$

By aggregating inverters of the same type, their admittances are summed, yielding the following:

$$\begin{gathered}
\sum_{k=1}^{i} i_{\mathrm{p1_equGFMk}}(s) = -u_{\mathrm{p}}(s)\sum_{k=1}^{i} Y_{\mathrm{11_equGFMk}}(s) \\
\sum_{k=1}^{i} i_{\mathrm{p2_equGFMk}}(s) = -u_{\mathrm{n}}(s)\sum_{k=1}^{i} Y_{\mathrm{12_equGFMk}}(s-2\mathrm{j}\omega) \\
\sum_{k=1}^{i} i_{\mathrm{n1_equGFMk}}(s) = -u_{\mathrm{p}}(s)\sum_{k=1}^{i} Y_{\mathrm{21_equGFMk}}(s) \\
\sum_{k=1}^{i} i_{\mathrm{n2_equGFMk}}(s) = -u_{\mathrm{n}}(s)\sum_{k=1}^{i} Y_{\mathrm{22_equGFMk}}(s-2\mathrm{j}\omega) \\
\sum_{k=1}^{j} i_{\mathrm{p1_equGFLk}}(s) = -u_{\mathrm{p}}(s)\sum_{k=1}^{j} Y_{\mathrm{11_equGFLk}}(s) \\
\sum_{k=1}^{j} i_{\mathrm{p2_equGFLk}}(s) = -u_{\mathrm{n}}(s)\sum_{k=1}^{j} Y_{\mathrm{12_equGFLk}}(s-2\mathrm{j}\omega) \\
\sum_{k=1}^{j} i_{\mathrm{n1_equGFLk}}(s) = -u_{\mathrm{p}}(s)\sum_{k=1}^{j} Y_{\mathrm{21_equGFLk}}(s) \\
\sum_{k=1}^{j} i_{\mathrm{n2_equGFLk}}(s) = -u_{\mathrm{n}}(s)\sum_{k=1}^{j} Y_{\mathrm{22_equGFLk}}(s-2\mathrm{j}\omega) \\
u_{\mathrm{p}}(s) = Z_{\mathrm{g}}(s)[\sum_{k=1}^{i} i_{\mathrm{p1_equGFMk}}(s) + \sum_{k=1}^{i} i_{\mathrm{p2_equGFMk}}(s) + \sum_{k=1}^{j} i_{\mathrm{p1_equGFLk}}(s) + \sum_{k=1}^{j} i_{\mathrm{p2_equGFLk}}(s)] \\
u_{\mathrm{n}}(s) = Z_{\mathrm{g}}(s-2\mathrm{j}\omega)[\sum_{k=1}^{i} i_{\mathrm{n1_equGFMk}}(s) + \sum_{k=1}^{i} i_{\mathrm{n2_equGFMk}}(s) + \sum_{k=1}^{j} i_{\mathrm{n1_equGFLk}}(s) + \sum_{k=1}^{j} i_{\mathrm{n2_equGFLk}}(s)]
\end{gathered} \tag{13}$$

According to Figure 6, the calculation formulas for the equivalent grid-following and grid-forming models can be expressed as follows:

$$\begin{aligned}
Y_{\mathrm{ep_GFM}}(s) &= -\frac{i_{\mathrm{p_GFM}}(s)}{u_{\mathrm{p}}(s)} = -\frac{\sum_{k=1}^{i} i_{\mathrm{p1_equGFMk}}(s) + \sum_{k=1}^{i} i_{\mathrm{p2_equGFMk}}(s)}{u_{\mathrm{p}}(s)} \\
Y_{\mathrm{en_GFM}}(s) &= -\frac{i_{\mathrm{n_GFM}}(s)}{u_{\mathrm{n}}(s)} = -\frac{\sum_{k=1}^{i} i_{\mathrm{n1_equGFMk}}(s) + \sum_{k=1}^{i} i_{\mathrm{n2_equGFMk}}(s)}{u_{\mathrm{n}}(s)} \\
Y_{\mathrm{ep_GFL}}(s) &= -\frac{i_{\mathrm{p_GFL}}(s)}{u_{\mathrm{p}}(s)} = -\frac{\sum_{k=1}^{i} i_{\mathrm{p1_equGFLk}}(s) + \sum_{k=1}^{i} i_{\mathrm{p2_equGFLk}}(s)}{u_{\mathrm{p}}(s)} \\
Y_{\mathrm{en_GFL}}(s) &= -\frac{i_{\mathrm{n_GFL}}(s)}{u_{\mathrm{n}}(s)} = -\frac{\sum_{k=1}^{i} i_{\mathrm{n1_equGFLk}}(s) + \sum_{k=1}^{i} i_{\mathrm{n2_equGFLk}}(s)}{u_{\mathrm{n}}(s)}
\end{aligned} \tag{14}$$

Based on Equations (13) and (14), the admittance matrix expressions for the equivalent grid-following and grid-forming inverters can be derived as follows:

$$
\begin{aligned}
Y_{ep\_GFL}(s) &= \sum_{k=1}^{j} Y_{11\_equGFLk} - \sum_{k=1}^{j} Y_{12\_equGFLk}(s-2j\omega)\cdot\frac{\sum_{k=1}^{j} Y_{21\_equGFLk}(s)+\sum_{k=1}^{i} Y_{21\_equGFMk}(s)}{\sum_{k=1}^{j} Y_{22\_equGFLk}(s-2j\omega)+\sum_{k=1}^{i} Y_{22\_equGFMk}(s-2j\omega)+Y_g(s-2j\omega)} \\
Y_{en\_GFL}(s) &= \sum_{k=1}^{j} Y_{22\_equGFLk} - \sum_{k=1}^{j} Y_{21\_equGFLk}(s+2j\omega)\cdot\frac{\sum_{k=1}^{j} Y_{12\_equGFLk}(s)+\sum_{k=1}^{i} Y_{12\_equGFMk}(s)}{\sum_{k=1}^{j} Y_{11\_equGFLk}(s+2j\omega)+\sum_{k=1}^{i} Y_{11\_equGFMk}(s+2j\omega)+Y_g(s+2j\omega)} \\
Y_{ep\_GFM}(s) &= \sum_{k=1}^{i} Y_{11\_equGFMk} - \sum_{k=1}^{i} Y_{12\_equGFMk}(s-2j\omega)\cdot\frac{\sum_{k=1}^{j} Y_{21\_equGFLk}(s)+\sum_{k=1}^{i} Y_{21\_equGFMk}(s)}{\sum_{k=1}^{j} Y_{22\_equGFLk}(s-2j\omega)+\sum_{k=1}^{i} Y_{22\_equGFMk}(s-2j\omega)+Y_g(s-2j\omega)} \\
Y_{en\_GFM}(s) &= \sum_{k=1}^{i} Y_{22\_equGFMk} - \sum_{k=1}^{i} Y_{21\_equGFMk}(s+2j\omega)\cdot\frac{\sum_{k=1}^{j} Y_{12\_equGFLk}(s)+\sum_{k=1}^{i} Y_{12\_equGFMk}(s)}{\sum_{k=1}^{j} Y_{11\_equGFLk}(s+2j\omega)+\sum_{k=1}^{i} Y_{11\_equGFMk}(s+2j\omega)+Y_g(s+2j\omega)}
\end{aligned} \quad (15)
$$

In Equation (15), $Y_{ep\_GFL}(s)$, $Y_{en\_GFL}(s)$, $Y_{ep\_GFM}(s)$, and $Y_{en\_GFM}(s)$ represent the admittance matrices for the equivalent grid-following and grid-forming systems. After a reduction in order, the grid-following and grid-forming hybrid renewable energy plant can be simplified to a simple system containing only the equivalent grid-following and grid-forming systems. The topology is shown in Figure 7.

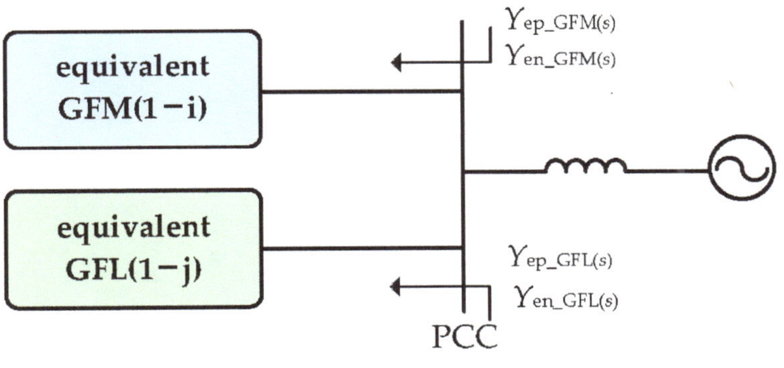

Figure 7. Reduced-order circuit diagram of the grid-following and grid-forming hybrid renewable energy plant.

3.4. Model Validation

3.4.1. Inverters Parameter Design

The admittance matrix is measured using the harmonic injection method. A hybrid renewable energy plant consisting of two grid-following inverters and two grid-forming inverters was built in Matlab/Simulink2023a, where harmonic signals of different frequencies are injected at the point of common coupling (PCC), and the resulting harmonic currents are measured. The admittance at each frequency is calculated accordingly. The admittance at each frequency is calculated using the following equation: $Y[f] = u[f]/i[f]$, where $u[f]$ and $i[f]$ are the measured harmonic current and the voltage at the PCC. By comparing the measured admittance values with the calculated results, the accuracy of the model order reduction method is verified.

Next, the parameter design of the grid-following and grid-forming converters is introduced. The grid-following converter design refers to [25], while the main circuit parameters of the grid-forming converter, including the LCL filter, remain the same. For the grid-forming inverter, the most critical aspect lies in the parameter regulation of the active power loop, which can be simplified into a typical second-order system. According to the control block diagram of the active power loop and the power angle equation, the closed-loop transfer function of the active power loop can be obtained as follows:

$$G_p(s) = \frac{1}{Js^2 + Ds + U_g E_0 / X_0} \tag{16}$$

In the Equation (16), U_g represents the grid voltage, E_0 is the initial voltage of the GFM inverter, and X_0 is the impedance between the initial voltage and the grid voltage. By combining the typical control parameters of a second-order system, the inertia J and damping coefficient D of the grid-forming inverter can be derived.

The parameters for the grid-following and grid-forming inverters are provided in Tables 1 and 2.

Table 1. Grid-forming inverter parameters.

Parameter	Value	Parameter	Value
P_n/MW	1	U_{dc}/V	1000
U_g/V	220	L_1/mH	0.04125
L_2/mH	0.04125	C_1/μF	982.44
L_s/mH	0.004125	L_x/mH	0.2475
D_p	202.5079	J/(kg.m^2)	10.6583
D_q	10	K	0.1
L_v/mH	0.0016	R_v/Ω	0.05

Table 2. Grid-following inverter parameters.

Parameter	Value	Parameter	Value
P_n/MW	1	U_{dc}/V	1000
U_g/V	220	L_1/mH	0.04125
L_2/mH	0.04125	C_1/μF	982.44
L_s/mH	0.004125	L_x/mH	0.2475
H_i	0.00044	K_{pPLL}	253
H_u	0.0034	K_{iPLL}	16,016

3.4.2. Admittance Curve Comparison

During the modeling process, some studies simplify the calculation by neglecting the coupling effects between the positive and negative sequences [26]. However, as discussed earlier, this approach does not accurately represent the real coupling paths within the system. Ignoring these effects and using the single inverter impedance as the actual impedance of the inverter in the renewable energy plant leads to discrepancies between the calculated and measured values, ultimately impacting the model's accuracy and reliability. A comparison of the calculated values of the equivalent grid-following and grid-forming admittances, with and without considering coupling effects, demonstrates the importance of frequency decoupling. As shown in Figure 8, the calculated values align perfectly with the simulation results when coupling effects are considered. By contrast, ignoring coupling effects leads to significant discrepancies between the calculated and simulated values. This demonstrates the importance of incorporating coupling effects in impedance modeling.

The simulation validation of the proposed reduced-order modeling method confirms that a grid-following and grid-forming hybrid renewable energy plant can be effectively simplified into a system comprising a grid-following inverter, a grid-forming inverter, and a grid. By following the reduction steps outlined in Figure 4, this simplified model accurately captures the impedance characteristics of the inverters. This reduced-order method categorizes and aggregates different inverters, reducing the admittance matrix of order (2i + 2j) × (2i + 2j) in Equation (4) to a 4th-order equivalent grid-following/grid-forming admittance matrix. Additionally, the order of this equivalent admittance matrix remains constant regardless of the number of inverters. The modeling process incorporates coupling effects to ensure accuracy. By utilizing this reduced-order system, the stability characteristics of the hybrid inverter system can be analyzed, significantly simplifying the stability analysis. The following section will apply this reduced-order modeling approach to assess the stability of grid-following and grid-forming hybrid renewable energy plants.

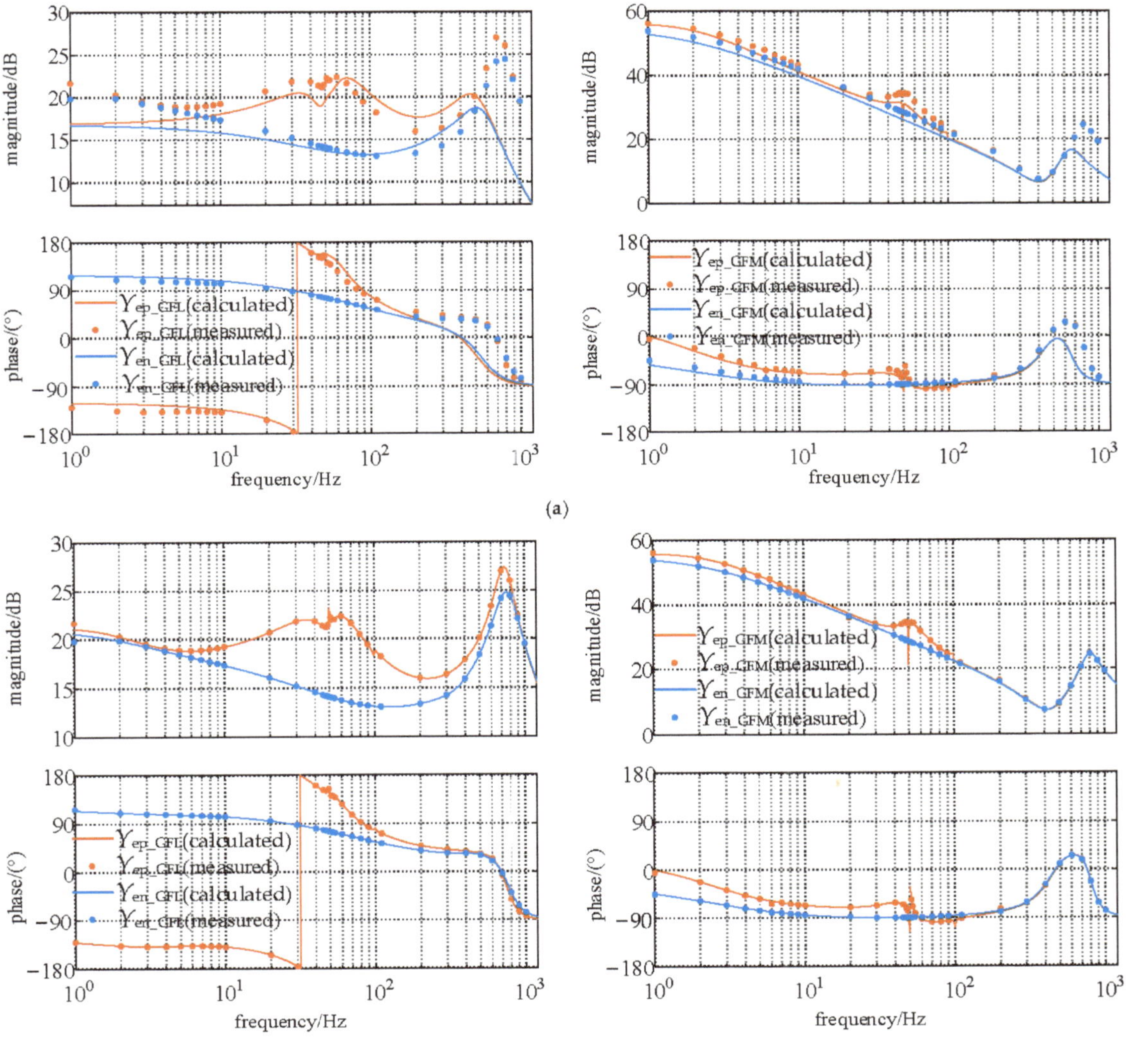

Figure 8. Comparison plot of calculated and simulated admittance matrix values: (**a**) admittance matrix ignoring frequency coupling effects; (**b**) admittance matrix considering frequency coupling effects.

4. Stability Analysis of Grid-Following and Grid-Forming Hybrid New Energy Power Plants

Grid-following and grid-forming hybrid renewable energy plants exhibit distinct impedance characteristics compared to single-inverter systems, resulting in increased complexity in their stability analysis and necessitating further in-depth investigation. The model order reduction method presented earlier establishes a solid theoretical foundation for the stability studies of such systems. In this section, representative case studies are conducted by integrating the reduced-order model with various stability analysis approaches to systematically examine the underlying mechanisms through which grid-forming inverters enhance the stability of renewable energy plants.

4.1. Stability Analysis Methods

4.1.1. Harmonic Characteristic Analysis Method

As power electronic devices, inverters are susceptible to harmonic distortion. The harmonic characteristic analysis method, based on the impedance approach, enables the examination of an inverter's amplification or attenuation effects on harmonics, facilitating an analysis of the system's harmonic behavior across different frequencies.

The harmonic equivalent circuit for a grid-following inverter operating in isolation is shown in Figure 9a, while the harmonic equivalent circuit for a multi-inverter system is illustrated in Figure 9b. An oscillation source, $U_{\mathrm{p}}(s)$, exists on the grid side. When the inverter is disconnected, the harmonics at the grid connection point match those of the grid. However, when the inverter is connected, the harmonics at the connection point change. A lower harmonic content at the connection point is more favorable for the stable operation of the inverter.

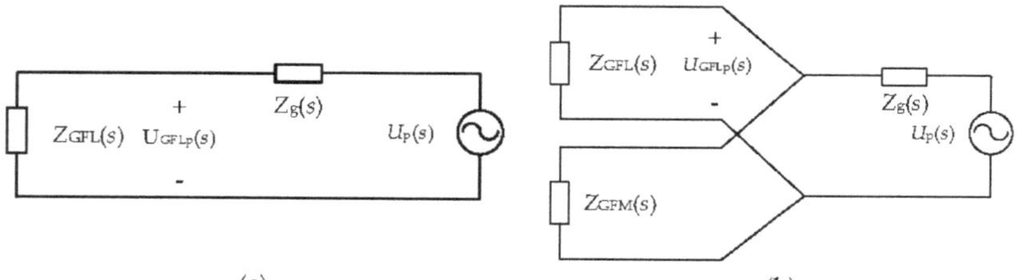

Figure 9. (**a**) Harmonic equivalent circuit diagram of the grid-following system. (**b**) Harmonic equivalent circuit diagram of the grid-following and grid-forming hybrid renewable energy plant.

To investigate the suppression or amplification of harmonics when grid-side oscillations are transmitted to the point of connection, the transfer function from the grid-side harmonic voltage to the connection point can be calculated. As shown in Figure 9a, the transfer function expression for a system consisting only of a grid-following inverter is as follows:

$$G_{u_{\mathrm{p}}}^{u_{\mathrm{GFLp}}}(s) = \frac{Z_{\mathrm{ep\_GFL}}(s)}{Z_{\mathrm{g}}(s) + Z_{\mathrm{ep\_GFL}}(s)} \tag{17}$$

$$G_{u_{n}}^{u_{\mathrm{GFLn}}}(s) = \frac{Z_{\mathrm{en\_GFL}}(s)}{Z_{\mathrm{g}}(s) + Z_{\mathrm{en\_GFL}}(s)} \tag{18}$$

Based on Figure 9b, the transfer function for the harmonic voltage from the grid to the point of connection in a grid-following and grid-forming hybrid system can be calculated as follows:

$$G_{u_{\mathrm{p}}}^{u_{\mathrm{GFLp}}}(s) = \frac{Z_{\mathrm{ep\_GFM}}(s)||Z_{\mathrm{ep\_GFL}}(s)}{Z_{\mathrm{g}}(s) + Z_{\mathrm{ep\_GFM}}(s)||Z_{\mathrm{ep\_GFL}}(s)} \tag{19}$$

$$G_{u_n}^{u_{\mathrm{GFLn}}}(s) = \frac{Z_{\mathrm{en\_GFM}}(s)||Z_{\mathrm{en\_GFL}}(s)}{Z_g(s) + Z_{\mathrm{en\_GFM}}(s)||Z_{\mathrm{en\_GFL}}(s)} \tag{20}$$

The harmonic characteristic curve of the system is plotted based on Equations (17)–(20). The amplitude–frequency curve illustrates how grid-side disturbances are transmitted to the point of connection across different frequency ranges. An amplitude greater than 1 indicates that the disturbances are amplified at the point of connection, posing a risk of instability, while an amplitude less than 1 indicates that the disturbances are attenuated. If attenuation occurs over a wider frequency range, the control strategy demonstrates stronger disturbance rejection capability, ensuring reliable operation even when grid-side disturbances are present.

4.1.2. Nyquist Stability Criterion of Grid-Following and Grid-Forming Hybrid New Energy Power Plants

Due to the voltage source characteristics of the grid-forming inverter, the three-port network comprising the grid-following inverter, the grid-forming inverter, and the grid can be simplified into a two-port network. In this configuration, the grid-forming inverter operates in parallel with the grid, while the grid-following inverter is connected in series. Figure 10 illustrates the equivalent circuit of the grid-following and grid-forming hybrid renewable energy plant.

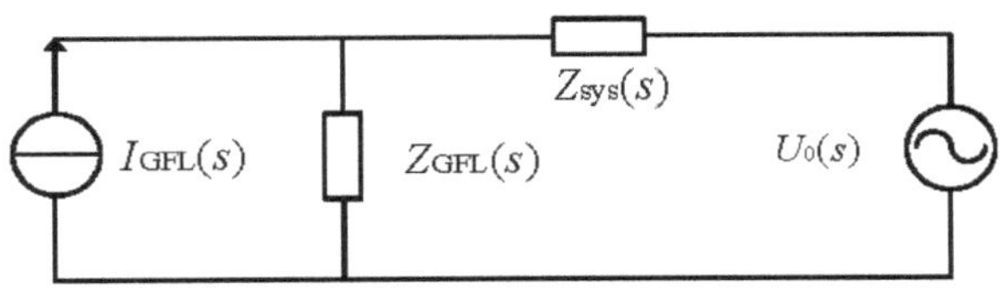

Figure 10. Equivalent circuit diagram of the grid-following and grid-forming hybrid new energy power plant.

The equivalent grid impedance, denoted as $Z_{sys}(s) = Z_{GFM}(s) || Z_g(s)$, represents the parallel combination of the grid-forming inverter and the power grid. The equivalent voltage after this parallel connection is represented as $U_0(s)$.

Based on the above analysis, the impedance ratio of the grid-following and grid-forming hybrid renewable energy plant is given by the following:

$$IR_{\mathrm{mixp}}(s) = \frac{Z_{\mathrm{sysp}}(s)}{Z_{\mathrm{ep\_GFL}}(s)} = \frac{Z_{\mathrm{ep\_GFM}}(s)||Z_g(s)}{Z_{\mathrm{ep\_GFL}}(s)} \tag{21}$$

$$IR_{\mathrm{mixn}}(s) = \frac{Z_{\mathrm{sysn}}(s)}{Z_{\mathrm{en\_GFL}}(s)} = \frac{Z_{\mathrm{en\_GFM}}(s)||Z_g(s)}{Z_{\mathrm{en\_GFL}}(s)} \tag{22}$$

In Equations (21) and (22), $Z_{\mathrm{ep\_GFL}}(s)$ and $Z_{\mathrm{ep\_GFM}}(s)$ represent the positive sequence impedances of the equivalent grid-following and grid-forming inverters, while $Z_{\mathrm{en\_GFL}}(s)$ and $Z_{\mathrm{en\_GFM}}(s)$ represent their negative sequence impedances, and $Z_g(s)$ is the grid impedance. The stability of the system can be assessed by analyzing whether the Nyquist plot of the open loop transfer functions, as shown in Equations (21) and (22), encircles the point $(-1, j0)$.

4.2. Capacity Ratio Analysis of the Grid-Forming and Grid-Following Hybrid New Energy Power Plants

Grid-following inverters can effectively control harmonic content within a small range under strong grid conditions, resulting in a higher stability margin for the system. However, in weak grid conditions, the stability margin of the inverters decreases, and the risk of harmonic generation increases. To address this, increasing the capacity of the grid-forming inverter in a dual-inverter system can improve the system's stability [27]. In renewable energy plants, the number of inverters is typically large, and interactions between inverters are likely to occur. This conclusion may require further investigation.

The following section employs the reduced-order modeling method for grid-following and grid-forming inverters to investigate how grid-forming inverters enhance the stability of small grid-following inverters in weak grids. Four scenarios are analyzed, each with a short-circuit ratio of 1.5. To provide a comprehensive assessment of the impact of grid-forming inverters on grid-following systems, Scenario 4 includes two grid-forming inverters with different parameter sets, both of which ensure stable operation in a weak grid with a short-circuit ratio of 1.5. Table 3 outlines the inverter configurations for each scenario, while detailed inverter parameters are provided in Appendix A.

Table 3. Parameters of four scenarios with different inverters.

	Inverter 1	Inverter 2	Inverter 3	Inverter 4	Total Inverter Capacity
Scenario 1	GFL/1 MW	GFL/1 MW	GFL/0.5 MW	GFL/1 MW	3.5 MW
Scenario 2	GFM/1 MW	GFL/1 MW	GFL/0.5 MW	GFL/1 MW	3.5 MW
Scenario 3	GFM/1 MW	GFM/1 MW	GFL/0.5 MW	GFL/1 MW	3.5 MW
Scenario 4	GFM/1 MW	GFM/1 MW	GFM/0.5 MW	GFL/1 MW	3.5 MW

To gain a deeper understanding of harmonic suppression in the four scenarios, harmonic characteristic curves were plotted for each case. As shown in Figure 11, when the amplitude of the harmonic characteristic curve exceeds 1, it indicates that disturbances from the power source side are amplified as they propagate to the grid connection point. Conversely, an amplitude below 1 signifies that disturbances are attenuated during transmission. In scenarios where only the grid-following inverters are present, the harmonic characteristic curve exceeds 1 within the 1–176.3 Hz range, indicating that the grid-following inverters amplify harmonics in this frequency band. The harmonic characteristic curves from Scenario 1 to Scenario 3 reveal that, as the grid-forming inverters are introduced, disturbances in the low and medium frequency range are suppressed to varying degrees. Moreover, the suppression effect becomes more pronounced as the number of grid-forming inverters increases. Nevertheless, the harmonic characteristic curve of Scenario 4 shows that, when the grid-forming inverters with different parameters are present in the system, harmonics around 50 Hz and 100 Hz increase significantly, highlighting emerging stability issues among the grid-forming inverters. Both grid-following and grid-forming control loops incorporate feedback mechanisms. Improved harmonic suppression at the point of interconnection results in lower harmonic content in the feedback signals, which in turn enhances system stability. In conclusion, from the perspective of harmonic characteristics, the integration of grid-forming inverters improves the harmonic behavior in the low-to-mid frequency range at the point of common coupling. In a grid-following and grid-forming hybrid renewable energy plant, increasing the number of grid-forming inverters under the same operating conditions significantly reduces harmonic content in these frequency ranges. However, when grid-forming inverters with different operating conditions are added, interactions between the inverters can lead to stability issues around the power frequency.

To comprehensively analyze the impact of inverter configuration on a grid-following and grid-forming hybrid renewable energy plant, impedance curves for the equivalent grid-following and grid-forming systems were plotted for four different scenarios with the same parameters. As shown in Figure 12a, with the increase in grid-forming inverters, the magnitude of the equivalent grid-following impedance gradually increases. In Figure 12b, the impedance of the equivalent grid decreases. The phase margin at the intersection of the equivalent grid-following and grid-forming impedances reflects the stability of the system: PM = $180° - |\angle(\varphi_{sys} - \varphi_{GFL})|$. As seen in Figure 13, from Scenario 1 to Scenario 3, with an increase in the grid-forming inverters, the intersection of the magnitude of the equivalent grid-following impedance and the equivalent grid impedance shifts to the right. The phase angle of the equivalent grid increases, while the phase angle of the equivalent grid-following impedance decreases, causing the phase difference to decrease and the phase margin to gradually increase. As seen in Scenario 4, with the addition of a grid-forming inverter with different parameters from the previous ones, the intersection of the equivalent grid-following and grid-forming impedances shifts to the left near 50 Hz. At this point, the equivalent grid-following impedance lies within the negative damping region, leading to system instability. In conclusion, from the perspective of impedance curves, the addition of grid-forming inverters increases the magnitude of the equivalent grid-following impedance and reduces the magnitude of the equivalent grid impedance. This causes the intersection of the two impedance curves to shift to the right, thereby increasing the phase margin and enhancing system stability. The stability improvement is more significant when grid-forming inverters with the same operating conditions are added. However, when grid-forming inverters with different operating conditions are introduced, the impedance intersection shifts earlier, leading to a risk of system instability.

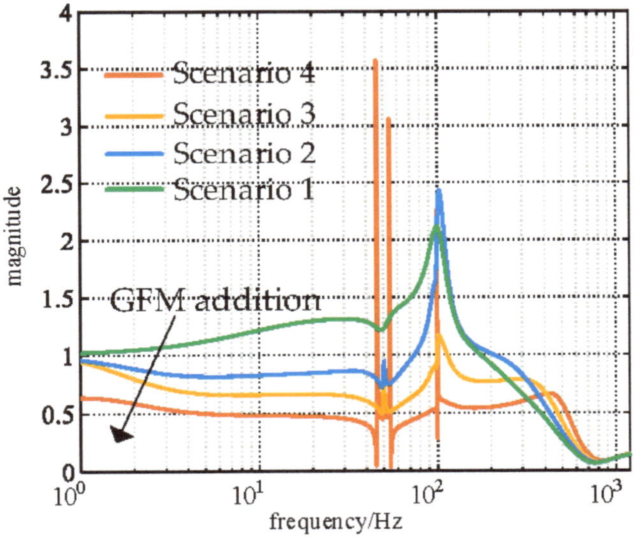

Figure 11. Harmonic characteristics of different scenarios.

To analyze the stability of the four scenarios, the solid and dashed lines in Figure 14 represent the positive sequence and negative sequence Nyquist curves. In scenarios one and four, the Nyquist curves pass through $(-1, j0)$, indicating system stability. By contrast, the Nyquist curves in scenarios two and three do not pass through $(-1, j0)$, indicating system instability. This result aligns with the previous stability analysis, which suggests that adding grid-forming inverters with the same operating conditions to a grid-following system improves system stability, while introducing different inverters poses a risk of instability.

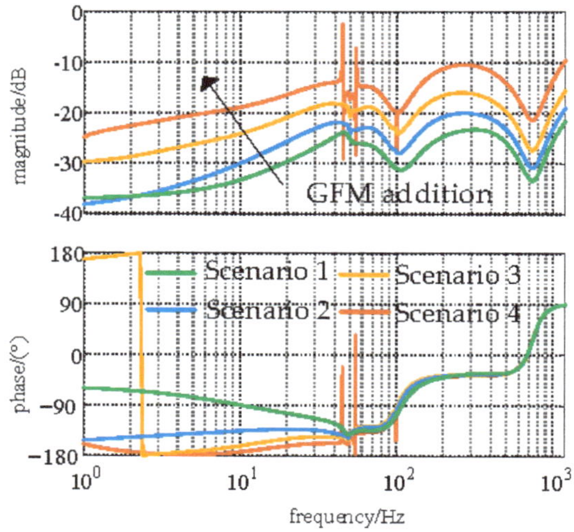

(a)

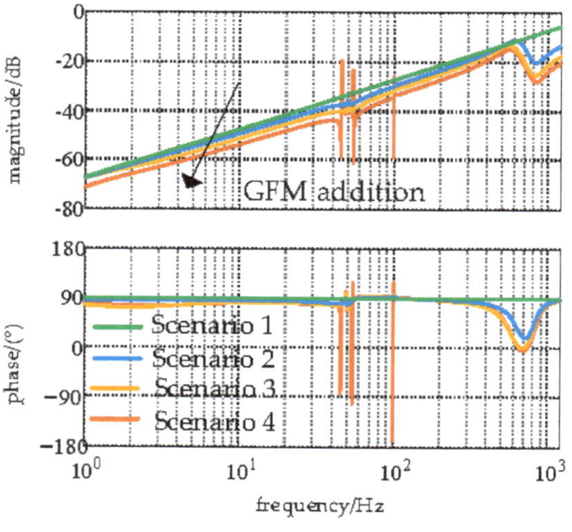

(b)

Figure 12. Impedance curves of different scenarios: (**a**) equivalent grid-following impedance curves in the four scenarios; (**b**) equivalent grid impedance curves in the four scenarios.

To conclude, the addition of grid-forming inverters improves the harmonic content at the grid connection point, reduces the control complexity of both the grid-following and grid-forming inverters, and enhances the system's disturbance rejection capability. From the perspective of system impedance, the inclusion of the grid-forming inverters increases the impedance of the equivalent grid-following inverter and decreases the impedance of the equivalent power grid, shifting the impedance intersection to the right, while increasing the stability margin. However, when multiple grid-forming inverters with different operating conditions are present in the system, their interactions may introduce new stability issues.

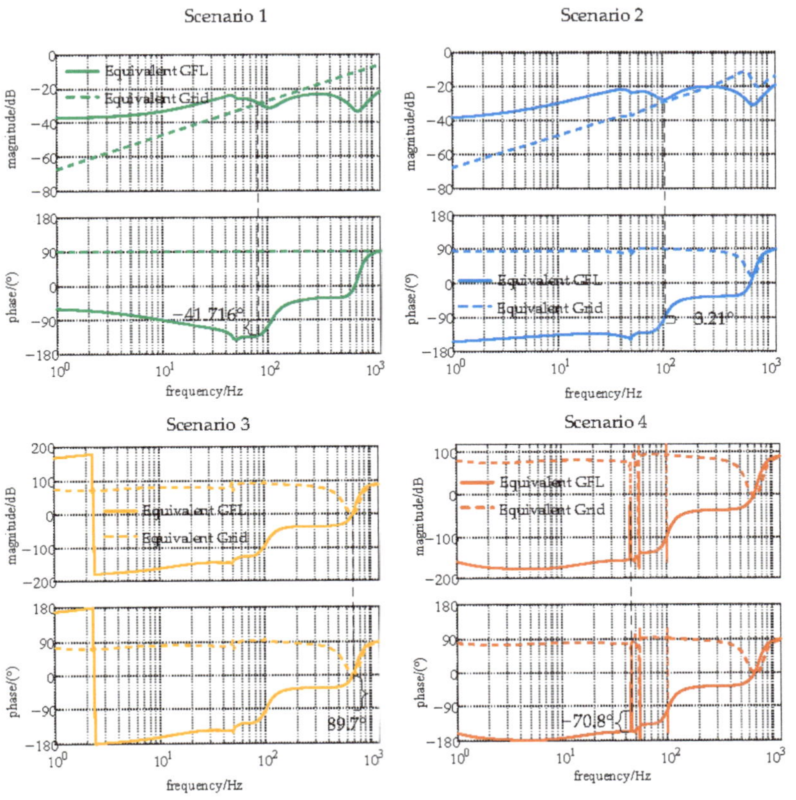

Figure 13. Phase margin under the different scenarios.

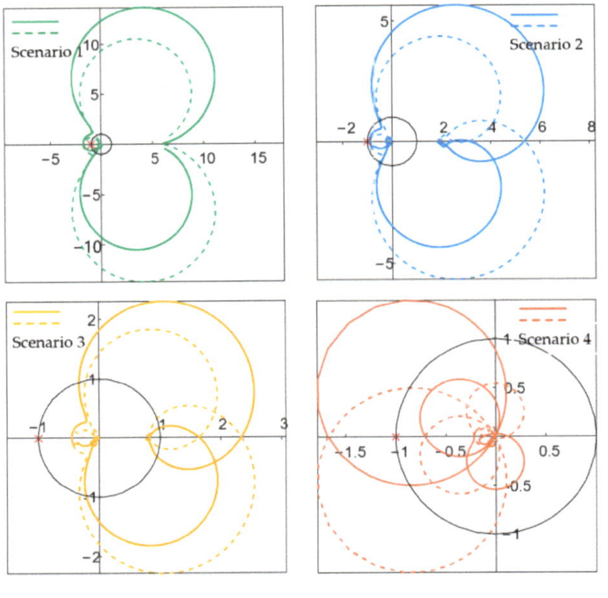

Figure 14. Nyquist curves in different scenarios.

5. Simulation Verification

To validate the conclusions in Section 4, simulation models for the four scenarios were constructed under the condition of SCR = 1.5. The inverter current waveforms are shown in Figure 15a. As depicted, the current waveforms in Scenario 1 and Scenario 4 exhibit significant oscillations, whereas those in Scenarios 2 and 3 remain stable. This corresponds with the stability analysis presented in Section 4.2. Figure 15b analyzes the harmonic content of the current. In Scenario 1, the dominant harmonics are 13.8 Hz and 86.2 Hz, while in Scenario 4, the dominant harmonics are 49 Hz and 51 Hz, which aligns with the harmonic characteristic analysis from Section 4.2.

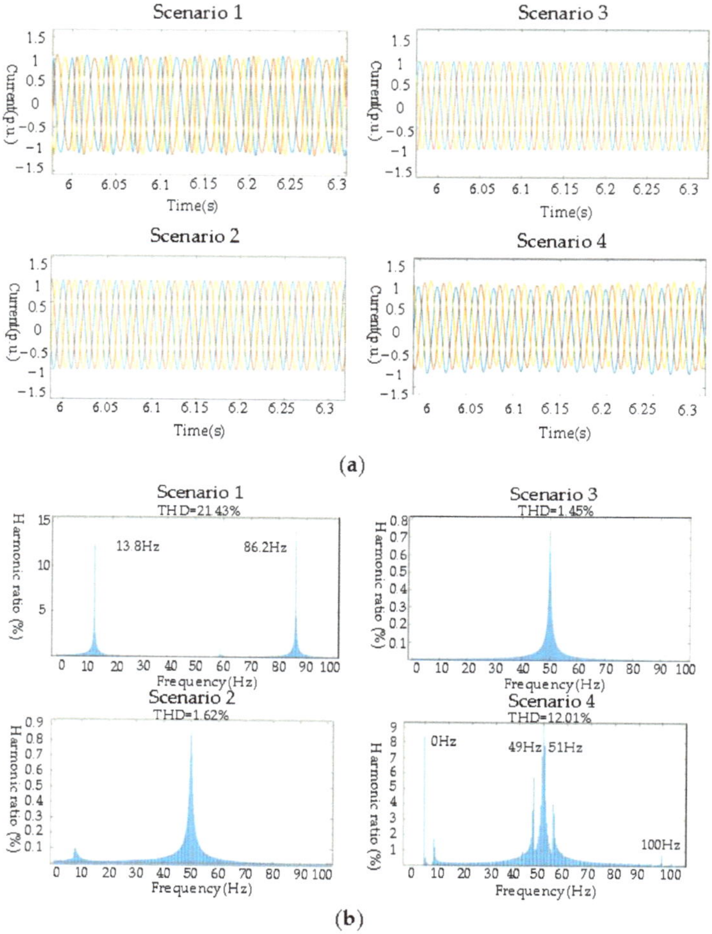

Figure 15. Simulation results of four scenarios: (**a**) current curves of the four scenarios; (**b**) harmonic analysis of the four scenarios.

Overall, the consistency between simulation experiments and theoretical analysis demonstrates that the impedance matrix calculated via the reduced-order modeling method effectively reflects the impedance characteristics of the grid-following and grid-forming hybrid renewable energy plants, thereby enabling accurate stability assessments. Simulation results across various scenarios further indicate that incorporating grid-forming inverters into renewable energy plants with multiple grid-following inverters can suppress harmonics in the low and medium frequency ranges at the connection point, and reduce the harmonic content in the feedback control loops of both inverter types. Adjusting the impedance magnitudes of the equivalent grid-following inverter and the equivalent grid indirectly enhances the system's phase margin, thereby improving its overall stability. However, if a grid-following and grid-forming hybrid renewable energy plant includes multiple grid-forming inverters under different operating conditions, interactions among these units may introduce new stability challenges.

6. Conclusions

This paper presents a reduced-order modeling approach for grid-following and grid-forming hybrid renewable energy plants. By employing harmonic characteristic analysis and the Nyquist stability criterion, this study examines how grid-forming inverters enhance the stability of grid-following inverters in weak grid conditions. Additionally, this study explores the capacity ratio between grid-following and grid-forming inverters in the plant. The main contributions and conclusions are summarized as follows:

(1) For grid-following and grid-forming hybrid renewable energy plants, the proposed reduced-order modeling method employs the aggregation of inverters with similar control strategies and the decoupling of positive and negative frequency among different components. This approach simplifies the complex renewable energy plant into a straightforward system comprising an equivalent grid-following system, an equivalent grid-forming system, and the power grid, thereby combining simplicity and generality with accuracy.

(2) The incorporation of grid-forming inverters markedly reduces the harmonic content in the low-to-medium frequency range at the grid connection point. In addition, by increasing the magnitude of the equivalent grid-following impedance while decreasing that of the equivalent grid impedance, the phase margin of the renewable energy plant is indirectly enhanced, thereby improving the stability of grid-following inverters under weak grid conditions. However, when multiple grid-forming inverters are present, caution must be exercised, since interactions among them may give rise to new stability challenges.

In addition, one small limitation of this paper is that none of the content has been experimentally validated in the laboratory. We will address this limitation in future work.

Author Contributions: Conceptualization, Y.M. and N.C.; methodology, L.G.; validation, Y.M., N.C. and L.G.; formal analysis, N.C.; investigation, L.G.; resources, Y.M.; data curation, Y.M.; writing—original draft preparation, Y.M.; writing—review and editing, Y.M.; visualization, L.G.; supervision, N.C.; project administration, L.G.; funding acquisition, N.C. All authors have read and agreed to the published version of the manuscript.

Funding: The research was funded by the National Key Research and Development Program of China under grant 2024YFB2408900 and supported by State Grid Jiangsu Electric Power Co., Ltd.

Data Availability Statement: The original contributions presented in this study are included in the article; further inquiries can be directed to the corresponding author.

Conflicts of Interest: The authors declare no conflicts of interest.

Abbreviations

GFM	Grid-forming
GFL	Grid-following
PLL	Phase-locked Loop
VSG	Virtual Synchronous Generator

L_1	Filter Inductor
L_2	Filter Inductor
C_1	Filter Capacitor
L_g	Grid Inductor
u_{abc}	Point of Common Coupling Voltage
i_{abc}	Point of Common Coupling Current
i_{dref}	*d*-axis Reference Current
i_{qref}	*q*-axis Reference Current
u_{dref}	*d*-axis Reference Voltage
u_{qref}	*q*-axis Reference Voltage
H_u	Voltage Conversion Coefficient
H_i	Current Conversion Coefficient
Kpwm	Pulse Width Modulation Coefficient
u_{mabc}	Bridge Arm Voltage
PWM	Pulse Width Modulation
θ	Phase of the Point of Common Coupling
PI	Proportional-Integral Controller
J	Inertia of the Active Power Loop
D_p	Droop Coefficient in the Active Power Loop
P_{set}	Active Power Setpoint
P	Active Power
Q_{set}	Reactive Power Setpoint
Q	Reactive Power
D_q	Droop Coefficient in the Reactive Power Loop
R_v	Virtual Resistor
L_v	Virtual Inductor
L_s	Transformer Impedance
L_x	Transmission Line Impedance
P_n	Rated Power
ω_0	Power Frequency
THD	Total Harmonic Distortion

Appendix A

$$
\begin{aligned}
&A = \{\{(\boldsymbol{U}_1 + \mathrm{j}\omega_0 L_1 \boldsymbol{I}_1) + [\boldsymbol{I}_1 + (\boldsymbol{U}_1 + \mathrm{j}\omega_0 L_1 \boldsymbol{I}_1)\mathrm{j}\omega_0 C_1]\mathrm{j}\omega_0 L_2\}/K_{\mathrm{PWM}} + k_c H_i[(\boldsymbol{U}_1 + \mathrm{j}\omega_0 L_1 \boldsymbol{I}_1)\mathrm{j}\omega_0 C_1]\}\\
&U_{\mathrm{md}}[0] = (\tfrac{1}{2}e^{-\mathrm{j}\varphi_v})A + (\tfrac{1}{2}e^{+\mathrm{j}\varphi_v})A^*\\
&U_{\mathrm{mq}}[0] = -[(\mathrm{j}\tfrac{1}{2}e^{-\mathrm{j}\varphi_v})A + (-\mathrm{j}\tfrac{1}{2}e^{+\mathrm{j}\varphi_v})A^*]\\
&H_{\mathrm{PLL}}(s) = (K_{\mathrm{pPLL}} + \tfrac{K_{\mathrm{iPLL}}}{s})\tfrac{1}{s}\\
&G_p(s) = \tfrac{-jH_{\mathrm{PLL}}(s)}{(1+H_u \boldsymbol{U}_1 e^{-\mathrm{j}\varphi_v} H_{\mathrm{PLL}}(s))}\\
&G_n(s) = \tfrac{+jH_{\mathrm{PLL}}(s)}{(1+H_u \boldsymbol{U}_1 e^{-\mathrm{j}\varphi_v} H_{\mathrm{PLL}}(s))}\\
&G_i(s) = \mathrm{k}_{\mathrm{pPLL}} + \tfrac{\mathrm{k}_{\mathrm{iPLL}}}{s}
\end{aligned}
\tag{A1}
$$

$$
Y_{11\_\mathrm{GFL}}(s) = -\frac{\begin{array}{c}[\mathrm{j}\tfrac{1}{2}H_u G_p(s)]U_{\mathrm{md}}[0]K_{\mathrm{pwm}} - [\tfrac{1}{2}H_u G_p(s)]U_{\mathrm{mq}}[0]K_{\mathrm{pwm}}\\ +(\tfrac{1}{2}e^{+\mathrm{j}\varphi_u})[G_i(s-\mathrm{j}\omega_0)-\mathrm{jk}_0]K_{\mathrm{pwm}}[\mathrm{j}I_1 H_i H_u G_p(s-\mathrm{j}\omega_0)e^{-2\mathrm{j}\varphi_u}] - sk_c H_i C_1 K_{\mathrm{pwm}} - s^2 L_2 C_1 - 1\end{array}}{\begin{array}{c}(\tfrac{1}{2}e^{+\mathrm{j}\varphi_u})[G_i(s-\mathrm{j}\omega_0)-\mathrm{jk}_0]K_{\mathrm{pwm}}(2H_i e^{-\mathrm{j}\varphi_u})\\ +sL_1 + sL_2 + s^3 L_2 C_1 L_1 + s^2 k_c H_i C_1 L_1 K_{\mathrm{pwm}}\end{array}}
\tag{A2}
$$

$$
Y_{21\_\mathrm{GFL}}(s) = -\frac{\begin{array}{c}(-\mathrm{j}\tfrac{1}{2}H_u G_p e^{-2\mathrm{j}\varphi_u})K_{\mathrm{pwm}}U_{\mathrm{md}}[0] - (\tfrac{1}{2}H_u G_p e^{-2\mathrm{j}\varphi_u})K_{\mathrm{pwm}}U_{\mathrm{mq}}[0]\\ +(\tfrac{1}{2}e^{-\mathrm{j}\varphi_u})(G_i(s-\mathrm{j}\omega_0)+\mathrm{j}K_0)K_{\mathrm{pwm}}(-\mathrm{j}I_1 H_i H_u G_p(s-\mathrm{j}\omega_0)e^{-2\mathrm{j}\varphi_i})\end{array}}{\begin{array}{c}-(\tfrac{1}{2}e^{-\mathrm{j}\varphi_u})(G_i(s-\mathrm{j}\omega_0)+\mathrm{j}K_0)K_{\mathrm{pwm}}(-2H_i e^{+\mathrm{j}\varphi_u}) + (s-2\mathrm{j}\omega_0)L_1\\ +(s-2\mathrm{j}\omega_0)L_2 + (s-2\mathrm{j}\omega_0)^3 L_2 L_1 C_1 + (s-2\mathrm{j}\omega_0)^2 k_c L_1 H_i C_1 K_{\mathrm{pwm}}\end{array}}
\tag{A3}
$$

$$Y_{22\_\mathrm{GFL}}(s) = -\frac{\begin{array}{c}(-\mathrm{j}\tfrac{1}{2}H_{\mathrm{u}}G_{\mathrm{n}})U^{*}{}_{\mathrm{md}}[0]K_{\mathrm{pwm}} - (\tfrac{1}{2}H_{u}G_{\mathrm{n}})U^{*}{}_{\mathrm{mq}}[0]K_{\mathrm{pwm}} \\ +(\tfrac{1}{2}e^{-\mathrm{j}\varphi_{\mathrm{u}}})[G_{\mathrm{i}}(s+\mathrm{j}\omega_0)+\mathrm{j}k_0]K_{\mathrm{pwm}}[-\mathrm{j}I_1H_{\mathrm{i}}H_{u}G_{\mathrm{p}}(s+\mathrm{j}\omega_0)e^{+2\mathrm{j}\varphi_{\mathrm{u}}}] - sk_{\mathrm{c}}H_{\mathrm{i}}C_1K_{\mathrm{pwm}} - s^2L_2C_1 - 1\end{array}}{\begin{array}{c}(\tfrac{1}{2}e^{-\mathrm{j}\varphi_{\mathrm{u}}})[G_{\mathrm{i}}(s+\mathrm{j}\omega_0)+\mathrm{j}k_0]K_{\mathrm{pwm}}(2H_{\mathrm{i}}e^{-\mathrm{j}\varphi_{\mathrm{u}}}) \\ +sL_1+sL_2+s^3L_2C_1L_1+s^2k_{\mathrm{c}}H_{\mathrm{i}}C_1L_1K_{\mathrm{pwm}}\end{array}} \tag{A4}$$

$$Y_{12\_\mathrm{GFL}}(s) = -\frac{\begin{array}{c}(+\mathrm{j}\tfrac{1}{2}H_{u}G_{n}e^{+2\mathrm{j}\varphi_{u}})K_{\mathrm{pwm}}U^{*}{}_{\mathrm{md}}[0] - (\tfrac{1}{2}H_{u}G_{n}e^{+2\mathrm{j}\varphi_{u}})K_{\mathrm{pwm}}U^{*}{}_{\mathrm{mq}}[0] \\ +(\tfrac{1}{2}e^{+\mathrm{j}\varphi_{u}})[G_{\mathrm{i}}(s+\mathrm{j}\omega_0)-\mathrm{j}K_0]K_{\mathrm{pwm}}[\mathrm{j}I_1H_{\mathrm{i}}H_{u}G_{\mathrm{p}}(s+\mathrm{j}\omega_0)e^{+2\mathrm{j}\varphi_{i}}]\end{array}}{\begin{array}{c}-(\tfrac{1}{2}e^{+\mathrm{j}\varphi_{u}})[G_{\mathrm{i}}(s+\mathrm{j}\omega_0)-\mathrm{j}K_0]K_{\mathrm{pwm}}(-2H_{\mathrm{i}}e^{-\mathrm{j}\varphi_{u}}) + (s+\mathrm{j}\omega_0)L_1 \\ +(s+\mathrm{j}\omega_0)L_2 + (s+2\mathrm{j}\omega_0)^3L_2L_1C_1 + (s+2\mathrm{j}\omega_0)^2k_{\mathrm{c}}L_1H_{\mathrm{i}}C_1K_{\mathrm{pwm}}\end{array}} \tag{A5}$$

$$C = [(K_{\mathrm{PWM}}H_{\mathrm{u}}E_0e^{\pm j\varphi_{vir}} - H_{\mathrm{u}}\boldsymbol{U}_1)/(R_{\mathrm{v}}+\mathrm{j}\omega_0L_{\mathrm{v}})] \tag{A6}$$

$$B = \{\{(\boldsymbol{U}_1+\mathrm{j}\omega_0L_1I_1) + [I_1 + (\boldsymbol{U}_1+\mathrm{j}\omega_0L_1I_1)\mathrm{j}\omega_0C_1]\mathrm{j}\omega_0L_2\}/K_{\mathrm{PWM}} + k_{\mathrm{c}}H_{\mathrm{i}}(\boldsymbol{U}_1+\mathrm{j}\omega_0L_1I_1)\mathrm{j}\omega_0C_1\} \tag{A7}$$

$$\begin{aligned}D = \{&-(\tfrac{1}{2}\mathrm{j})H_{\omega}H_{\mathrm{u}}H_{\mathrm{i}}M(s-\mathrm{j}\omega_0)/(s-\mathrm{j}\omega_0)K_{\mathrm{PWM}}T_sB+ \\ &(\mathrm{j}\tfrac{1}{2})G_{\mathrm{i}}(s_1)H_{\omega}H_{\mathrm{u}}H_{\mathrm{i}}M(s-\mathrm{j}\omega_0)/(s-\mathrm{j}\omega_0)K_{\mathrm{PWM}}T_sC- \\ &(\mathrm{j}\tfrac{1}{2}e^{+\mathrm{j}\varphi_{vir}})G_{\mathrm{i}}(s-\mathrm{j}\omega_0)E_0K_{\mathrm{PWM}}H_{\mathrm{u}}/(R_{\mathrm{v}}-L_{\mathrm{v}}s)H_{\omega}H_{\mathrm{u}}H_{\mathrm{i}}M(s-\mathrm{j}\omega_0)/(s-\mathrm{j}\omega_0)K_{\mathrm{PWM}}T_s- \\ &(\mathrm{j}\tfrac{1}{2})G_{\mathrm{i}}(s-\mathrm{j}\omega_0)H_{\mathrm{i}}I_1H_{\omega}H_{\mathrm{u}}H_{\mathrm{i}}M(s-\mathrm{j}\omega_0)/(s-\mathrm{j}\omega_0)K_{\mathrm{PWM}}T_s\}\end{aligned} \tag{A8}$$

$$\begin{aligned}E = \{&(\mathrm{j}\tfrac{1}{2})\{H_{\omega}H_{\mathrm{u}}H_{\mathrm{i}}M(s-\mathrm{j}\omega_0)/(s-\mathrm{j}\omega_0)T_sK_{\mathrm{PWM}}B^{*}- \\ &(\mathrm{j}\tfrac{1}{2})G_{\mathrm{i}}(s-\mathrm{j}\omega_0)H_{\omega}H_{\mathrm{u}}H_{\mathrm{i}}M(s-\mathrm{j}\omega_0)/(s-\mathrm{j}\omega_0)T_sK_{\mathrm{PWM}}C^{*}+ \\ &(\mathrm{j}\tfrac{1}{2}e^{-\mathrm{j}\varphi_{vir}})G_{\mathrm{i}}(s-\mathrm{j}\omega_0)E_0K_{\mathrm{PWM}}H_{\mathrm{u}}/(R_{\mathrm{v}}+L_{\mathrm{v}}s_2)H_{\omega}H_{\mathrm{u}}H_{\mathrm{i}}M(s-\mathrm{j}\omega_0)/(s-\mathrm{j}\omega_0)T_sK_{\mathrm{PWM}}+ \\ &(\mathrm{j}\tfrac{1}{2})G_{\mathrm{i}}(s-\mathrm{j}\omega_0)H_{\mathrm{i}}I_1H_{\omega}H_{\mathrm{u}}H_{\mathrm{i}}M(s-\mathrm{j}\omega_0)/(s-\mathrm{j}\omega_0)T_sK_{\mathrm{PWM}}\}\end{aligned} \tag{A9}$$

$$F = \{(U_1+\mathrm{j}\omega_0L_1I_1) + [I_1 + (U_1+\mathrm{j}\omega_0L_1I_1)\mathrm{j}\omega_0C_1]\mathrm{j}\omega_0L_2\}/K_{\mathrm{PWM}} + k_{\mathrm{c}}H_{\mathrm{i}}(U_1+\mathrm{j}\omega_0L_1I_1)\mathrm{j}\omega_0C_1 \tag{A10}$$

$$H = [(K_{\mathrm{PWM}}H_{\mathrm{u}}E_0e^{\pm j\varphi_{vir}} - H_{\mathrm{u}}U_1)/(R_{\mathrm{v}}+L_{\mathrm{v}}\mathrm{j}\omega_0)] \tag{A11}$$

$$\begin{aligned}K = \{&(\tfrac{1}{2}\mathrm{j})K_{\mathrm{PWM}}T_sH_{\omega}H_{\mathrm{u}}H_{\mathrm{i}}M(s+\mathrm{j}\omega_0)/(s+\mathrm{j}\omega_0)F^{*}- \\ &(\mathrm{j}\tfrac{1}{2})G_{\mathrm{i}}(s_3)K_{\mathrm{PWM}}T_sH_{\omega}H_{\mathrm{u}}H_{\mathrm{i}}M(s+\mathrm{j}\omega_0)/(s+\mathrm{j}\omega_0)H^{*}+ \\ &(\mathrm{j}\tfrac{1}{2}e^{-\mathrm{j}\varphi_{vir}})G_{\mathrm{i}}(s+\mathrm{j}\omega_0)E_0K_{\mathrm{PWM}}H_{\mathrm{u}}/(R_{\mathrm{v}}+L_{\mathrm{v}}s)K_{\mathrm{PWM}}T_sH_{\omega}H_{\mathrm{u}}H_{\mathrm{i}}M(s+\mathrm{j}\omega_0)/(s+\mathrm{j}\omega_0)+ \\ &(\mathrm{j}\tfrac{1}{2})G_{\mathrm{i}}(s+\mathrm{j}\omega_0)H_{\mathrm{i}}K_{\mathrm{PWM}}T_sH_{\omega}H_{\mathrm{u}}H_{\mathrm{i}}M(s+\mathrm{j}\omega_0)/(s+\mathrm{j}\omega_0)\boldsymbol{I}_1\}\end{aligned} \tag{A12}$$

$$\begin{aligned}M = \{&(-\mathrm{j}\tfrac{1}{2})H_{\omega}H_{\mathrm{u}}H_{\mathrm{i}}M(s+\mathrm{j}\omega_0)/(s+\mathrm{j}\omega_0)T_sK_{\mathrm{PWM}}F- \\ &(-\mathrm{j}\tfrac{1}{2})G_{\mathrm{i}}(s_3)H_{\omega}H_{\mathrm{u}}H_{\mathrm{i}}M(s+\mathrm{j}\omega_0)/(s+\mathrm{j}\omega_0)T_sK_{\mathrm{PWM}}H+ \\ &(-\mathrm{j}\tfrac{1}{2}e^{+j\varphi_{vir}})G_{\mathrm{i}}(s+\mathrm{j}\omega_0)E_0K_{\mathrm{PWM}}H_{\mathrm{u}}/[R_{\mathrm{v}}+L_{\mathrm{v}}(s+2\mathrm{j}\omega_0)]H_{\omega}H_{\mathrm{u}}H_{\mathrm{i}}M(s+\mathrm{j}\omega_0)/(s+\mathrm{j}\omega_0)T_sK_{\mathrm{PWM}}+ \\ &(-\mathrm{j}\tfrac{1}{2})G_{\mathrm{i}}(s+\mathrm{j}\omega_0)H_{\mathrm{i}}T_sK_{\mathrm{PWM}}H_{\omega}H_{\mathrm{u}}H_{\mathrm{i}}M(s+\mathrm{j}\omega_0)/(s+\mathrm{j}\omega_0)\boldsymbol{I}_1\}\end{aligned} \tag{A13}$$

$$\begin{aligned}&Y_{11\_\mathrm{GFM}} = \\ &-\frac{\begin{array}{l}\{-\{De^{-2\mathrm{j}\varphi_{\mathrm{i}}}\boldsymbol{I}_1 + (\tfrac{1}{2}e^{+\mathrm{j}\varphi_{vir}})G_{\mathrm{i}}(s-\mathrm{j}\omega_0)/[K(s-\mathrm{j}\omega_0)]K_{\mathrm{PWM}}H_{\mathrm{u}}/(R_{\mathrm{v}}+L_{\mathrm{v}}s)[-e^{\mp\mathrm{j}\varphi_{u}}H_{\mathrm{u}}(D_{\mathrm{q}}-\mathrm{j}H_{\mathrm{i}}e^{-2\mathrm{j}\varphi_{\mathrm{i}}}e^{+\mathrm{j}\varphi_{u}}\boldsymbol{I}_1)]K_{\mathrm{PWM}}T_s- \\ G_{\mathrm{i}}(s-\mathrm{j}\omega_0)H_{\mathrm{u}}/(R_{\mathrm{v}}+L_{\mathrm{v}}s)K_{\mathrm{PWM}}T_s - sk_{\mathrm{c}}H_{\mathrm{i}}C_1K_{\mathrm{PWM}}T_s - s^2L_2C_1 - 1\}+ \\ \{Ee^{-2\mathrm{j}\varphi_{\mathrm{i}}}\boldsymbol{I}_1 + (\tfrac{1}{2}e^{-\mathrm{j}\varphi_{vir}})G_{\mathrm{i}}(s-\mathrm{j}\omega_0)/[K(s-\mathrm{j}\omega_0)]K_{\mathrm{PWM}}H_{\mathrm{u}}/ \\ [R_{\mathrm{v}}+L_{\mathrm{v}}(s-2\mathrm{j}\omega_0)]T_sK_{\mathrm{PWM}}[-e^{-\mathrm{j}\varphi_{u}}H_{\mathrm{u}}(D_{\mathrm{q}}-\mathrm{j}H_{\mathrm{i}}e^{-2\mathrm{j}\varphi_{\mathrm{i}}}e^{+\mathrm{j}\varphi_{u}}\boldsymbol{I}_1)]\}/ \\ \{E\boldsymbol{U}_1 + (\tfrac{1}{2}e^{-\mathrm{j}\varphi_{vir}})G_{\mathrm{i}}(s-\mathrm{j}\omega_0)/[K(s-\mathrm{j}\omega_0)]K_{\mathrm{PWM}}H_{\mathrm{u}}/[R_{\mathrm{v}}+L_{\mathrm{v}}(s-2\mathrm{j}\omega_0)]T_sK_{\mathrm{PWM}}(\mathrm{j}H_{\mathrm{u}}H_{\mathrm{i}}\boldsymbol{U}_1)- \\ G_{\mathrm{i}}(s-\mathrm{j}\omega_0)H_{\mathrm{i}}T_sK_{\mathrm{PWM}} - k_{\mathrm{c}}L_1H_{\mathrm{i}}(s-2\mathrm{j}\omega_0)^2C_1T_sK_{\mathrm{PWM}} - (s-2\mathrm{j}\omega_0)L_1 - (s-2\mathrm{j}\omega_0)L_2 - (s-2\mathrm{j}\omega_0)^3L_2L_1C_1\} \\ \{D\boldsymbol{U}_1 + (\tfrac{1}{2}e^{+j\varphi_{vir}})G_{\mathrm{i}}(s-\mathrm{j}\omega_0)/[K(s-\mathrm{j}\omega_0)]K_{\mathrm{PWM}}H_{\mathrm{u}}/(R_{\mathrm{v}}+L_{\mathrm{v}}s)K_{\mathrm{PWM}}T_s(\mathrm{j}H_{\mathrm{u}}H_{\mathrm{i}}\boldsymbol{U}_1)\}\}\end{array}}{\begin{array}{l}\{-\{E\boldsymbol{U}_1e^{-2\mathrm{j}\varphi_{\mathrm{u}}} + (\tfrac{1}{2}e^{-\mathrm{j}\varphi_{vir}})G_{\mathrm{i}}(s-\mathrm{j}\omega_0)/[K(s-\mathrm{j}\omega_0)]K_{\mathrm{PWM}}H_{\mathrm{u}}/[R_{\mathrm{v}}+L_{\mathrm{v}}(s-2\mathrm{j}\omega_0)]T_sK_{\mathrm{PWM}}(-\mathrm{j}H_{\mathrm{u}}H_{\mathrm{i}}e^{-2\mathrm{j}\varphi_{\mathrm{u}}}\boldsymbol{U}_1)\}/ \\ \{E\boldsymbol{U}_1 + (\tfrac{1}{2}e^{-j\varphi_{vir}})G_{\mathrm{i}}(s-\mathrm{j}\omega_0)/[K(s-\mathrm{j}\omega_0)]K_{\mathrm{PWM}}H_{\mathrm{u}}/[R_{\mathrm{v}}+L_{\mathrm{v}}(s-2\mathrm{j}\omega_0)]T_sK_{\mathrm{PWM}}(\mathrm{j}H_{\mathrm{u}}H_{\mathrm{i}}\boldsymbol{U}_1)- \\ G_{\mathrm{i}}(s-\mathrm{j}\omega_0)H_{\mathrm{i}}T_sK_{\mathrm{PWM}} - k_{\mathrm{c}}L_1H_{\mathrm{i}}(s-2\mathrm{j}\omega_0)^2C_1T_sK_{\mathrm{PWM}} - (s-2\mathrm{j}\omega_0)L_1 - (s-2\mathrm{j}\omega_0)L_2 - (s-2\mathrm{j}\omega_0)^3L_2L_1C_1\} \\ \{D\boldsymbol{U}_1 + (\tfrac{1}{2}e^{+j\varphi_{vir}})G_{\mathrm{i}}(s-\mathrm{j}\omega_0)/[K(s-\mathrm{j}\omega_0)]K_{\mathrm{PWM}}H_{\mathrm{u}}/(R_{\mathrm{v}}+L_{\mathrm{v}}s)K_{\mathrm{PWM}}T_s(\mathrm{j}H_{\mathrm{u}}H_{\mathrm{i}}\boldsymbol{U}_1)\}+ \\ \{D\boldsymbol{U}_1e^{-2\mathrm{j}\varphi_{\mathrm{u}}} + (\tfrac{1}{2}e^{+\mathrm{j}\varphi_{vir}})G_{\mathrm{i}}(s-\mathrm{j}\omega_0)/[K(s-\mathrm{j}\omega_0)]K_{\mathrm{PWM}}H_{\mathrm{u}}/(R_{\mathrm{v}}+L_{\mathrm{v}}s)K_{\mathrm{PWM}}T_s(-\mathrm{j}H_{\mathrm{u}}H_{\mathrm{i}}e^{-2\mathrm{j}\varphi_{u}}\boldsymbol{U}_1)- \\ G_{\mathrm{i}}(s-\mathrm{j}\omega_0)H_{\mathrm{i}}K_{\mathrm{PWM}}T_s - s^2k_{\mathrm{c}}L_1H_{\mathrm{i}}C_1K_{\mathrm{PWM}}T_s - sL_1 - sL_2 - s^3L_2C_1L_1\}\}\end{array}}\end{aligned} \tag{A14}$$

$$Y_{21\_GFM} = -\frac{\begin{array}{l} -\{De^{-2j\varphi_i}\boldsymbol{I}_1 + (\frac{1}{2}e^{+j\varphi_{vir}})G_i(s-j\omega_0)/[K(s-j\omega_0)]K_{PWM}H_u/(R_v+L_vs)K_{PWM}T_s[-e^{-j\varphi_v}H_u(D_q-jH_ie^{-2j\varphi_i}e^{+j\varphi_u}\boldsymbol{I}_1)]- \\ G_i(s-j\omega_0)H_u/(R_v+L_vs)K_{PWM}T_s - sk_cH_iC_1K_{PWM}T_s - s^2L_2C_1 - 1\}+ \\ \{Ee^{-2j\varphi_i}\boldsymbol{I}_1 + (\frac{1}{2}e^{-j\varphi_{vir}})G_i(s-j\omega_0)/[K(s-j\omega_0)]K_{PWM}H_u/ \\ [R_v+L_v(s-2j\omega_0)]T_sK_{PWM}[-e^{-j\varphi_u}H_u(D_q-jH_ie^{-2j\varphi_i}e^{+j\varphi_u}\boldsymbol{I}_1)]\}/ \\ \{Ee^{-2j\varphi_v}\boldsymbol{U}_1 + (\frac{1}{2}e^{-j\varphi_{vir}})G_i(s-j\omega_0)/[K(s-j\omega_0)]K_{PWM}H_u/[R_v+L_v(s-2j\omega_0)]T_sK_{PWM}(-jH_uH_ie^{-2j\varphi_u}\boldsymbol{U}_1)\} \\ \{De^{-2j\varphi_v}\boldsymbol{U}_1 + (\frac{1}{2}e^{+j\varphi_{vir}})G_i(s-j\omega_0)/[K(s-j\omega_0)]K_{PWM}H_u/(R_v+L_vs)K_{PWM}T_s(-jH_uH_ie^{-2j\varphi_u}\boldsymbol{U}_1)- \\ G_i(s-j\omega_0)H_iK_{PWM}T_s - s^2k_cL_1H_iC_1K_{PWM}T_s - sL_1 - sL_2 - s^3L_2C_1L_1\} \end{array}}{\begin{array}{l} \{D\boldsymbol{U}_1 + (\frac{1}{2}e^{+j\varphi_{vir}})G_i(s-j\omega_0)/[K(s-j\omega_0)]K_{PWM}H_u/(R_v+L_vs)K_{PWM}T_s(jH_uH_i\boldsymbol{U}_1)\}- \\ \{E\boldsymbol{U}_1 + (\frac{1}{2}e^{-j\varphi_{vir}})G_i(s-j\omega_0)/[K(s-j\omega_0)]K_{PWM}H_u/[R_v+L_v(s-2j\omega_0)]T_sK_{PWM}(jH_uH_i\boldsymbol{U}_1)- \\ G(s-j\omega_0)H_iT_sK_{PWM} - (s-2j\omega_0)^2k_cL_1H_iL_vC_1T_sK_{PWM} - (s-2j\omega_0)L_1 - (s-2j\omega_0)L_2 - (s-2j\omega_0)^3L_2L_1C_1\}/ \\ \{Ee^{-2j\varphi_v}\boldsymbol{U}_1 + (\frac{1}{2}e^{-j\varphi_{vir}})G_i(s-j\omega_0)/[K(s-j\omega_0)]K_{PWM}H_u/[R_v+L_v(s-2j\omega_0)]T_sK_{PWM}(-jH_uH_ie^{-2j\varphi_u}\boldsymbol{U}_1)\} \\ \{De^{-2j\varphi_v}\boldsymbol{U}_1 + (\frac{1}{2}e^{+j\varphi_{vir}})G_i(s-j\omega_0)/[K(s-j\omega_0)]K_{PWM}H_u/(R_v+L_vs)K_{PWM}T_s(-jH_uH_ie^{-2j\varphi_u}\boldsymbol{U}_1)- \\ G_i(s-j\omega_0)H_iK_{PWM}T_s - s^2k_cL_1H_iC_1K_{PWM}T_s - sL_1 - sL_2 - s^3L_2C_1L_1\} \end{array}} \quad \text{(A15)}$$

$$Y_{22\_GFM} = -\frac{\begin{array}{l} -\{Ke^{+2j\varphi_i}\boldsymbol{I}_1{}^* + (\frac{1}{2}e^{-j\varphi_{vir}})G_i(s+j\omega_0)/[K(s+j\omega_0)]K_{PWM}H_u/(R_v+L_vs)K_{PWM}T_s[-e^{+j\varphi_u}H_u(D_q+jH_ie^{+2j\varphi_i}e^{-j\varphi_u}\boldsymbol{I}_1{}^*)] \\ -G_i(s+j\omega_0)H_u/(R_v+L_vs)K_{PWM}T_s - k_cH_iC_1sK_{PWM}T_s - L_2C_1s^2 - 1\} \\ +\{Me^{+2j\varphi_i}\boldsymbol{I}_1{}^* + (\frac{1}{2}e^{+j\varphi_{vir}})G_i(s+j\omega_0)/[K(s+j\omega_0)]K_{PWM}H_u \\ /[R_v+L_v(s+2j\omega_0)]T_sK_{PWM}[-e^{+j\varphi_u}H_u(D_q+jH_ie^{+2j\varphi_i}e^{-j\varphi_u}\boldsymbol{I}_1{}^*)]\}/ \\ \{M\boldsymbol{U}_1{}^* + (\frac{1}{2}e^{+j\varphi_{vir}})G_i(s+j\omega_0)/[K(s+j\omega_0)]K_{PWM}H_u/[R_v+L_v(s+2j\omega_0)]T_sK_{PWM}(-jH_uH_i\boldsymbol{U}_1{}^*) \\ -G(s+j\omega_0)H_iT_sK_{PWM} - (s+2j\omega_0)^2k_cL_1H_iC_1T_sK_{PWM} - (s+2j\omega_0)L_1 - (s+2j\omega_0)L_2 - (s+2j\omega_0)^3L_2L_1C_1\} \\ \{K\boldsymbol{U}_1{}^* + (\frac{1}{2}e^{-j\varphi_{vir}})G_i(s+j\omega_0)/[K(s+j\omega_0)]K_{PWM}H_u/(R_v+L_vs)K_{PWM}T_s(-jH_uH_i\boldsymbol{U}_1{}^*)\} \end{array}}{\begin{array}{l} -\{M\boldsymbol{U}_1{}^*e^{\pm 2j\varphi_u} + (\frac{1}{2}e^{+j\varphi_{vir}})G_i(s+j\omega_0)/[K(s+j\omega_0)]K_{PWM}H_u/[R_v+L_v(s+2j\omega_0)]T_sK_{PWM}(jH_uH_ie^{+2j\varphi_u}\boldsymbol{U}_1{}^*)\}/ \\ \{M\boldsymbol{U}_1{}^* + (\frac{1}{2}e^{+j\varphi_{vir}})G_i(s+j\omega_0)/[K(s+j\omega_0)]K_{PWM}H_u/[R_v+L_v(s+2j\omega_0)]T_sK_{PWM}(-jH_uH_i\boldsymbol{U}_1{}^*) \\ -G_i(s+j\omega_0)H_iT_sK_{PWM} - (s+2j\omega_0)^2k_cL_1H_iC_1T_sK_{PWM} - (s+2j\omega_0)L_1 - (s+2j\omega_0)L_2 - (s+2j\omega_0)^3L_2L_1C_1\} \\ \{K\boldsymbol{U}_1{}^* + (\frac{1}{2}e^{-j\varphi_{vir}})G_i(s+j\omega_0)/[K(s+j\omega_0)]K_{PWM}H_u/(R_v+L_vs)K_{PWM}T_s(-jH_uH_i\boldsymbol{U}_1{}^*)\} \\ +\{Ke^{\pm 2j\varphi_v}\boldsymbol{U}_1{}^* + (\frac{1}{2}e^{-j\varphi_{vir}})G_i(s+j\omega_0)/[K(s+j\omega_0)]K_{PWM}H_u/(R_v+L_vs)K_{PWM}T_s(jH_uH_ie^{+2j\varphi_u}\boldsymbol{U}_1{}^*) \\ -G_i(s+j\omega_0)H_iK_{PWM}T_s - s^2k_cL_1H_iC_1K_{PWM}T_s - sL_1 - sL_2 - s^3L_2C_1L_1\} \end{array}} \quad \text{(A16)}$$

$$Y_{12\_GFM} = -\frac{\begin{array}{l} -\{Ke^{+2j\varphi_i}\boldsymbol{I}_1{}^* + (\frac{1}{2}e^{-j\varphi_{vir}})G_i(s+j\omega_0)/[K(s+j\omega_0)]K_{PWM}H_u/(R_v+L_vs)K_{PWM}T_s[-e^{+j\varphi_u}H_u(D_q+jH_ie^{+2j\varphi_i}e^{-j\varphi_u}\boldsymbol{I}_1{}^*)] \\ -G_i(s_3)H_u/(R_v+L_vs)K_{PWM}T_s - k_cH_iC_1sK_{PWM}T_s - L_2C_1s^2 - 1\} \\ +\{Me^{+2j\varphi_i}\boldsymbol{I}_1{}^* + (\frac{1}{2}e^{+j\varphi_{vir}})G_i(s+j\omega_0)/[K(s+j\omega_0)]K_{PWM}H_u \\ /[R_v+L_v(s+2j\omega_0)]T_sK_{PWM}[-e^{+j\varphi_u}H_u(D_q+jH_ie^{+2j\varphi_i}e^{-j\varphi_u}\boldsymbol{I}_1{}^*)]\}/ \\ \{Me^{+2j\varphi_u}\boldsymbol{U}_1{}^* + (\frac{1}{2}e^{+j\varphi_{vir}})G_i(s+j\omega_0)/[K(s+j\omega_0)]K_{PWM}H_u/[R_v+L_v(s+2j\omega_0)]T_sK_{PWM}(jH_uH_ie^{+2j\varphi_u}\boldsymbol{U}_1{}^*)\} \\ \{Ke^{+2j\varphi_u}\boldsymbol{U}_1{}^* + (\frac{1}{2}e^{-j\varphi_{vir}})G_i(s+j\omega_0)/[K(s+j\omega_0)]K_{PWM}H_u/(R_v+L_vs)K_{PWM}T_s(jH_uH_ie^{+2j\varphi_u}\boldsymbol{U}_1{}^*) \\ -G_i(s+j\omega_0)H_iK_{PWM}T_s - s^2k_cL_1H_iC_1K_{PWM}T_s - sL_1 - sL_2 - s^3L_2C_1L_1\} \end{array}}{\begin{array}{l} \{K\boldsymbol{U}_1{}^* + (\frac{1}{2}e^{-j\varphi_{vir}})G_i(s+j\omega_0)/[K(s+j\omega_0)]K_{PWM}H_u/(R_v+L_vs)K_{PWM}T_s(-jH_uH_i\boldsymbol{U}_1{}^*)\} \\ -\{M\boldsymbol{U}_1{}^* + (\frac{1}{2}e^{+j\varphi_{vir}})G_i(s+j\omega_0)/[K(s+j\omega_0)]K_{PWM}H_u/[R_v+L_v(s+2j\omega_0)]T_sK_{PWM}(-jH_uH_i\boldsymbol{U}_1{}^*) \\ -G(s_3)H_iT_sK_{PWM} - (s+2j\omega_0)^2k_cL_1H_iC_1T_sK_{PWM} - (s+2j\omega_0)L_1 - (s+2j\omega_0)L_2 - (s+2j\omega_0)^3L_2L_1C_1\}/ \\ \{Me^{+2j\varphi_u}\boldsymbol{U}_1{}^* + (\frac{1}{2}e^{+j\varphi_{vir}})G_i(s+j\omega_0)/[K(s+j\omega_0)]K_{PWM}H_u/[R_v+L_v(s+2j\omega_0)]T_sK_{PWM}(jH_uH_ie^{+2j\varphi_u}\boldsymbol{U}_1{}^*)\} \\ \{Ke^{+2j\varphi_u}\boldsymbol{U}_1{}^* + (\frac{1}{2}e^{-j\varphi_{vir}})G_i(s+j\omega_0)/[K(s+j\omega_0)]K_{PWM}H_u/(R_v+L_vs)K_{PWM}T_s(jH_uH_ie^{+2j\varphi_u}\boldsymbol{U}_1{}^*) \\ -G_i(s+j\omega_0)H_iK_{PWM}T_s - s^2k_cL_1H_iC_1K_{PWM}T_s - sL_1 - sL_2 - s^3L_2C_1L_1\} \end{array}} \quad \text{(A17)}$$

The grid-forming inverter parameters for the four scenarios are consistent with those in Table 1, and the grid-forming parameters are shown in Table A1.

Table A1. Grid-following inverter parameters.

Parameter	Value	Parameter	Value
P_n/MW	1	U_{dc}/V	1000
U_g/V	220	L_1/mH	0.04125
L_2/mH	0.04125	C_1/μF	982.44
L_s/mH	0.004125	L_x/mH	0.02475
H_i	0.00044	K_{pPLL}	253
H_u	0.0034	K_{iPLL}	104,104

In Scenario 4, inverters 1, 3, and 4 are as shown in the table above, and the parameters of inverter 2 are as follows.

Table A2. Parameters of inverter 3 in Scenario 4.

Parameter	Value	Parameter	Value
P_n/MW	1	U_{dc}/V	1000
U_g/V	220	L_1/mH	0.04125
L_2/mH	0.04125	C_1/μF	982.44
L_s/mH	0.004125	L_x/mH	0.02475
D_p	0.578594	J/(kg.m^2)	1.06583
D_q	10	K	0.1
L_v/mH	0.0008	R_v/Ω	0.0254

References

1. Dalala, Z.; Al-Omari, M.; Al-Addous, M.; Bdour, M.; Al-Khasawneh, Y.; Alkasrawi, M. Increased renewable energy penetration in national electrical grids constraints and solutions. *Energy* **2022**, *246*, 123361.
2. Ranjan, M.; Shankar, R. A literature survey on load frequency control considering renewable energy integration in power system: Recent trends and future prospects. *J. Energy Storage* **2022**, *45*, 103717.
3. Wen, B.; Boroyevich, D.; Burgos, R.; Mattavelli, P.; Shen, Z. Analysis of D-Q Small-Signal Impedance of Grid-Tied Inverters. *IEEE Trans. Power Electron.* **2016**, *31*, 675–687.
4. Huang, L.; Xin, H.; Li, Z.; Ju, P.; Yuan, H.; Lan, Z.; Wang, Z. Grid-Synchronization Stability Analysis and Loop Shaping for PLL-Based Power Converters With Different Reactive Power Control. *IEEE Trans. Smart Grid* **2020**, *11*, 501–516.
5. Li, G.; Chen, Y.; Luo, A.; Wang, H. An Enhancing Grid Stiffness Control Strategy of STATCOM/BESS for Damping Sub-Synchronous Resonance in Wind Farm Connected to Weak Grid. *IEEE Trans. Ind. Inf.* **2020**, *16*, 5835–5845. [CrossRef]
6. Liao, Y.; Wang, X.; Liu, F.; Xin, K.; Liu, Y. Sub-Synchronous Control Interaction in Grid-Forming VSCs with Droop Control. In Proceedings of the 2019 4th IEEE Workshop on the Electronic Grid (eGRID), Xiamen, China, 7–9 November 2019.
7. Shi, K.; Wang, Y.; Sun, Y.; Xu, P.; Gao, F. Frequency-Coupled Impedance Modeling of Virtual Synchronous Generators. *IEEE Trans. Power Syst.* **2021**, *36*, 3692–3700. [CrossRef]
8. Li, Y.; Gu, Y.; Green, T.C. Revisiting Grid-Forming and Grid-Following Inverters: A Duality Theory. *IEEE Trans. Power Syst.* **2022**, *37*, 4541–4554. [CrossRef]
9. Li, M.; Zhang, X.; Guo, Z.; Wang, J.; Li, F. The Dual-Mode Combined Control Strategy for Centralized Photovoltaic Grid-Connected Inverters Based on Double-Split Transformers. *IEEE Trans. Ind. Electron.* **2021**, *68*, 12322–12330. [CrossRef]
10. Lima, L.A.M.; Watanabe, E.H. Hybrid Control Scheme for VSC Presenting Both Grid-Forming and Grid-Following Capabilities. *IEEE Trans. Power Deliv.* **2022**, *37*, 4570–4581.
11. Han, F.; Zhang, X.; Li, M.; Li, F.; Zhao, W. Stability Control for Grid-Connected Inverters Based on Hybrid-Mode of Grid-Following and Grid-Forming. *IEEE Trans. Ind. Electron.* **2024**, *71*, 10750–10760.
12. Xu, Y.; Nian, H.; Kang, J.; Zhao, J.; Wang, Z.; Zhou, J. Impedance-based Analysis of Potential Stability Risk Between Grid-Forming and Grid-Following Wind Turbine Systems. In Proceedings of the 2021 6th Asia Conference on Power and Electrical Engineering (ACPEE), Chongqing, China, 8–11 April 2021; pp. 858–862.
13. Fang, Z.; Huang, Y.; Yan, W.; He, Z.; Wang, D.; Xiong, B.; He, Q.; Zhou, K. Stability Analysis of Grid-Forming and Grid-Following VSCS in Parallel Connected to Weak Grid. In Proceedings of the 2024 IEEE 10th International Power Electronics and Motion Control Conference, Chengdu, China, 17–20 May 2024; pp. 4992–4996.
14. Yang, C.; Huang, L.; Xin, H.; Ju, P. Placing Grid-Forming Converters to Enhance Small Signal Stability of PLL-Integrated Power Systems. *IEEE Trans. Power Syst.* **2021**, *36*, 3563–3573.
15. Sun, H.; Lin, X.; Chen, S.; Ji, X.; Liu, D.; Jiang, K. Transient Stability Analysis of the Standalone Solar-Storage AC Supply System Based on Grid-Forming and Grid-Following Converters During Sudden Load Variation. *IET Renew. Power Gener.* **2023**, *17*, 1286–1302.
16. Akwasi, A.M.; Chen, H.; Liu, J. Grid Forming Inverters Using Reduced Order-Based Luenberger Observer for Power Control. *Electr. Power Syst. Res.* **2025**, *241*, 111424.
17. Du, Y.; Lu, X.; Zhao, D. Model Reduction for Inverter-Dominated Networked Microgrids with Grid-Forming Inverters. In Proceedings of the 2021 47th Annual Conference of the IEEE Industrial Electronics Society (IECON), Toronto, ON, Canada, 13–16 October 2021; pp. 1–6.
18. Rosso, R.; Engelken, S.; Liserre, M. Robust Stability Investigation of the Interactions Among Grid-Forming and Grid-Following Converters. *IEEE J. Emerg. Sel. Topics Power Electron.* **2020**, *8*, 991–1003.

19. Wang, G.; Pei, W.; Xiong, J.; Li, Y. Analysis Method of Stability for Hybrid System of Grid-Following and Grid-Forming Converters. *Proc. CSEE* **2025**, *45*, 25–38.
20. Gu, Y.; Li, Y.; Zhu, Y.; Green, T.C. Impedance-Based Whole-System Modeling for a Composite Grid via Embedding of Frame Dynamics. *IEEE Trans. Power Syst.* **2021**, *36*, 336–345.
21. Xu, Y.; Cheng, Y. Reduced-Order Modeling of Grid-Forming Inverters Based on Frequency Domain Perturbation Operator. In Proceedings of the 2024 IEEE Power & Energy Society General Meeting (PESGM), Seattle, WA, USA, 21–25 July 2024; pp. 1–5.
22. Cespedes, M.; Sun, J. Impedance Modeling and Analysis of Grid-Connected Voltage-Source Converters. *IEEE Trans. Power Electron.* **2014**, *29*, 1254–1261. [CrossRef]
23. Rygg, A.; Molinas, M.; Zhang, C.; Cai, X. A Modified Sequence-Domain Impedance Definition and Its Equivalence to the dq-Domain Impedance Definition for the Stability Analysis of AC Power Electronic Systems. *IEEE J. Emerg. Sel. Top. Power Electron.* **2016**, *4*, 1383–1396.
24. Ding, L.; Men, Y.; Du, Y.; Lu, X.; Chen, B.; Tan, J.; Lin, Y. Region-Based Stability Analysis of Resilient Distribution Systems with Hybrid Grid-Forming and Grid-Following Inverters. In Proceedings of the 2020 IEEE Energy Conversion Congress and Exposition (ECCE), Detroit, MI, USA, 11–15 October 2020; pp. 3733–3740.
25. Ge, L.; Tang, Y.; Zhu, L.; Chen, N.; Qu, L. Virtual Node Voltage Feedback Control Strategy for Inverters in Weak Grids. *Proc. CSEE* **2024**, *44*, 8392–8405.
26. Zhao, Y.; Ye, H.; Yang, J. Interaction and Stability Analysis Between Grid-Forming and Grid-Following Converters. In Proceedings of the 2023 International Conference on Power System Technology (PowerCon), Jinan, China, 21–22 September 2023; pp. 1–6.
27. Wang, G.; Lyu, J.; Rao, Y. Small-Signal Stability Analysis of Hybrid Wind Farm with Grid-Following and Grid-Forming Wind Turbine Generators. In Proceedings of the 2023 IEEE PELS Students and Young Professionals Symposium (SYPS), Shanghai, China, 25–27 August 2023.

energies

MDPI

Review

Concentrated Solar Thermal Power Technology and Its Thermal Applications

Chunchao Wu [1,2], Yonghong Zhao [1,2], Wulin Li [1,2], Jianjun Fan [1,2], Haixiang Xu [3], Zhongqian Ling [3], Dingkun Yuan [3] and Xianyang Zeng [3,*]

1 Akesai Kazak Autonomous County Huidong New Energy Co., Ltd., Jiuquan 735000, China
2 SDIC Gansu New Energy Co., Ltd., Lanzhou 730070, China
3 College of Energy Environment and Safety Engineering, China Jiliang University, Hangzhou 310018, China
* Correspondence: 06a0204057@cjlu.edu.cn

Abstract: The industrial sector accounts for approximately 65% of global energy consumption, with projections indicating a steady annual increase of 1.2% in energy demand. In the context of growing concerns about climate change and the need for sustainable energy solutions, solar thermal energy has emerged as a promising technology for reducing reliance on fossil fuels. With its ability to provide high-efficiency heat for industrial processes at temperatures ranging from 150 °C to over 500 °C, solar thermal power generation offers significant potential for decarbonizing energy-intensive industries. This review provides a comprehensive analysis of various solar thermal technologies, including parabolic troughs, solar towers, and linear Fresnel reflectors, comparing their effectiveness across different industrial applications such as process heating, desalination, and combined heat and power (CHP) systems. For instance, parabolic trough systems have demonstrated optimal performance in high-temperature applications, achieving efficiency levels up to 80% for steam generation, while solar towers are particularly suitable for large-scale, high-temperature operations, reaching temperatures above 1000 °C. The paper also evaluates the economic feasibility of these technologies, showing that solar thermal systems can achieve a levelized cost of energy (LCOE) of USD 60–100 per MWh, making them competitive with conventional energy sources in many regions. However, challenges such as high initial investment, intermittency of solar resource, and integration into existing industrial infrastructure remain significant barriers. This review not only discusses the technical principles and economic aspects of solar thermal power generation but also outlines specific recommendations for enhancing the scalability and industrial applicability of these technologies in the near future.

Keywords: concentrated solar thermal (CST); solar thermal energy; industrial heat; thermal energy storage; renewable energy

1. Introduction

As the world pursues a low-carbon future, solar energy technologies are central to global clean energy strategies [1]. Concentrated solar thermal (CST) is a key solar technology that uses mirror-based optical systems to focus sunlight onto a small-area receiver, converting it into high-temperature heat. This high-grade thermal energy can then drive steam turbines for power generation or supply heat for industrial processes and heating/cooling applications [2]. Unlike photovoltaics, CST inherently supports thermal energy storage by storing hot fluids or materials, enabling a stable energy supply even when the sun is not shining. This ability to dispatch solar energy on demand makes CST especially suitable for large-scale, high-temperature applications that require reliability, such as around-the-clock power generation or continuous industrial heating [3].

Recent years have seen remarkable advancements in CST technology, driven by innovations in materials, optics, and thermal storage. Solar receiver materials have greatly improved; researchers have developed nano-engineered selective absorber coatings (including multilayer "tandem" structures) that significantly increase solar absorption while withstanding higher temperatures [4]. These advanced coatings reduce radiative heat losses and maintain efficiency at operating temperatures well above 600 °C. In addition, high-temperature alloys are being explored for receiver components; for example, high-entropy alloys (HEAs) offer excellent creep strength and corrosion resistance at elevated temperatures. A recent study demonstrated that a Fe-Mn-Ni-Al-Cr HEA can serve as a durable receiver tube material, with its oxide layer acting as a high-efficiency solar absorber coating. Such materials enable higher operating temperatures (≥700 °C), which in turn boost the thermal-to-electric conversion efficiency of CST systems [5].

Parallel progress in optical design and control has enhanced the solar collection efficiency. Modern CST plants employ improved concentrator configurations (e.g., refined heliostat field layouts and reflector designs) alongside smart control systems. Notably, artificial intelligence (AI) and advanced wireless communications are being integrated into heliostat control, dramatically improving solar tracking accuracy and reducing operation costs. AI-driven calibration and predictive tracking can increase the overall reflectance efficiency and reliability of heliostat fields [6]. These optical innovations ensure more sunlight is delivered to the receiver throughout the day, even under wind or alignment disturbances, thereby raising heat collection efficiency and system stability.

Advances in thermal energy storage (TES) are equally vital, since storage extends CST operation beyond sunny hours. State-of-the-art CST plants use sensible heat storage in molten salts (typically a nitrate salt mixture) to store several hours' worth of thermal energy at high efficiency [7]. Research continues to improve molten salt formulations for higher thermal stability and lower melting points. Beyond sensible heat, new latent heat storage systems using phase change materials (PCMs) are being developed to exploit the high latent heat during solid–liquid transitions, allowing compact storage of large energy quantities. Meanwhile, thermochemical storage is an emerging frontier; high-temperature reactions (reversible chemical cycles) can store solar heat in chemical bonds and release it on demand, offering very high energy densities. Each of these TES approaches—sensible, latent, and thermochemical—has seen significant R&D progress, and hybrid systems are being designed to achieve both short-term and long-duration storage for CST. Efficient TES integration has led to today's CST plants providing up to 10–15 h of dispatchable power after sunset. This progress in storage, coupled with incremental improvements in turbines and heat transfer fluids (e.g., adoption of supercritical CO_2 cycles), has steadily increased the overall efficiency and capacity factor of CST plants [8].

Crucially, CST is no longer limited to power stations in deserts; its application footprint has broadened. Industries are increasingly adopting concentrated solar thermal systems to provide process heat for manufacturing, mining, food processing, chemical production, and other sectors [9]. High-temperature solar heat can drive processes like steam generation, drying, and calcination, directly substituting for fossil-fuel boilers in some cases. This use of CST for industrial process heat (also known as "solar heat for industrial processes") not only helps decarbonize heavy industry but is opening new commercial markets for CST technologies [10]. The expansion into industrial heat and other thermal applications is expected to increase production volume and bring economies of scale, further reducing the costs of CST components. Additionally, CST systems have been demonstrated in building heating and cooling contexts, for example, in solar-driven air conditioning or district heating, underscoring their versatility beyond electricity generation [11]. The growing interest and deployments in these non-traditional sectors have accelerated CST's path toward wider commercialization and integration in the global energy mix.

This paper provides a comprehensive review of the current state of CST technology and its thermal applications, highlighting recent progress and persisting challenges. We begin by outlining the basic principles of CST and the main types of concentrator systems (parabolic trough, central tower, linear Fresnel, and parabolic dish). We then examine the latest developments in key technical areas—solar receivers and materials, heat transfer and energy storage techniques—and discuss how these innovations are enhancing performance and expanding CST capabilities. The current applications of CST in power generation, industrial processes, and other sectors are surveyed to evaluate its commercial status. Finally, we discuss future research directions and emerging opportunities, considering what advances are needed to fully unlock CST's potential. Through this review, we aim to clarify CST's role in a sustainable energy future and provide insights for researchers, industry practitioners, and policymakers. Ultimately, continued innovation and deployment of CST technology will help foster the wider use of solar thermal energy in the global energy transition.

2. Composition of Solar Thermal Power Generation Systems

CST technology focuses sunlight through reflectors, collectors convert light energy into high-temperature heat energy, thermal storage systems store heat to ensure a stable supply, and ultimately heat energy is converted into electricity through a power generation system. The four work in tandem to achieve efficient and sustainable energy output.

2.1. Reflector

In a solar thermal power generation system, reflectors play a crucial role in capturing and concentrating solar radiation onto the receiver, thereby improving the system's energy collection efficiency. Their primary function is to focus sunlight, generating high-density thermal energy on the receiver's surface. The efficiency and performance of the entire solar thermal system are directly influenced by factors such as the reflector material and the arrangement of the mirror field.

Reflectors can be categorized into four main types based on their material composition: silver-coated glass mirrors, silver-coated polymer mirrors, aluminum mirrors, and stainless steel mirrors. Figure 1 provides a visual representation of the classification of so-lar mirrors according to their fundamental materials.

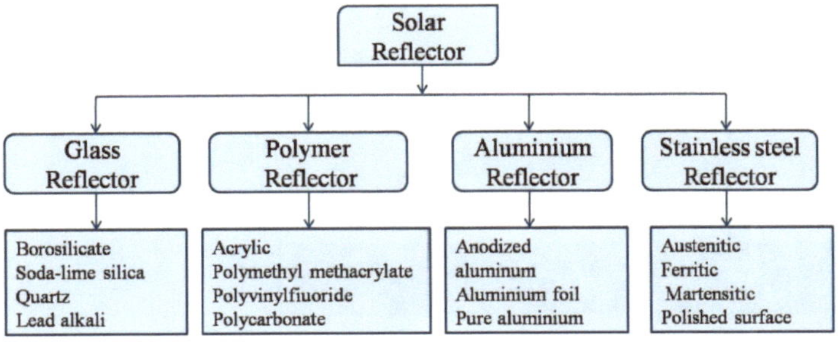

Figure 1. Classification of solar mirrors [12].

Glass mirrors are among the most commonly used reflectors in solar thermal power systems. When sunlight strikes the mirror, it reflects both heat and light to a designated location. The glass surface can function as both the substrate and the protective cover layer.

There are two types of glass mirrors based on the placement of the reflective coating. A first-surface mirror is created by applying a highly reflective coating directly onto the glass to enhance its optical performance. In contrast, a second-surface mirror has the reflective material applied to the backside of the glass, which is then protected by a paint coating [13]. An experimental study on silver-plating processes was conducted to enhance the mechanical properties and environmental durability of glass mirrors. The study found that a tantalum oxide coating provided excellent protection for silver reflectors. In 1979, the glass division of the Ford Motor Company developed thin glass for heliostats in concentrated solar power plants, where the silver-plated backside achieved a solar reflectivity of 89.3% [14].

Silver-coated polymer reflectors are lightweight and offer greater design flexibility. However, they have a shorter lifespan due to poor adhesion with silver when exposed to water [15]. An aluminum-coated first surface polymer mirror was fabricated by depositing an aluminum layer on a polycarbonate substrate. The focus of the research has been on creating lightweight and durable reflectors. The study also demonstrated that these plastic mirrors are resistant to extreme outdoor conditions, with a long-term reflectivity of up to 88%.

Aluminum oxide is a commonly used reflective material in concentrated solar power plants. Aluminum is the most abundant metal, relatively inexpensive, and widely used as a non-ferrous metal. The solar reflectivity of aluminum reflectors typically ranges from 85% to 91% [16]. A composite mirror for parabolic solar concentrators was developed using a magnetron sputtering technique. Efforts were made to enhance the reflectivity and longevity of the mirror by applying a bilayer reflective coating of aluminum (Al) and silicon dioxide (SiO_2). The study also demonstrated high mirror-like solar reflectivity and excellent protection against harsh environmental conditions by providing multiple protective layers.

Stainless steel is tough and does not require substantial thickness; however, its main limitation is its low solar reflectivity. Although aluminum has a higher reflectivity than stainless steel, aluminum mirrors require a minimum thickness of 4 mm to be self-supporting [17]. A reflector material was designed and fabricated that combined the beneficial properties of steel and aluminum. The reflective layer was created by directly evaporating aluminum onto the backside of a polyethylene terephthalate (PET) protective layer. A 9 μm aluminum foil substrate was then provided to support the reflectivity. To prevent aluminum damage, an additional PET layer was applied, and a stainless steel plate was pressed onto the back to enhance the strength of the reflector. A schematic diagram of the aluminum-steel reflector manufacturing process is shown in Figure 2.

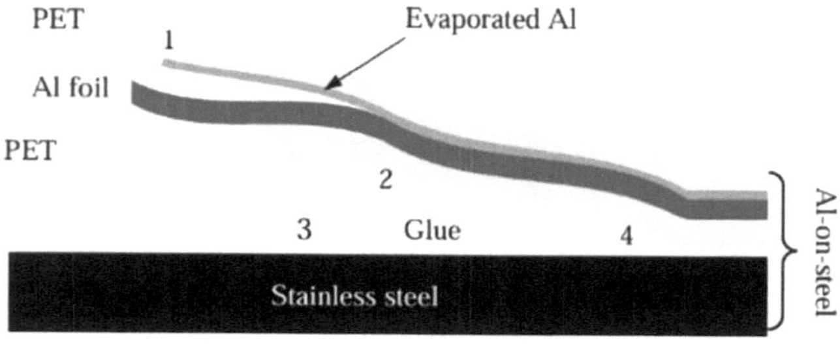

Figure 2. Manufacture of aluminum-steel reflectors. A 20 nm layer of aluminium is evaporated on a 25 μm PET foil (1), the PET foil with evaporated aluminium, a 9 μm rolled aluminium foil, and a 20 μm PET foil are laminated together (2), a 4 μm layer of glue is applied on a 0.5 mm thick sheet of stainless steel (3), the PET/Al/Al/PET sandwich and the steel sheet is hot pressed together to form the reflector laminate (4) [17].

Reflectors are key components in solar thermal power stations, responsible for reflecting sunlight and concentrating it onto the receiver. The design of the reflector must adapt to the position of the sun, ensuring that the solar rays are reflected as vertically as possible onto the receiver. Additionally, to enhance reflection efficiency, the surface of the reflector should be appropriately treated for reflectivity. This can be achieved through the choice of materials for the reflector, such as glass mirrors, ceramic mirrors, metal mirrors, self-cleaning high-reflectivity solar film materials, and sputtering targets.

2.2. Collector

A solar collector is a special type of heat exchanger that converts solar radiation into the internal energy of a transport medium. Solar collectors are the primary components of any solar energy system. They absorb incoming solar radiation, convert it into heat, and transfer this heat to a fluid (usually air, water, or oil) that flows through the collector. The collected solar energy is then directly transferred to hot water or space conditioning systems or stored in a thermal storage tank, from which it can be extracted for use during the night and/or on cloudy days. This chapter will introduce flat-plate collectors, compound parabolic concentrators, and vacuum tube collectors, starting with their types.

A typical flat-plate solar collector is illustrated in Figure 3. When solar radiation passes through the transparent cover and reaches the high-absorption blackened surface of the absorber, most of the energy is absorbed by the plate and then transferred to the heat transfer medium within the fluid pipes. This medium carries the absorbed heat away for storage or direct use. To minimize conductive heat losses, the bottom surface of the absorber plate and the sides of the housing are well insulated. The fluid pipes, responsible for transporting the heat transfer medium, can either be welded to the absorber plate or integrated directly into it. At both ends of the fluid pipes, a large-diameter manifold is in-stalled to ensure efficient fluid circulation [18].

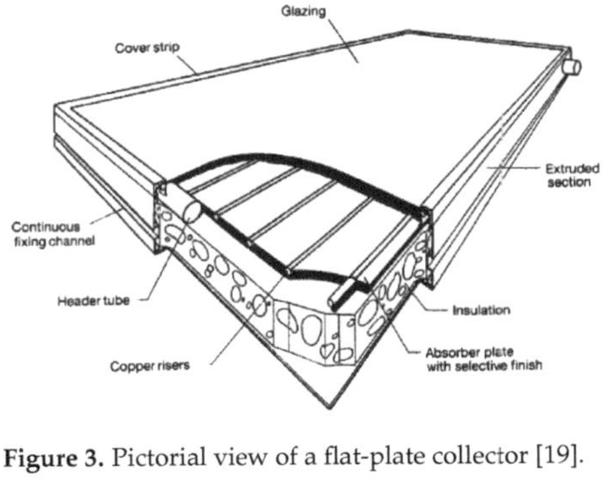

Figure 3. Pictorial view of a flat-plate collector [19].

The transparent cover plate serves to reduce convective losses from the absorber plate by restricting the stagnant air layer between the absorber plate and the glass. It also minimizes radiation losses from the collector, as the glass is transparent to the short-wave solar radiation received by the sun, while being almost opaque to the long-wave thermal radiation emitted by the absorber plate.

Flat-plate collectors are made with various designs and different materials. They have been used to heat fluids such as water, water with antifreeze additives, or air. Their primary goal is to collect as much solar energy as possible at the lowest total cost.

Glass has been widely used as a glazing material for solar collectors because it can transmit up to 90% of incident short-wave solar radiation, while being almost opaque to the long-wave radiation emitted by the absorber plate. Standard window glass and greenhouse glass have normal transmission rates of about 0.87 and 0.85, respectively. For direct radiation, the transmission rate varies significantly with the angle of incidence. Reflective coatings and surface texturing can also significantly improve the transmission rate. Dirt and dust have minimal impact on the collector glass, and occasional rainfall is usually sufficient to keep the transmission rate within 2–4% of the maximum.

Compound parabolic concentrators (CPCs) are non-imaging concentrators. They are capable of reflecting all incident radiation onto an absorber over a wide range [20]. Winston [21] noted their potential as solar collectors. The use of two parabolically orientated slots reduces the need to move the concentrator to accommodate changes in solar orientation, as shown in Figure 4. Composite parabolic concentrators can accept incoming radiation over a relatively wide range of angles. Through multiple internal reflections, any radiation entering the aperture within the collector's acceptance angle reaches the surface of the absorber located at the bottom of the collector [22]. The absorber is available in a variety of configurations. It can be cylindrical or flat, as shown in Figure 4. In the CPC shown in Figure 4, the lower part of the reflector (AB and AC) is circular while the upper part (BD and CE) is parabolic. Since the upper part of the CPC has little effect on the radiation reaching the absorber, it is often truncated, resulting in a shorter CPC, which is also cheaper. The CPC is usually covered with glass to avoid dust and other substances from entering the collector, which reduces the reflectivity of the collector walls. These collectors are more useful as linear or trough concentrators. The acceptance angle is defined as the angle at which the light source can move and still converge on the absorber. The orientation of a CPC collector is related to its acceptance angle (θc; in Figure 4) [19]. The way the collector moves depends on its acceptance angle, which determines whether it can be stationary or needs to track the sun. The long axis of a CPC collector can be orientated either in a north–south direction or an east–west direction, with the aperture tilted at an angle equal to the local latitude, pointing towards the equator [19]. When the long axis is orientated in a north–south direction, the collector must face the sun continuously by rotating its axis, as the collector receives a larger angle along the long axis and therefore does not require seasonal tilt adjustment. If kept stationary, the collector will only receive radiation during the periods when the sun is within the receiving angle. When the collector's long axis is orientated east–west, a slight adjustment of the seasonal tilt angle will allow the sun to be captured efficiently through the larger acceptance angle. In this case, the minimum acceptance angle should be equal to the maximum angle of incidence in the north–south vertical plane when the collector needs to output.

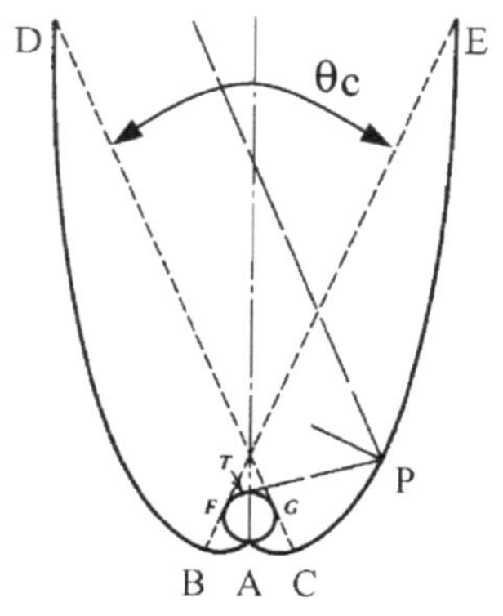

Figure 4. Schematic diagram of a compound parabolic collector. (AB and AC) are the lower part of the reflector in a circular configuration, and (BD and CE) are the upper part of the reflector in a parabolic configuration [19].

Traditional simple flat-plate solar collectors were developed for use in sunny and warm climates. However, their benefits are greatly diminished under conditions of cold, cloudy, and windy days. Additionally, weathering effects such as condensation and moisture can lead to the premature degradation of internal materials, resulting in performance deterioration and system failure. Vacuum tube solar collectors (tubes) operate differently from other collectors on the market. These collectors consist of heat pipes sealed in vacuum tubes, as shown in Figure 5.

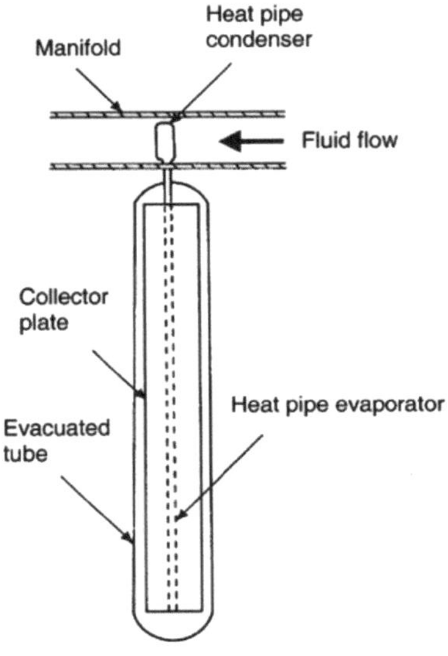

Figure 5. Schematic diagram of an evacuated tube collector [19].

The vacuum tube collector primarily consists of heat pipes housed within vacuum-sealed tubes. There are numerous variations of vacuum tube absorbers available on the market [23]. Recently, a manufacturer has introduced a fully glass evacuated tube collector (ETC), which could represent a significant advancement in reducing costs and extending the system's lifespan. Another variation of this collector is the Dewar tube, which uses two concentric glass tubes, with the space between them evacuated to create a vacuum jacket. A key advantage of this design is that it is entirely constructed from glass, allowing heat to be extracted from the tube without the need to penetrate the glass shell, thereby eliminating the potential for leakage losses. This design is also more cost effective compared to single-shell systems [24].

In addition to the flat-plate, evacuated tube, and compound parabolic collectors discussed earlier, concentrated solar collectors form the core of modern concentrated solar thermal (CST) and concentrated solar power (CSP) technologies. These collectors use optical devices to focus direct normal irradiance (DNI) onto a small receiver area, thereby achieving very high operating temperatures (typically between 250 °C and 1000 °C). Depending on the optical geometry and tracking configuration, concentrated collectors are categorized into four main types (Figure 6): Parabolic trough collectors (Figure 6a) use linearly curved mirrors that focus sunlight onto an absorber tube located at the focal line. A heat transfer fluid (HTF), typically synthetic oil or molten salt, circulates through this tube to absorb the concentrated thermal energy. PTC systems usually operate at temperatures between 300 °C and 400 °C, although advanced molten salt designs can reach 550 °C [25]. Central receiver (tower) systems (Figure 6b) employ a field of heliostats (flat, sun-tracking mirrors) that reflect sunlight to a central receiver atop a tower. The concentrated energy heats a working fluid—such as molten salt or air—circulated through the receiver. CRS systems are known for achieving very high operating temperatures up to 1000 °C, suitable for advanced thermodynamic cycles like supercritical CO_2 [26]. Regarding parabolic dishes (Figure 6c), these collectors are composed of a parabolic dish that focuses sunlight onto a receiver at its focal point, often integrated with a Stirling engine or Brayton cycle for direct power generation. Parabolic dishes offer the highest optical efficiency among all CSP technologies and are ideal for modular and distributed applications [27]. And regarding linear Fresnel reflectors (Figure 6d), LFR systems consist of long rows of flat or slightly curved mirrors that focus sunlight onto a fixed receiver positioned above the mirror field. Although the concentration ratio and thermal efficiency are generally lower than PTCs, the simpler design and reduced structural requirements make LFR systems more cost effective [28]. They are often used where land constraints and cost optimization are critical. These systems are key enablers for high-efficiency solar thermal power generation and high-temperature industrial heat supply.

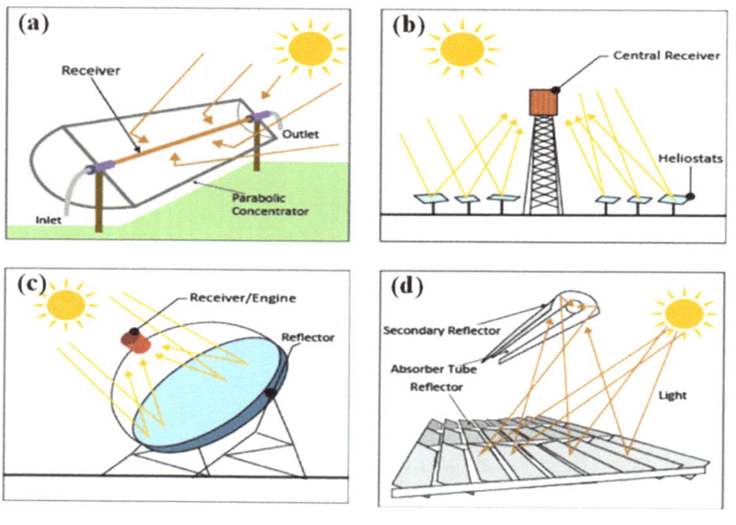

Figure 6. Several main types of concentrating solar power generation [29]. (**a**) Parabolic trough collectors, (**b**) Central receiver (tower) systems, (**c**) parabolic dishes collectors, (**d**) linear Fresnel reflectors.

2.3. Heat Storage System

To address the need for on-demand power generation under variable solar conditions, energy storage is essential. It enhances the reliability of renewable energy and helps manage peak demand for solar power generation [30]. Thermal energy storage (TES) is one of the key advantages of concentrated solar power (CSP); compared to battery storage, TES is more cost effective and better suited for integration into the electrical grid [31]. Studies have shown that CSP systems with energy storage can offer over twice the value of photovoltaic power generation in certain contexts [32].

TES technologies include sensible heat storage, which raises the temperature of a medium to store heat but has lower density, latent heat storage (PCM), which stores heat through phase changes and offers higher density, and thermochemical storage, which stores heat via reversible chemical reactions and has the highest potential for heat density, although it is still in early development (Figure 7).

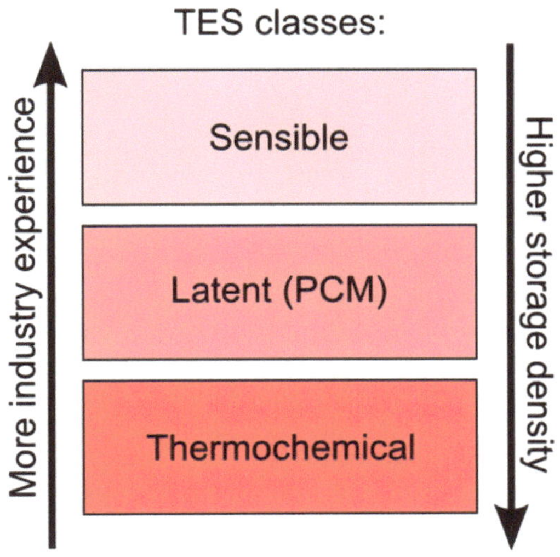

Figure 7. The different categories of thermal energy storage include sensible heat, latent heat (or phase change materials, PCM), and thermochemical storage [33].

With technological advancements, CSP plants are expected to transition from sensible heat storage to latent heat storage, and eventually to thermochemical storage, in order to achieve higher efficiency and greater energy storage capacity.

2.3.1. Sensible Heat Storage

Sensible heat exchangers can generally be divided into two types: solid and liquid. There are also systems that combine both solid and liquid, such as fluid flowing through a packed bed made of solid particles. The working principle of both solid and liquid sensible heat exchangers is the same; heat is stored by raising the temperature of the material.

Molten salts are an ideal choice for high-temperature energy storage due to their high energy density and low cost. The operational temperature of molten salts is limited by their freezing point and corrosion issues, with freezing being the primary concern as it can damage pipes and pumps. Therefore, auxiliary heaters are typically required. Common molten salts include nitrates, nitrites, and carbonates, with mixed salts improving performance, particularly by lowering the melting point. For example, $NaNO_3$-$NaNO_2$-KNO_3 (Hitec) and $Ca(NO_3)_2$-$NaNO_3$-KNO_3 (HitecXL) are common commercial salt mixtures [34]. Ongoing research is focused on developing better molten salts, such as the quaternary salt mixture (K, $NaNO_2$, Cl, and NO_3) designed by Peng et al., and low-melting-point salts (80 °C) presented by Zhao and Wu [35].

When using solid materials, packed bed structures are typically employed, with heat storage and discharge occurring via a heat transfer fluid (HTF). If the HTF is gas (e.g., air), it is used solely for heat transfer; if it is a fluid, its heat capacity is comparable to that of the solid material and can act as an auxiliary storage medium. Inexpensive options include sand, gravel, or concrete. Sand and gravel are generally used as packing materials, with some proposals for using a fluidized bed structure [36]. Concrete can be formed into bricks to create a packed bed or can be used with heat transfer tubes running through large concrete blocks [37]. To enhance the thermal performance of concrete and to be compatible with finned HTF heat exchangers, concrete and castable ceramics with volumetric heat capacities greater than 0.7 kWhth/m^3/K have been developed [38]. For high-temperature applications, silica bricks, alumina bricks, and magnesia bricks are recommended, often paired with molten salts as HTFs [39]. Graphite has also been proposed as a sensible heat storage material due to its excellent stability at high temperatures and high specific heat capacity.

2.3.2. Latent Heat Storage

In latent heat storage systems, thermal energy is stored by inducing phase changes in the storage material. The most common phase change is from solid to liquid, while sol-id-to-solid transitions are occasionally used. Liquid-to-gas phase changes are rarely employed due to the significant expansion and the need for strict pressure containment.

Metals and metal alloys are often proposed as Phase Change Materials (PCMs) due to their high thermal conductivity and latent heat. While metals exhibit excellent properties, their high cost can be a drawback. Alloying metals with cheaper materials, such as aluminum–silicon alloys, can reduce costs while maintaining good thermal performance [40].

To enable the widespread use of PCM-based thermal storage, cost-effective and high-performance materials must be developed. Additionally, low-cost containers, heat exchangers, and encapsulating materials compatible with PCM are needed to support efficient charging and discharging processes. High-performance PCMs, suitable materials, and efficient system design are crucial for the practical application of latent heat storage.

2.3.3. Thermochemical Heat Storage

Thermochemical TES stores thermal energy via reversible chemical reactions, where heat input drives internal reactions, storing energy in chemical bonds [41]. The reverse exothermic reactions are used for heat recovery and power generation. Common energy storage processes include the following:

Metal hydroxide dehydration: energy is stored by dehydration, but the low thermal conductivity of the materials limits the reaction rate [15].

Metal hydride dehydrogenation: Energy is stored by removing hydrogen, although issues with poor heat transfer and slow reaction kinetics exist. However, doping elements such as iron or nickel can increase the reaction rate [42].

Metal carbonate decarbonation: Energy is stored by removing carbon dioxide, typically requiring high temperatures (>450 °C), making it suitable for future high-temperature power plants [43].

Metal oxide redox: Energy is stored by reducing metal oxides. Although this process is less studied, it holds significant potential for high energy density.

Thermochemical storage materials offer high energy density and, in theory, virtually unlimited storage duration. However, challenges related to reaction reversibility, rates, conversion efficiency, and economic feasibility must be addressed. Overcoming these obstacles will improve CSP system performance and advance their integration into global renewable energy networks.

2.4. Power Generation Systems

Power generation systems convert thermal energy into electricity, with traditional plants using steam generators to produce high-temperature steam that drives turbines. A key area of research is the development of technologies that directly convert thermal energy into electricity, eliminating the need for intermediate mechanical conversion. Efficient, low-cost direct conversion technologies could improve the scalability of solar thermal systems.

2.4.1. Thermal Engine Systems

A thermal engine is the system in a CSP plant that converts collected heat into electrical energy. The traditional approach involves converting thermal energy into mechanical energy via a thermodynamic cycle, which is then used to drive a generator for electricity production. The main characteristics, advantages, and challenges of various traditional power generation systems are summarized in Table 1.

Table 1. Key features, advantages, and challenges of various traditional power generation systems.

Technology Type	Description	Advantages	Challenges
Parabolic Trough Solar Power	Mirrors focus sunlight onto pipes, heating a fluid to produce steam that drives a turbine for power generation.	Mature technology, suitable for large-scale power generation.	Requires large land area, highly weather dependent.
Solar Power Tower	Mirrors focus sunlight onto pipes, heating a fluid to produce steam that drives a turbine for power generation.	High efficiency, capable of generating power around the clock.	High initial construction cost, large land requirements.
Parabolic Dish Solar Power	Flat mirrors concentrate sunlight onto a collector to generate steam that drives a turbine for power generation.	Low cost, suitable for medium- to small-scale projects.	Lower efficiency, performance not as good as trough and tower systems.
Carbide Concentrated System	Parabolic dish mirrors concentrate sunlight to drive a Stirling engine for power generation.	High efficiency, suitable for distributed generation.	High maintenance, large initial investment.
Solar Multi-Receiver Collection System	Multiple heliostats focus sunlight onto a high tower receiver, and heat transfer drives a turbine for power generation.	High efficiency, suitable for large-scale generation, and peak load management.	High cost, complex technology, requires precise tracking.

2.4.2. Direct Thermoelectric Conversion

The most active research areas in direct solar thermoelectric conversion are Solar Thermoelectric Generators (STEGs) and Solar Thermophotovoltaic (STPV). A brief overview of these two fields is provided below.

STEGs convert thermal energy into electrical energy through thermoelectric generators. Due to the Seebeck effect, thermoelectric materials generate a voltage gradient under the influence of a temperature gradient, which in turn produces a current and generates electricity [44]. Thermoelectric generators are typically used for waste heat recovery, but when coupled with solar energy absorbers, they can generate power from solar energy. A typical configuration of a STEG is illustrated in Figure 8.

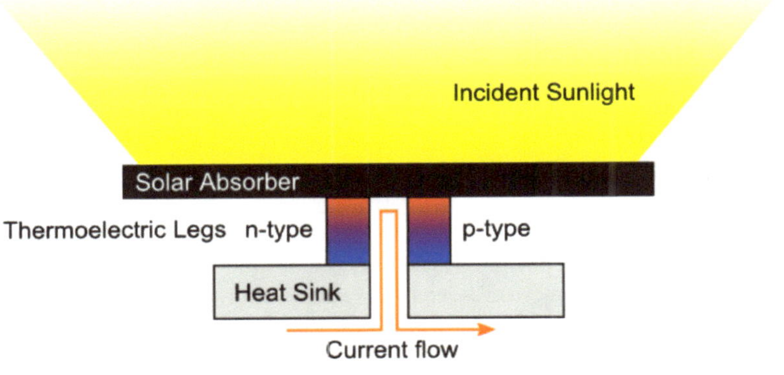

Figure 8. Schematic of a solar thermoelectric generator (STEG) [33].

Solar Thermophotovoltaic (STPV) systems absorb solar radiation to generate heat, which is then converted into thermal radiation. This radiation is directed to a single-junction thermophotovoltaic (TPV) cell, where it is converted into electricity. STPV's advantage lies in using frequency-dependent thermoelectric converters to achieve an ideal spectral irradiation, resulting in higher conversion efficiency than direct solar irradiation. The challenge is ensuring an effective intermediate absorption/re-emission process, so that the spectral enhancement outweighs its negative impacts. A typical STPV system configuration is shown in Figure 9.

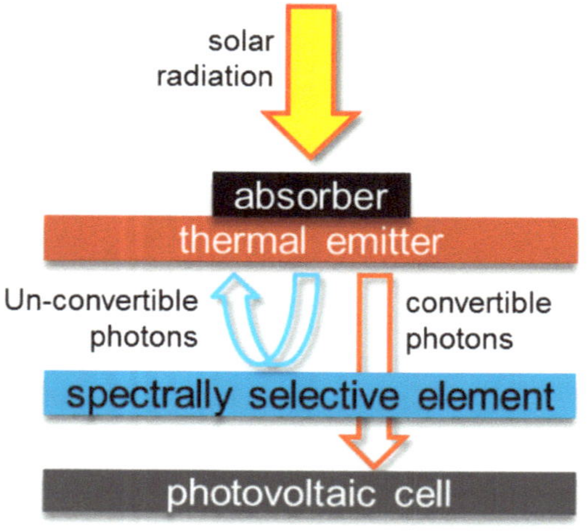

Figure 9. Schematic of an STPV system [33].

The main challenge faced by STPV systems is the high-temperature requirement, as they rely on converting thermal radiation into electrical energy. Even with the semiconductors that have the lowest bandgap (around 0.5 eV), a blackbody spectrum temperature of over 1000 °C is needed to achieve efficient conversion. Although 1000 °C is below the melting point of many materials, constructing a highly efficient and reliable STPV system remains a significant engineering challenge [33].

3. Types of Solar Thermal Power Generation

3.1. Trough Solar Power System

Among the various concentrated solar power (CSP) technologies, parabolic trough solar power systems stand out as the most widely utilized solution. While CSP systems encompass different types of concentrating collectors, the parabolic trough configuration has garnered significant attention due to its proven effectiveness in large-scale power generation. In this section, we focus on the parabolic trough solar power system, which represents a key approach to harnessing solar energy in an efficient and scalable manner.

Parabolic trough solar power systems are currently the most widely utilized concentrated solar power (CSP) technology. The full name of this system is the parabolic trough solar thermal power system, which typically consists of a concentrating collector, a heat storage unit, a heat engine power generation device, and auxiliary energy systems (such as boilers). It primarily comprises several parabolic trough-type concentrating collectors arranged in series and parallel. These collectors harness solar energy to heat the working fluid within the heat pipe, generating high-temperature steam that drives a turbine, which in turn powers a generator to produce electricity [45]. The medium used (typically heat transfer oil or a nitrate mixture) is enclosed within a vacuum glass cover in the heating tube. The concentration ratio generally ranges from 10 to 100. When oil is employed as the heat transfer fluid, the maximum heat collection temperature can reach 400 °C. Conversely, when a nitrate mixture is used, the maximum heat collection temperature can reach 700 °C. The latter maintains superior thermal efficiency at high temperatures, resulting in a higher power generation efficiency [46–48].

Parabolic trough concentrating collectors (Figure 10) are linear focusing collectors. A large number of trough-type concentrators are typically connected in series and parallel to form a concentrating collector array. One-dimensional tracking of solar radiation is employed. The advantages of the trough system include its relatively simple structure, low maintenance costs, potential for mass production, and compact size, which enables easy installation and maintenance when arranged on the ground. However, due to the low geometric concentration ratio of the parabolic concentrator and the relatively low collector temperature, the thermal-to-power efficiency of the power subsystem in the parabolic trough solar thermal power generation system remains relatively low. Consequently, it is challenging for a pure parabolic trough solar thermal power generation system to further enhance thermal efficiency and reduce power generation costs.

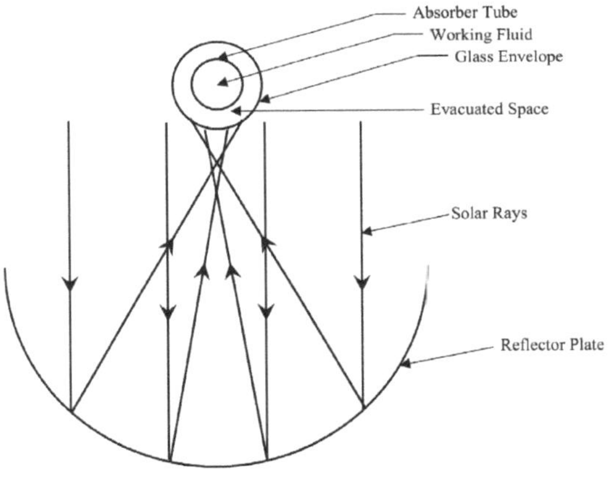

Figure 10. Schematic diagram of a parabolic trough concentrating collector [49].

To enhance the efficiency of the system and reduce costs, new materials and technologies are continuously being explored and implemented. In parabolic trough concentrated solar power (CSP), the selection of collector materials is crucial. In recent years, studies [50,51] have highlighted the development of highly efficient nano-functionalized absorber materials that can effectively utilize most of the radiation in sunlight. Despite the nanoscale manipulation of the material, its photothermal conversion efficiency can be significantly enhanced, thereby improving the overall system performance. Due to the variability of light during the day and night, as well as under sunny and cloudy conditions, existing parabolic trough CSP systems are increasingly being integrated with thermal energy storage technology. By employing molten salt energy storage and other technologies [52], excess thermal energy collected during periods of optimal sunlight can be stored for use during the night or on cloudy days, ensuring a continuous supply of electricity.

3.2. Tower Solar Power System

Applying solar power generation technology in the form of a tower provides a novel method for generating electricity. Not only can it efficiently store thermal energy, but it can also operate at high temperatures. A solar power generation system primarily consists of a power generation unit, a main control system, a heat storage tank, a receiver, and a heliostat array. A specific number of heliostats (spherical mirror groups that automatically track the sun) are positioned on the ground, with a solar collector tower constructed at an optimal location within the heliostat array. The heliostats reflect solar energy into the collector at the top of the tower, heating the working fluid to produce high-temperature steam that drives a turbine for electricity generation [47]. The SDIC Gansu Akesai Huidong "Solar Thermal + Photovoltaic" Project has a total installed capacity of 750 MW, with 110 MW allocated to solar thermal power generation and 640 MW to photovoltaic power generation [29]. As illustrated in Figure 11, the solar thermal section includes 11,960 pentagonal heliostats and a collection tower approximately 200 m high. Each heliostat is capable of rotating to track the sun and reflect solar radiation onto the absorber at the top of the heat collection tower.

Figure 11. The tower system in Akesai Kazak Autonomous County Jiuquan, Gansu Province, China.

Due to the large number of heliostats in a tower power generation system, its concentration ratio can reach 1500. Compared to a parabolic trough solar power generation system, the concentration ratio of a tower solar power generation system is significantly higher. The unique design allows the cavity temperature of its collector to exceed 1000 °C [53,54]. Due to its advantages in high concentration and high thermal conversion efficiency, a tower solar power system can achieve high power generation capacity. This makes it highly suitable for large-scale, high-capacity commercial applications. However, a tower solar thermal power system requires a significant one-time investment, has a complex device structure and control system, and incurs high maintenance costs [55].

Similarly, several emerging technologies are anticipated to be applied to tower power generation systems. Göttsche et al. proposed a novel design utilizing a mini-mirror array (10 × 10 cm), with each mirror mounted on a ball joint driven by a stepper motor (Figure 12) [56]. The purpose of this design is to mitigate the adverse effects that strong winds may have on the mirrors while reducing costs. However, the final optical performance was not satisfactory.

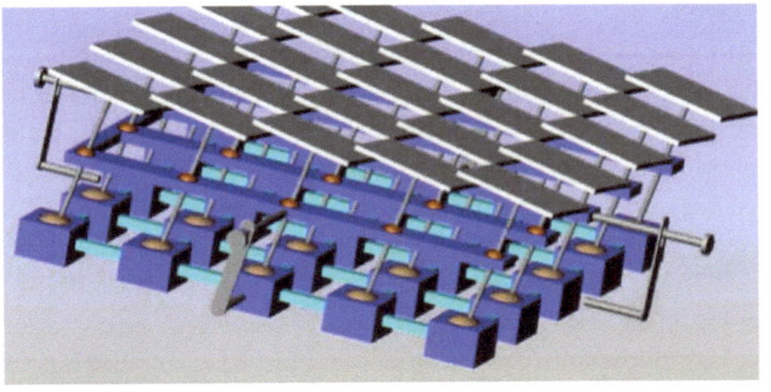

Figure 12. Schematic of mini-mirror array [56,57].

The hydraulic tracking system [58] utilizes a hydraulic structure to track the movement of the sun, ensuring that the device is always positioned to receive the maximum amount of sunlight. Additionally, the incorporation of a dual-axis tracking system and a high-precision solar tracking algorithm will enhance both the efficiency and reliability of the CSP system. Due to the higher temperatures in tower power generation systems, the application of materials with high thermal conductivity and thermal stability has become essential. High-entropy (HE) materials [59] have garnered widespread attention in recent years. These materials possess excellent light-to-heat conversion capabilities and broadband light absorption characteristics, enabling selective absorption of the solar spectrum, reducing infrared light absorption, and minimizing heat loss.

3.3. Dish Stirling Solar Power Generation System

The dish Stirling solar power generation system is a solar energy system that uses parabolic mirrors to concentrate sunlight for power generation. It primarily consists of a parabolic dish, a receiver, a Stirling engine, and a generator. When sunlight hits the parabolic mirror, the light is focused onto the receiver, heating the working fluid inside the receiver to a high temperature. This, in turn, drives the Stirling engine to generate electricity. This system can operate independently, serving as a small power source for off-grid remote areas, or multiple units can be connected in parallel to form a small-scale solar thermal power plant. The dish Stirling system is known for its high efficiency, modularity, and ability to operate independently.

This technology converts solar thermal energy into rotational motion via the Stirling cycle, with AC generators transforming mechanical energy into electricity [60]. After accounting for power losses, the system can convert nearly 30% of the direct normal solar radiation into electricity, achieving an efficiency of about 30% [61].

As early as 1988, Roelf [62] optimized the solar application of the STM4-120 Stirling engine. The engine's power output at 25 rpm was 1800 kW, with an efficiency of 40% from heat input to shaft output. The use of a simple variable slotted piston stroke control eliminated sealing and power control issues that previously plagued Stirling engines.

Singh et al. [63] designed, simulated, and optimized a solar Stirling engine. The study found that the maximum thermal efficiency of the dish Stirling system at an absorber temperature of 850 K was 32%. The energy and exergy efficiency of the dish Stirling system were 17% and 19%, respectively, with the majority of the losses occurring at the receiver. Mari et al. [64] addressed gaps in the mathematical modeling of the parameters of the parabolic solar dish Stirling engine system, providing numerical equations and simulations related to solar dish applications. Pheng et al. [65] reviewed the parabolic dish Stirling system based on CSP technology, considering aspects such as performance, overall efficiency, dish position, and levelized cost of electricity. They also analyzed the technical and high capital cost barriers to the practical application of parabolic dish systems.

Zayed et al. [66] developed a theoretical model based on optical geometry and thermodynamic analysis for the solar dish Stirling system, considering the optical configurations and energy balance of different system components. An optimization algorithm was implemented using multi-objective particle swarm optimization to simultaneously maximize output power and total efficiency, achieving a maximum power output of 23.46 kW and an optimal overall efficiency of 30.15%. Subsequent research [67] conducted comprehensive assessments of the following: design standards, geometric parameters' optimization, thermal performance analysis, thermodynamic optimization, and techno-economic aspects.

Typically, a solar thermal receiver is added to a parabolic solar dish to capture the maximum concentrated solar radiation from the dish. Senthil et al. [68] experimentally analyzed the improvement of the thermal inertia balance and heat storage capacity of a solar thermal receiver using phase change materials (PCMs). The average energy and exergy efficiency of the receiver reached 66.7% and 13.8%, respectively.

3.4. Fresnel Solar Thermal Power Generation System

The Fresnel solar thermal power generation system employs arrays of flat mirrors to reflect sunlight onto a collector tube. It has a modest concentration ratio but is cost effective and suitable for large-scale applications. The advantages of the Fresnel system include its simple structure, high land utilization, ease of cleaning, and low cost.

In 1957, Baum et al. [69] first proposed modularizing parabolic trough mirrors into flat mirror arrays. A tracking system with flat glass mirrors can reflect and concentrate sunlight onto a fixed secondary concentrator. Francia et al. [70] implemented this concept practically in a linear Fresnel concentrator system, achieving pioneering work with a prototype featuring a dual-axis tracking system, laying the foundation for large-scale applications of linear Fresnel reflectors (LFRs). Since then, researchers have extensively studied the Fresnel solar thermal power generation system.

Feuermann et al. [71] introduced the concept of secondary concentrators by adding secondary mirrors to the receiver to enhance concentration efficiency and reduce thermal losses. The advent of compound parabolic secondary concentrators has significantly advanced the linear Fresnel solar thermal collection technology. Grena et al. [72] used molten nitrate salt as the heat transfer fluid in a solar Fresnel linear concentrator. They introduced a system designed to handle molten nitrate salts more effectively, with vacuum tube collectors that can tolerate temperatures up to 550 °C. Through optical and thermodynamic simulations, they found that, with an adequately precise tracking system, the efficiency loss of the molten salt Fresnel system was about 10% to 20% compared to the trough system. Taramona et al. [73] overcame the main limitations of previously proposed hyperbolic secondary concentrators by proposing a new type of secondary concentrator made up of several fixed flat mirrors at the same height. This setup established a new solar field model and achieved a concentration ratio of up to 31% and optical efficiency up to 60%. Grena [74] discussed the impact of optical geometry on linear Fresnel concentrators, the parameters required to determine the geometry, and the main optical concepts. He also summarized a ray-tracking program to simulate the mirror field and a fast calibration analysis method for optimization and real-time calculation.

An important subsystem in LFR systems is the collector system. Currently, research on the heat collection performance of linear Fresnel collectors is relatively limited, focusing mainly on two areas: new heat transfer fluids and collector tube designs. Zhai [75] conducted experimental and simulation studies on a Fresnel lens solar collector with vacuum tubes and found that for water heating, when the inlet water temperature was 80–90 °C, the heat efficiency of the collector reached 9.80%, which was 90% higher than the commonly used evacuated tube collectors. However, optical losses were the primary source of energy loss, accounting for nearly 40% of solar radiation. Khan et al. [76] studied the effect of new corrugated collector tubes with nanofluids on the energy efficiency of a novel dual-fluid heat collection system and presented an optimization model for the corrugated collector tube through simulation. Alamdari et al. [77] reviewed research on linear Fresnel solar power generation systems over the past decade, emphasizing that various types of thermal losses are key parameters affecting efficiency.

Each of the four CSP technologies mentioned above has its own set of advantages and disadvantages, which can be evaluated based on the factors in Table 2. The table presents the key technical characteristics of each technology, including their capacity range, associated costs, efficiency percentages, and electricity costs (calculated using the Levelized Cost of Electricity model). These factors play a crucial role in determining the economic viability and performance of each CSP system, and they vary depending on the system design, installation location, and solar radiation levels. For instance, systems with higher efficiency tend to have lower electricity costs, but they might come with higher initial installation costs. The comparison of these attributes allows for a comprehensive assessment of the strengths and limitations of each CSP technology.

Table 2. The technical characteristics for CSP [78–86].

	Capacity (MW)	Installation Cost	Power Generation Efficiency (%)	Cost of Electricity (USD/kWh)	Thermal Loss Coefficient	Optical Efficiency Attenuation Rate
Trough solar power system	10–200	Low	13–18	0.17	High (significant radiation and conduction losses)	Relatively high (affected by radiation and weather variations)
Tower solar power system	10–200	High	16–17	0.14	Low (highly efficient light concentration)	Low (precise optical design)
Dish Stirling Solar Power Generation System	0.01–0.4	High	20–30	0.15	High (large collector area and Stirling engine thermal losses)	High (significantly affected by external environmental changes)
Fresnel Solar Thermal Power Generation System	10–200	Low	8–11	0.14	Moderate (relatively high optical component efficiency)	Relatively low (influenced by dust accumulation on mirrors)

4. Application of Solar Thermal Systems in Industrial Process Heating

4.1. Energy Demand and Challenges in Industrial Process Heating

Industrial process heating accounts for a significant portion of global energy consumption, with a major share of energy allocated to thermal applications across various industries. These demands range from low to high temperatures, with the majority concentrated in the low- to medium-temperature range, particularly in sectors such as food processing, textiles, chemicals, and paper manufacturing. However, the predominant re-liance on fossil fuels, especially natural gas and oil, raises concerns about environmental pollution, greenhouse gas emissions, and resource depletion. As the world shifts toward more sustainable energy sources, renewable options, such as solar thermal energy, offer substantial potential to meet industrial heating needs, particularly in regions with abundant solar resources. However, the widespread adoption of solar thermal systems faces challenges, including high initial costs and the intermittent nature of solar energy, which necessitates the integration of energy storage solutions and hybrid systems for greater economic feasibility and stability.

4.1.1. Energy Demand for Industrial Process Heating

Industrial process heating constitutes a major share of global energy consumption, representing over 35% of total energy demand, with nearly 54% of industrial energy allocated to thermal applications [19]. This demand spans a wide range of applications, from low-temperature (<100 °C) to medium-temperature (100–400 °C) and high-temperature (>400 °C) needs. However, the majority of the industrial heating demand is concentrated in the low- to medium-temperature range, particularly in industries such as food processing, textiles, chemicals, and paper manufacturing. In these sectors, thermal energy is typically used for processes such as pasteurization, drying, cooking, bleaching, and dyeing. For example, the temperature range for cleaning, pasteurization, and drying in the food processing industry generally falls between 60 °C and 250 °C, while the thermal energy requirements for cooking and drying in the paper industry typically range from 90 °C to 200 °C. In the chemical and pharmaceutical industries, distillation, evaporation, and reaction processes often require higher temperatures, typically in the range of 100 °C to 400 °C [87].

At present, industrial process heating is predominantly dependent on fossil fuels, particularly natural gas and oil. While these energy sources can meet the high-temperature requirements of industrial processes, their usage is associated with environmental pollution, greenhouse gas emissions, and resource depletion [88]. In many developing countries, coal remains the primary energy source for industrial process heating due to technological and economic limitations, while more industrialized regions tend to rely on natural gas [89]. With the increasing global focus on energy efficiency, fluctuating fossil fuel prices, and the tightening of environmental regulations, governments and businesses are progressively shifting towards renewable energy sources as alternatives to conventional energy. This transition is not only aimed at alleviating environmental pressures but also at addressing the potential crisis arising from the gradual depletion of fossil fuels.

4.1.2. Potential for Application in the Industrial Sector

Solar thermal energy is the process of converting solar radiation into heat energy. As shown in Figure 13, solar thermal systems utilize solar collectors to capture solar radiation. The potential of solar thermal energy systems in industrial process heating is primarily reflected in their ability to fulfill low- and medium-temperature demands while providing substantial economic and environmental advantages. These systems employ various types of solar collectors, including flat plate collectors, evacuated tube collectors, and parabolic trough collectors, to provide thermal energy for multiple industrial applications such as food processing, textiles, chemicals, and pharmaceuticals [90]. The application of solar thermal technology is not confined to specific industries but also demonstrates notable benefits in particular regions. For example, in regions with abundant solar resources, such as India, the Middle East, and Africa, the strong solar radiation provides crucial support for the economic viability and efficiency of solar thermal systems. Industrial enterprises in these regions can substantially reduce their dependence on traditional fossil fuels, while simultaneously lowering energy costs and carbon emissions [87]. Despite the extensive potential of solar thermal systems in the industrial sector, their large-scale implementation faces considerable obstacles. High initial investment cost represent a significant barrier to widespread adoption, while the intermittent nature of solar radiation impacts system stability. As a result, the incorporation of energy storage technologies, such as phase change materials (PCMs) and molten salt storage, is of critical importance to maintaining a stable energy supply. Moreover, to enhance adaptability and economic feasibility, hybrid heating technologies that integrate solar thermal technology with conventional fossil fuel systems have emerged as a crucial area of research.

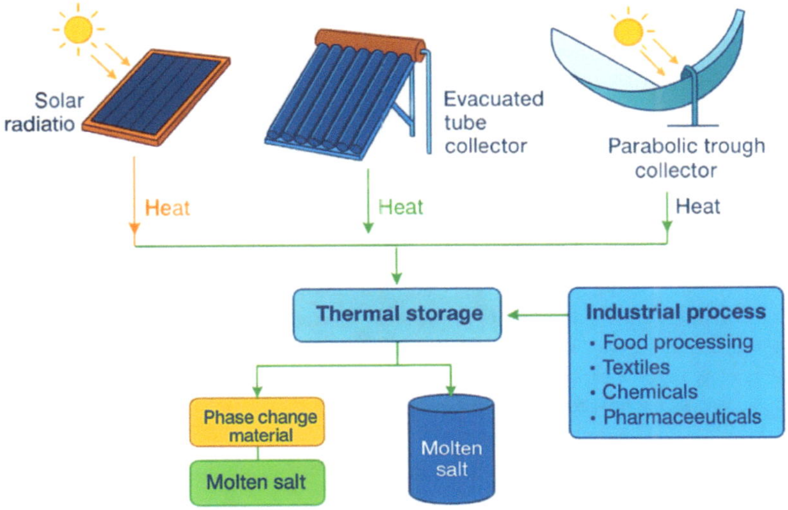

Figure 13. Solar thermal system in industry.

4.2. Application Scenarios of Solar Thermal Systems in Industrial Processes

Solar thermal systems have a wide range of applications in industrial processes, from low-temperature hot water to high-temperature steam, meeting the diverse needs of industries such as food processing, textiles, chemicals, and metalworking (Figure 14). These systems provide clean and sustainable energy solutions, significantly reducing dependence on fossil fuels and lowering carbon emissions. The application scenarios of solar thermal systems can be broadly classified into low-temperature, medium-temperature, and high-temperature categories, depending on the specific thermal requirements of industrial processes.

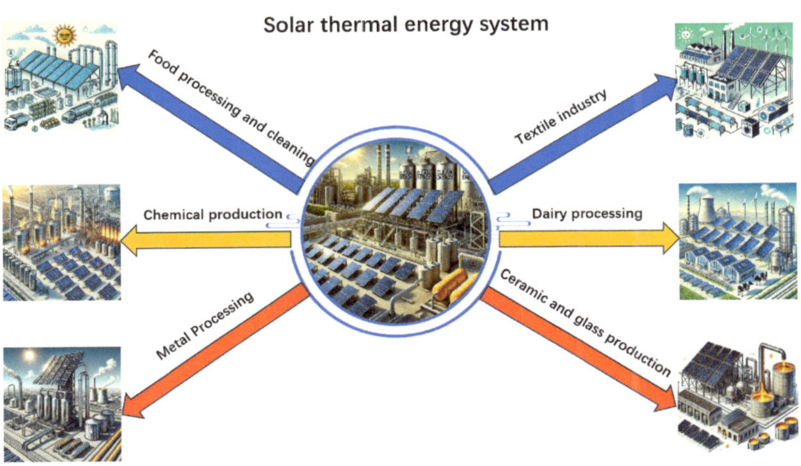

Figure 14. Application scenarios of solar thermal systems in different industries.

4.2.1. Low-Temperature Application Scenarios

Solar thermal systems exhibit broad applicability in low-temperature industrial processes, particularly in industries requiring low-temperature hot water or thermal energy. Low-temperature solar thermal systems, typically used for applications below 100 °C, mainly convert solar radiation into heat through solar collectors, providing sustainable energy alternatives for industrial production. Key application scenarios include food and beverage processing, textile industries, and chemical processes.

In the food industry, solar thermal systems can be utilized for cleaning, pasteurization, and disinfection, all of which require low-temperature heat. Pasteurization, a widely used technique, necessitates a constant supply of hot water at temperatures between 60 and 70 °C, a requirement efficiently met by solar thermal systems. This approach not only reduces reliance on fossil fuels but also significantly cuts energy costs and carbon emissions [91]. Additionally, the beverage industry often uses solar energy to provide the required low-temperature hot water for cleaning in bottling production lines.

In the textile industry, low-temperature solar thermal systems play an essential role. Dyeing, rinsing, and washing processes in textile manufacturing demand substantial quantities of warm water, typically within the 50–80 °C range. Solar thermal systems supply the necessary heat for these processes, reducing dependence on coal and natural gas while enhancing energy independence for enterprises [92].

Low-temperature applications in the chemical industry demonstrate considerable potential, finding extensive utilization in processes such as low-temperature evaporation, dissolution, and precision cleaning. This system reliably supplies the necessary thermal energy while mitigating cost pressures associated with fluctuations in fossil fuel prices. In industrial production processes with a continuous yet relatively low heat demand, the integration of solar thermal systems with thermal energy storage technologies significantly improves the system's flexibility and operational stability.

Consequently, in low-temperature industrial applications, solar thermal systems serve as a low-carbon, cost-effective, and sustainable energy solution, positioning them as an essential enabler in advancing industrial energy efficiency and emission reduction objectives.

4.2.2. Medium-Temperature Application Scenarios

Solar thermal systems are also widely applied in medium-temperature industrial processes (100–250 °C), representing an important pathway for achieving industrial energy savings and sustainable development. This temperature range is commonly required in industries such as paper and pulp, dairy processing, chemicals, and pharmaceuticals. In these fields, solar thermal systems use solar collectors and moderately concentrated parabolic trough systems to provide hot water or steam to meet medium-temperature energy demands.

In the paper industry, pulp drying and thermal treatment are the main energy-consuming stages. Solar thermal systems can provide a stable heat source, reducing reliance on traditional fossil fuels. In dairy processing, such as milk powder production, a significant amount of medium-temperature steam is required for evaporation and drying. Empirical studies conducted in dairy processing facilities indicate that solar thermal systems can fulfill up to 76–94% of the thermal load demand [93], leading to a substantial reduction in both energy expenditures and carbon emissions. In the chemical and pharmaceutical industries, solar thermal energy is used for solvent evaporation, chemical reactions, and distillation, all of which require stable temperatures. By integrating solar thermal systems with thermal energy storage devices (such as phase change materials or thermal storage tanks), a continuous and stable heat supply can be ensured, overcoming the intermittent nature of solar energy [87].

Additionally, the application of solar thermal systems in medium-temperature scenarios in beverage industries, such as breweries, has been highly successful, where they are used to heat water and steam for cleaning, pasteurization, and production processes. Through the implementation of advanced solar thermal technologies, industrial facilities not only achieve significant reductions in operating costs but also substantially cut their carbon emissions, thus playing a crucial role in advancing carbon neutrality objectives [91]. To clearly illustrate the correlation between solar thermal applications and specific industrial process temperature ranges, Table 3 summarizes the typical industrial sectors and their representative thermal processes categorized by low-, medium-, and high-temperature requirements.

Table 3. Solar systems at different temperatures' generation.

Temperature Divisione	Industrial Type	Application
Low (<100 °C)	Food industry Textile industry Chemical industry	Wash, pasteurize, and disinfect Dyeing, rinsing, and washing Low-temperature evaporation, dissolution, and cleaning
Medium (100–250 °C)	Paper industry Beverage industry Chemical and pharmaceutical industries	Pulp drying, heat treatment Wash and pasteurize Solvent evaporation, chemical reaction, and distillation
High (<250 °C)	Metalworking industry Chemical industry	Metal melting, heat treatment Pyrolysis, gasification, and synthesis at high temperature

Table 3 shows typical industrial sectors and thermal process examples at different temperature levels for solar thermal systems.

Solar thermal systems demonstrate immense potential in medium-temperature industrial applications by significantly reducing fossil fuel usage and enhancing energy efficiency. When integrated with thermal energy storage technologies, they further improve reliability and adaptability. The effective deployment of these technologies offers practical and scalable solutions to enhancing industrial energy efficiency and promoting environmental sustainability.

4.2.3. High-Temperature Application Scenarios

Solar thermal systems in industrial processes are primarily applied in fields that require high-temperature heat sources, such as metal processing, ceramic manufacturing, glass production, and high-temperature chemical reactions. High-temperature solar thermal systems primarily rely on concentrated solar power (CSP) technologies, including parabolic trough collectors, solar power towers, and Fresnel lens collectors, to achieve temperatures of approximately 250 °C or higher. These technologies focus solar radiation onto specific areas using mirrors, achieving the required high temperatures. In the metal processing industry, solar thermal systems can provide sustainable high-temperature heat sources for metal smelting and heat treatment, replacing traditional coal or gas energy sources [94]. In ceramic and glass manufacturing, where high-temperature kilns frequently operate at temperatures exceeding 1000 °C, solar power tower systems can provide the necessary thermal energy, thereby lowering production costs and mitigating environmental pollution [92].

The chemical industry presents considerable potential for high-temperature applications. In high-temperature cracking, gasification, and synthesis processes, solar thermal systems can provide high-temperature steam and heat support. These processes are inherently energy-intensive, and integrating solar thermal systems can substantially decrease fossil fuel consumption while enhancing overall energy efficiency. High-temperature solar systems can be coupled with thermal energy storage technologies to ensure a continuous supply of high-temperature heat, even during periods of insufficient sunlight. Molten salt storage technology, for example, stores solar thermal energy during the day and releases it at night, providing a stable heat source for chemical reactions or high-temperature manufacturing.

In summary, solar thermal systems hold great promise for high-temperature industrial applications. These systems serve as clean and efficient heat sources, reduce industrial dependence on conventional energy, and play a pivotal role in enhancing energy efficiency and reducing emissions.

4.3. Economic Effects of Solar Thermal Systems in Industrial Processes

As the global energy transition accelerates and industrial sustainability becomes a priority, solar thermal systems, a clean and efficient renewable energy technology, are increasingly being integrated into the industrial sector. These systems provide a sustainable source of thermal energy for industrial processes, reducing reliance on fossil fuels and mitigating greenhouse gas emissions. While the adoption of solar thermal systems in the industry is still evolving, numerous studies highlight their substantial potential to lower operational costs, meet carbon reduction targets, and foster long-term economic stability.

4.3.1. Cost Savings and Investment Return Analysis

The economic benefits of solar thermal systems in industry are primarily evident in long-term fuel cost savings and rapid return on returns. Although the initial capital investment for solar thermal systems—comprising equipment, installation, and integration—can be high, these systems significantly reduce reliance on traditional fossil fuels, resulting in notable fuel cost reductions. For example, a dairy plant in Austria installed 1000 m^2 of solar collectors, which saves 85,000 m^3 of natural gas annually and reduces CO_2 emissions by 170 tons. This not only generates considerable environmental benefits but also shortens the payback period to under three years [91].

Life cycle cost (LCC) analysis further underscores the economic viability of solar thermal systems. Despite higher initial investments, energy cost savings during operation, combined with low maintenance costs, render these systems economically advantageous in the medium to long term [95,96]. Furthermore, the monetization of carbon reduction benefits, supported by policy incentives, further enhances the economic appeal of these systems. For example, under the European Union Emissions Trading System (EU ETS), the value of carbon emission reductions is integrated into company revenues, providing additional incentives for investors. This dual benefit—direct fuel cost savings and policy incentives—makes solar thermal systems a powerful tool for reducing industrial operating costs.

4.3.2. Regional and Industry-Specific Differential Benefits

The economic benefits of solar thermal systems vary significantly across different geographical regions and industries. The key factors influencing these benefits include local climate conditions and fossil fuel prices. In regions with abundant solar and high fossil fuel prices, such as Southern Europe and Australia, the payback period for solar thermal systems is typically shorter, and the economic benefits are particularly pronounced [96]. In contrast, regions with lower solar radiation or fuel prices, such as some developing countries, experience more limited economic benefits. These areas often require additional policy support and technological optimization to fully unlock the potential of solar thermal energy [97].

At the industry level, low-temperature processes (40–120 °C) generally show higher compatibility with solar thermal systems. Industries such as food, beverage, and textiles, where thermal energy demand is concentrated within a low-temperature range, can effectively meet significant thermal needs using flat-plate collectors or vacuum tube collectors. For instance, a dairy plant in New Zealand implemented a solar thermal system to meet part of its thermal demand, achieving additional economic benefits through heat storage and system optimization [98]. The characteristics of these industries allow for a faster realization of high investment returns.

For industries requiring medium-temperature heat (150–400 °C), such as chemical and plastic processing, economic benefits are more dependent on technological support. These sectors typically employ highly efficient collectors, such as parabolic troughs or linear Fresnel reflectors, often coupled with high-temperature storage systems to meet thermal demands. While the initial investment is higher, the gradual optimization of these technologies and their increasing deployment are demonstrating their economic feasibility [97]. However, for high-temperature processes exceeding 400 °C, such as steelmaking and glass manufacturing, the economic benefits of solar thermal systems remain relatively low due to technological and efficiency limitations. Further research and breakthroughs in these technologies are essential for future development [96].

4.3.3. Long-Term Economic Stability and Intangible Value

In addition to direct cost savings, solar thermal systems contribute to long-term economic stability and provide intangible asset value to businesses. As a renewable energy technology, they reduce a company's reliance on fossil fuels, thereby mitigating the financial risks associated with energy price volatility. This risk avoidance benefit is especially crucial for regions dependent on imported fuels [95].

Moreover, the intangible value of carbon reduction is increasingly recognized as a key component of corporate competitiveness. As the global focus on sustainable development intensifies, the adoption of solar thermal systems can significantly enhance a company's environmental reputation and social responsibility image [91]. These intangible assets not only help attract more investments but also increase market competitiveness, particularly in industries that place a high emphasis on green production.

Although the promotion of solar thermal systems in high-temperature industrial sectors still faces technological and economic challenges, the economic benefits in low- and medium-temperature industrial processes have been widely recognized. Future research directions should include further optimization of system designs, reducing initial investment costs and driving the global adoption of solar thermal systems through policy support and business model innovation (Figure 15). Additionally, improvements in energy storage technologies and the development of new, efficient collectors will further enhance the economic potential of solar thermal systems in high-temperature applications, providing new possibilities for achieving sustainable industrial development.

4.4. Technical Advantages and Challenges in Industrial Applications

4.4.1. Technical Advantages

Solar thermal systems represent a clean, renewable energy source that can significantly reduce carbon emissions in industrial production. This is especially true in energy-intensive industrial sectors such as metal smelting, chemical production, and food processing, where solar thermal energy can effectively replace traditional fossil fuels, thereby reducing carbon dioxide and other greenhouse gas emissions. This contributes to addressing the global challenge of climate change [99]. Moreover, with the continuous development of concentrated solar thermal technologies, such as tower systems, parabolic trough collectors, and concentrating collectors, solar energy can be efficiently converted into thermal energy. These technologies generally exhibit high heat collection efficiency, meeting industrial heat demands ranging from low to high temperatures. This is particularly beneficial in industries such as metal processing and ceramic manufacturing, which require high-temperature heat sources [100]. Solar thermal systems can also be integrated with traditional industrial energy systems to form hybrid energy supply systems. For instance, solar thermal energy can be combined with natural gas boilers or electricity systems to ensure the stability and reliability of heat supply. Especially in regions with weaker solar radiation, the combination of solar energy and traditional energy sources can provide around-the-clock energy supply [94]. With the advancement of energy storage technologies, solar thermal systems can integrate advanced thermal storage technologies (e.g., molten salt storage and phase change material storage), which allow excess thermal energy to be stored during the day and used during the night or on cloudy days. This resolves the intermittency issue associated with solar energy. The application of such storage systems makes solar thermal systems more widely applicable, particularly in industrial processes that require a stable heat source [101].

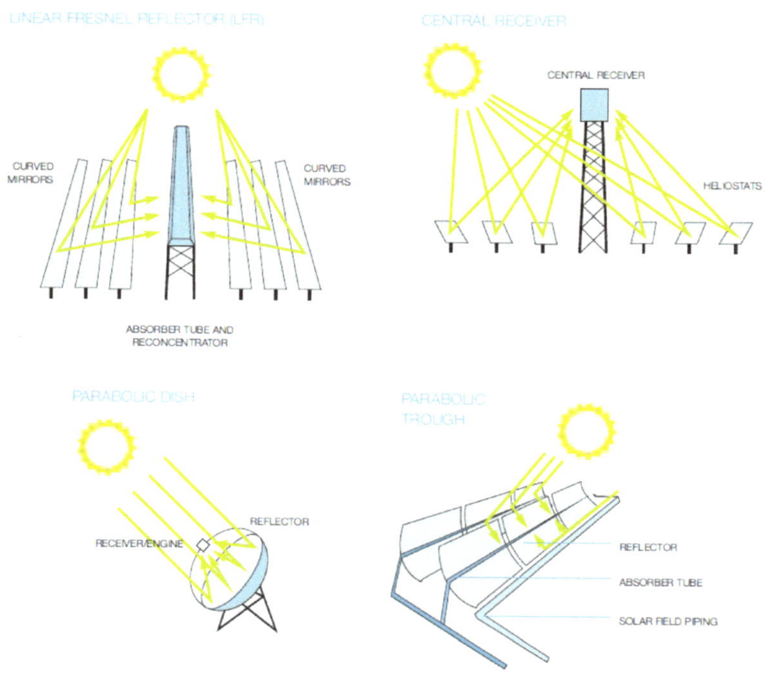

Figure 15. An illustration of commonly used CST technologies [34].

Despite the higher initial investment in solar thermal systems, their operational costs are much lower than those of fossil fuels, particularly in the context of rising energy prices. Solar thermal energy thus becomes a competitive alternative energy source. Many studies show that in some regions, the payback period for solar thermal systems typically ranges from 3 to 5 years, effectively reducing energy costs and improving economic efficiency [94].

The main technical advantages of solar thermal systems in industrial applications are their clean, renewable energy characteristics, high energy conversion efficiency, low operating costs, good system integration, and ability to integrate with energy storage technologies. With continuous technological innovation and the gradual maturation of markets, solar thermal systems are expected to be more widely applied in industrial fields, providing sustainable solutions for reducing carbon emissions and lowering energy consumption.

4.4.2. Technical Challenges

Despite the significant advantages of solar thermal systems in industrial applications, there remain numerous technical challenges to their widespread implementation.

First, the initial investment for solar thermal systems is typically high, especially in large-scale industrial applications such as chemical processing, food manufacturing, and metal smelting. While the operational costs of these systems are relatively low, the initial costs for equipment purchase, installation, and system integration require substantial capital investment. This can be a major challenge for smaller and medium-sized enterprises looking to adopt solar thermal systems [101].

Second, the efficiency of solar thermal systems is heavily influenced by climate conditions and geographic location, particularly in regions with insufficient sunlight. In such areas, the heat supply capability of the system may be compromised. Although thermal storage technologies, such as molten salt storage or phase change material storage, can partially address this issue, challenges remain with the widespread adoption of efficient storage technologies and the high cost of storage equipment. Particularly in regions with long winters or frequent cloudy weather, solar thermal systems may not meet the continuous industrial heat demand [100].

Furthermore, integrating solar thermal systems into existing industrial energy systems poses another challenge. The integration design, installation, and maintenance of these systems require a high level of technical expertise, adding complexity to their deployment. Determining the optimal configuration and scale of collectors for different industrial applications and effectively storing and supplying solar heat when needed are significant technical challenges [99].

Currently, although solar thermal technology has made significant progress in certain areas, it is still in the process of gradual refinement and broader adoption. This is especially true for high-temperature industrial applications, where the maturity and reliability of the technology still need to be improved. For example, solar tower collector technology and concentrating systems require higher costs and precision, and their reliability and efficiency in actual industrial applications are still under ongoing optimization [94].

Finally, despite being seen as a clean energy solution in many countries, the promotion of solar thermal technology faces challenges in some regions due to insufficient policy support and low market demand. The lack of adequate government incentives, subsidy policies, and a general lack of corporate awareness about green development often leads to the slow adoption of solar thermal systems in industrial applications. Without policy guidance, enterprises may be reluctant to bear the high initial costs, even though solar thermal systems may offer higher economic and environmental benefits in the long term [96].

In conclusion, the application of solar thermal systems in the industry faces challenges related to technical costs, system integration complexity, weather dependency, technology maturity, and insufficient policy support. To further advance the application of solar thermal technology in industrial settings, technological innovation, policy support, and an improved market environment are urgently needed to lower costs, enhance system stability and reliability, and promote the widespread use of renewable energy in the industrial sector.

5. Emerging Technologies in Concentrating Solar Thermal Systems

5.1. Innovative Solar Collector Materials

Advances in nano-functionalized absorber materials and high-performance optical systems are expected to enhance the heat collection efficiency of CST systems. Nano-functionalized materials, such as nanostructured coatings and multi-layered absorbers, offer superior light absorption and reduced thermal losses. These innovations help to significantly improve the conversion of sunlight into thermal energy, particularly for high-temperature applications where traditional materials might not perform as well. For example, the development of selective absorbers with tailored optical properties allows CST systems to operate efficiently at temperatures exceeding 500 °C. A schematic representation of the nanocomposite-based single-layer absorber coating developed on SS 304 substrate is shown in Figure 16, which highlights the fabrication process using SiO_2 nanoparticles and the improvement in solar absorptance through nano-void textured surfaces [102].

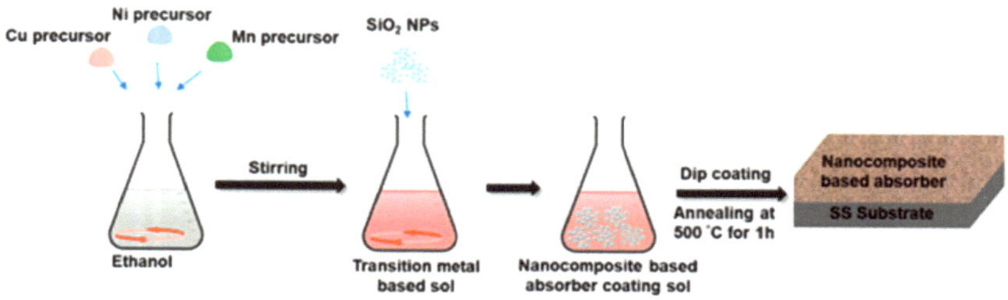

Figure 16. Schematic represents the development of nanocomposite-based single-layer absorber coating on SS 304 substrate [102].

5.2. Thermal Energy Storage (TES) Innovations

Emerging TES technologies are playing a critical role in enhancing the performance of CST systems by enabling them to provide continuous power even when sunlight is unavailable. Technologies like phase change materials (PCMs) and thermochemical storage systems offer higher energy density and cost effectiveness compared to traditional methods such as sensible heat storage. PCMs, which store thermal energy through phase transitions (solid to liquid), allow for more compact and efficient thermal storage. Thermo-chemical storage systems, which store energy via reversible chemical reactions, provide even higher energy densities and longer storage durations, making them especially useful for large-scale applications. The integration of TES with CST systems has been well explained through Figure 17, which illustrates the indirect configuration using super-critical carbon dioxide (s-CO_2) as the heat transfer fluid (HTF). The integrated system in this study, which utilizes supercritical carbon dioxide (s-CO_2) as the heat transfer fluid (HTF) in a parabolic trough collector (PTC) system, operates as follows. Solar radiation is focused by the parabolic mirrors onto a receiver tube, where it heats the circulating HTF (Points 1–5). The heated HTF is transported through the PTC loop, passing through a heat exchanger (Point 8) to transfer thermal energy to a thermal energy storage (TES) system. The system includes both hot and cold TES tanks (Points 11–12), where energy is stored during the day and discharged during non-sunny periods (Point 13–15). The HTF, after being heated, flows into the hot tank for storage, while the cold tank helps maintain the temperature differential for energy efficiency. The stored thermal energy is later used to generate electricity in the power block (Point 17) through a conversion process. This process is facilitated by a secondary heat exchanger, which is specifically required for the s-CO_2 configuration (Point 9). The system's design ensures efficient energy storage and usage, with feedback mechanisms to regulate energy flow throughout the day and night (Points 16–19). This system design shows how s-CO_2 improves thermal energy storage efficiency and reduces system costs [52].

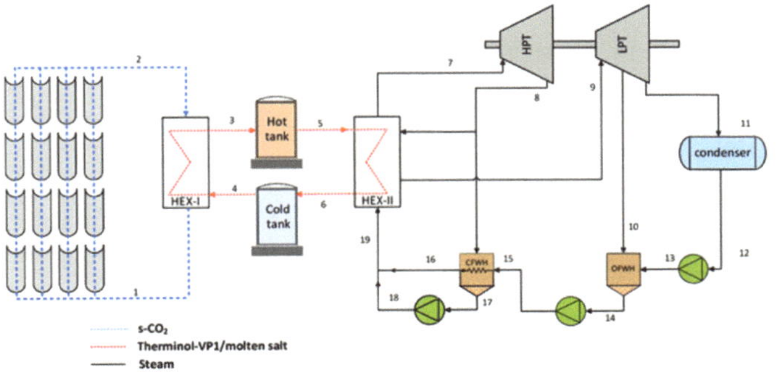

Figure 17. Schematic diagram for the indirect configuration of the integrated system using s-CO_2 as an HTF [52].

6. Conclusions and Outlook

This review comprehensively explored the technological evolution, thermal performance, and industrial applications of concentrated solar thermal (CST) systems, emphasizing their capabilities beyond power generation. CST systems, through optical concentration and efficient thermal conversion, can generate medium- to high-temperature heat suitable for a wide range of industrial sectors such as food processing, chemical manufacturing, textiles, and metal treatment. With growing global pressure to decarbonize industrial processes, CST technologies present a critical pathway for reducing fossil fuel reliance and greenhouse gas emissions in the heat domain.

Through an extensive survey of CST configurations—including parabolic troughs, central towers, linear Fresnel reflectors, and dish Stirling systems—we highlighted their respective thermal efficiencies, temperature capacities, land requirements, and application scopes. Recent research advancements in optical design, receiver materials, and heat transfer fluids are significantly enhancing the thermal efficiency and operational stability of CST systems. Especially, the development of high-absorptance, low-emissivity coatings and high-temperature-tolerant materials such as high-entropy alloys has elevated CST systems' ability to perform at temperatures exceeding 700 °C, opening new possibilities for energy-intensive processes.

Thermal energy storage (TES) technologies, such as molten salt storage, phase change materials, and thermochemical storage, have been shown to extend CST system performance during periods without solar radiation, allowing for continuous and dispatchable thermal energy delivery. This capability is crucial for ensuring the operational flexibility of CST systems in real industrial scenarios, especially where thermal loads are steady or require precise temperature control.

While the hybridization of CST with photovoltaic (PV) systems offers clear benefits for electricity dispatchability, it was not the core focus of this work. Our review intentionally centered on CST's role in heat production, not electrical power generation. As such, the discussion on PV-CSP hybrid systems—although important in the context of power systems—is outside the intended scope of this study, and should not constitute a key takeaway of the conclusion. Instead, future research and industrial implementation strategies should focus on enhancing CST's value as a sustainable heat supply solution, particularly in hard-to-electrify industrial sectors.

Looking forward, several challenges remain. These include high initial capital investment, integration complexity with existing heat infrastructure, and performance degradation in sub-optimal sunlight conditions. To overcome these, innovation should continue in areas such as smart solar tracking, AI-based operation control, durable receiver coatings, and cost-effective thermal storage materials. Meanwhile, policy incentives, industrial demonstration projects, and cost reduction through scale will play a decisive role in accelerating CST deployment in the thermal energy market.

In summary, CST systems are no longer limited to niche or experimental roles. Their ability to deliver reliable, high-grade heat makes them a promising and scalable solution for low-carbon industrial transformation. Continued advances in material science, thermal storage, and system integration will be key to realizing CST's full potential in achieving industrial decarbonization, thermal energy security, and sustainable development goals.

Author Contributions: Investigation, writing—original draft, C.W.; Formal analysis, Y.Z.; Data curation, W.L.; Investigation, J.F.; Methodology, H.X.; Resources, conceptualization, and supervision, Z.L.; Visualization, resources, and supervision, D.Y.; Project administration, supervision, and writing—review and editing, X.Z. All authors have read and agreed to the published version of the manuscript.

Funding: This research was supported by the research project of SDIC Power Holdings Corporation Limited (2024019).

Data Availability Statement: The data are available on reasonable request.

Conflicts of Interest: Authors Chunchao Wu, Yonghong Zhao, Wulin Li, and Jianjun Fan were employed by the Akesai Kazak Autonomous County Huidong New Energy Co., Ltd. and SDIC Gansu New Energy Co., Ltd. The remaining authors declare that the research was conducted in the absence of any commercial or financial relationships that could be construed as potential conflicts of interest.

References

1. Kabir, E.; Kumar, P.; Kumar, S.; Adelodun, A.A.; Kim, K.-H. Solar energy: Potential and future prospects. *Renew. Sustain. Energy Rev.* **2018**, *82*, 894–900. [CrossRef]
2. Bülow, M.; Pitz-Paal, R. Status and perspectives of concentrating solar technologies. In *Energy and Climate Change*; Elsevier: Amsterdam, The Netherlands, 2025; pp. 251–277.
3. Abdullah, M.; Corona, B.H.; Martins, M.; Ferber, N.L.; Armstrong, P.; Chiesa, M.; Calvet, N. Advancements, challenges, and opportunities in the measurement of high heat flux for concentrated solar thermal systems. *Sol. Energy* **2025**, *287*, 113252. [CrossRef]
4. Xu, B.; Ganesan, M.; Devi, R.K.; Ruan, X.; Chen, W.; Lin, C.C.; Chang, H.T.; Lizundia, E.; An, A.K.; Ravi, S.K. Hierarchically Promoted Light Harvesting and Management in Photothermal Solar Steam Generation. *Adv. Mater.* **2025**, *37*, 2406666. [CrossRef]
5. Zhao, S.-S.; Qiu, X.-L.; He, C.-Y.; Yu, D.-M.; Liu, G.; Gao, X.-H. Nanometer-thick high-entropy alloy nitride Al0. 4Hf0. 6NbTaTiZrN-based solar selective absorber coatings. *ACS Appl. Nano Mater.* **2021**, *4*, 4504–4512. [CrossRef]
6. Liu, Q.; Dai, Z.; Wei, Y.; Wang, D.; Xie, Y. Transformative Impacts of AI and Wireless Communication in CSP Heliostat Control Systems. *Energies* **2025**, *18*, 1069. [CrossRef]
7. Sarbu, I.; Dorca, A. Review on heat transfer analysis in thermal energy storage using latent heat storage systems and phase change materials. *Int. J. Energy Res.* **2019**, *43*, 29–64. [CrossRef]
8. Sarbu, I.; Sebarchievici, C. A comprehensive review of thermal energy storage. *Sustainability* **2018**, *10*, 191. [CrossRef]
9. Lovegrove, K.; Stein, W. *Concentrating Solar Power Technology: Principles, Developments and Applications*; Elsevier: Amsterdam, The Netherlands, 2012.
10. Mubarrat, M.; Mashfy, M.M.; Farhan, T.; Ehsan, M.M. Research advancement and potential prospects of thermal energy storage in concentrated solar power application. *Int. J. Thermofluids* **2023**, *20*, 100431. [CrossRef]
11. Liu, H.; Shuai, W.; Yao, Z.; Xuan, J.; Ni, M.; Xiao, G.; Xu, H. Optimization of solid oxide electrolysis cells using concentrated solar-thermal energy storage: A hybrid deep learning approach. *Appl. Energy* **2025**, *377*, 124610. [CrossRef]
12. Malwad, D.; Tungikar, V. Development and performance testing of reflector materials for concentrated solar power: A review. *Mater. Today Proc.* **2021**, *46*, 539–544. [CrossRef]

13. Viswanathan, M.; Nagendra, C.; Thutupalli, G. High efficiency front and rear surface silver reflectors. *Thin Solid Films* **1988**, *167*, 291–298. [CrossRef]
14. Goodyear, J.K.; Lindberg, V.L. Low absorption float glass for back surface solar reflectors. *Sol. Energy Mater.* **1980**, *3*, 57–67. [CrossRef]
15. Alder, F.A.; Charrault, E.; Zuber, K.; Fabretto, M.; Patil, A.; Murphy, P.; Llusca, M. Fabrication of robust solar mirrors on polymeric substrates by physical vapor deposition technique. *Sol. Energy Mater. Sol. Cells* **2020**, *209*, 110476. [CrossRef]
16. Hernández, P.; Cruz-Manjarrez, H.; Almanza, R. Multilayer reflective coating for solar energy concentrators. In *Proceedings of the ISES World Congress 2007 (Vol. I–Vol. V) Solar Energy and Human Settlement*; Springer: Berlin/Heidelberg, Germany, 2007; pp. 1929–1933.
17. Brogren, M.; Helgesson, A.; Karlsson, B.; Nilsson, J.; Roos, A. Optical properties, durability, and system aspects of a new aluminium-polymer-laminated steel reflector for solar concentrators. *Sol. Energy Mater. Sol. Cells* **2004**, *82*, 387–412. [CrossRef]
18. Tian, Y.; Zhao, C.-Y. A review of solar collectors and thermal energy storage in solar thermal applications. *Appl. Energy* **2013**, *104*, 538–553. [CrossRef]
19. Kalogirou, S.A. Solar thermal collectors and applications. *Prog. Energy Combust. Sci.* **2004**, *30*, 231–295. [CrossRef]
20. Daneshazarian, R.; Cuce, E.; Cuce, P.M.; Sher, F. Concentrating photovoltaic thermal (CPVT) collectors and systems: Theory, performance assessment and applications. *Renew. Sustain. Energy Rev.* **2018**, *81*, 473–492. [CrossRef]
21. Winston, R. Principles of solar concentrators of a novel design. *Sol. Energy* **1974**, *16*, 89–95. [CrossRef]
22. Bellos, E.; Tzivanidis, C. A review of concentrating solar thermal collectors with and without nanofluids. *J. Therm. Anal. Calorim.* **2018**, *135*, 763–786. [CrossRef]
23. Qin, L.; Furbo, S. Solar heating systems with evacuated tubular solar collector. In Proceedings of the EuroSun'98, Portoros, Slovenia, 14–17 September 1998.
24. Morrison, G.L. Solar collectors. In *Solar Energy*; Routledge: London, UK, 2013; pp. 163–240.
25. Wang, F.; Cheng, Z.; Tan, J.; Yuan, Y.; Shuai, Y.; Liu, L. Progress in concentrated solar power technology with parabolic trough collector system: A comprehensive review. *Renew. Sustain. Energy Rev.* **2017**, *79*, 1314–1328.
26. Vant-Hull, L.L. Central tower concentrating solar power systems. In *Concentrating Solar Power Technology*; Elsevier: Amsterdam, The Netherlands, 2021; pp. 267–310.
27. Nepveu, F.; Ferriere, A.; Bataille, F. Thermal model of a dish/Stirling systems. *Sol. Energy* **2009**, *83*, 81–89. [CrossRef]
28. Bellos, E. Progress in the design and the applications of linear Fresnel reflectors–A critical review. *Therm. Sci. Eng. Prog.* **2019**, *10*, 112–137. [CrossRef]
29. Wu, C.; Zhao, Y.; Li, W.; Fan, J.; Xu, H.; Yuan, D.; Ling, Z. Layered Operation Optimization Methods for Concentrated Solar Power (CSP) Technology and Multi-Energy Flow Coupling Systems. *Energies* **2024**, *17*, 6297. [CrossRef]
30. Barton, J.P.; Infield, D.G. Energy Storage and Its Use With Intermittent Renewable Energy. *IEEE Trans. Energy Convers.* **2004**, *19*, 441–448. [CrossRef]
31. Izquierdo, S.; Montañés, C.; Dopazo, C.; Fueyo, N. Analysis of CSP plants for the definition of energy policies: The influence on electricity cost of solar multiples, capacity factors and energy storage. *Energy Policy* **2010**, *38*, 6215–6221. [CrossRef]
32. Jorgenson, J.; Denholm, P.; Mehos, M. *Estimating the Value of Utility-Scale Solar Technologies in California Under a 40% Renewable Portfolio Standard*; Gold. CO: Calabasas, CA, USA, 2014.
33. Weinstein, L.A.; Loomis, J.; Bhatia, B.; Bierman, D.M.; Wang, E.N.; Chen, G. Concentrating Solar Power. *Chem. Rev.* **2015**, *115*, 12797–12838. [CrossRef] [PubMed]
34. Rafique, M.M.; Rehman, S.; Alhems, L.M. Recent Advancements in High-Temperature Solar Particle Receivers for Industrial Decarbonization. *Sustainability* **2024**, *16*, 103. [CrossRef]
35. Peng, Q.; Yang, X.; Ding, J.; Wei, X.; Yang, J. Design of new molten salt thermal energy storage material for solar thermal power plant. *Appl. Energy* **2013**, *112*, 682–689. [CrossRef]
36. Iniesta, A.C.; Diago, M.; Delclos, T.; Falcoz, Q.; Shamim, T.; Calvet, N. Gravity-fed Combined Solar Receiver/Storage System Using Sand Particles as Heat Collector, Heat Transfer and Thermal Energy Storage Media. *Energy Procedia* **2015**, *69*, 802–811. [CrossRef]
37. Bai, F.; Xu, C. Performance analysis of a two-stage thermal energy storage system using concrete and steam accumulator. *Appl. Therm. Eng.* **2011**, *31*, 2764–2771. [CrossRef]
38. Laing, D.; Steinmann, W.-D.; Tamme, R.; Richter, C. Solid media thermal storage for parabolic trough power plants. *Sol. Energy* **2006**, *80*, 1283–1289. [CrossRef]
39. Khare, S.; Dell'Amico, M.; Knight, C.; McGarry, S. Selection of materials for high temperature sensible energy storage. *Sol. Energy Mater. Sol. Cells* **2013**, *115*, 114–122. [CrossRef]
40. Wang, X.; Liu, J.; Zhang, Y.; Di, H.; Jiang, Y. Experimental research on a kind of novel high temperature phase change storage heater. *Energy Convers. Manag.* **2006**, *47*, 2211–2222. [CrossRef]

41. Petrasch, J.; Klausner, J. Integrated solar thermochemical cycles for energy storage and fuel production. *WIREs Energy Environ.* **2012**, *1*, 347–361. [CrossRef]
42. Felderhoff, M.; Bogdanovic, B. High temperature metal hydrides as heat storage materials for solar and related applications. *Int. J. Mol. Sci* **2009**, *10*, 325–344. [CrossRef]
43. Pardo, P.; Deydier, A.; Anxionnaz-Minvielle, Z.; Rougé, S.; Cabassud, M.; Cognet, P. A review on high temperature thermochemical heat energy storage. *Renew. Sustain. Energy Rev.* **2014**, *32*, 591–610. [CrossRef]
44. Goldsmid, H.J. A Review of: "Applications of Thermoelectricity". *J. Electron. Control* **1960**, *8*, 463. [CrossRef]
45. Islam, M.T.; Huda, N.; Abdullah, A.B.; Saidur, R. A comprehensive review of state-of-the-art concentrating solar power (CSP) technologies: Current status and research trends. *Renew. Sustain. Energy Rev.* **2018**, *91*, 987–1018. [CrossRef]
46. Fang, Y.; Zhu, Y.; Huang, R. Research on tower-type solar photothermal power generation technology. In *IOP Conference Series: Earth and Environmental Science*; IOP Publishing: Bristol, UK, 2021; p. 032028.
47. Kulkarni, K.; Havaldar, S.; Bhattacharya, N. Review on advance tubular receivers for central solar tower system. In *IOP Conference Series: Materials Science and Engineering*; IOP Publishing: Bristol, UK, 2021; p. 012018.
48. Zhang, H.L.; Baeyens, J.; Degrève, J.; Cacères, G. Concentrated solar power plants: Review and design methodology. *Renew. Sustain. Energy Rev.* **2013**, *22*, 466–481. [CrossRef]
49. Jebasingh, V.K.; Herbert, G.M.J. A review of solar parabolic trough collector. *Renew. Sustain. Energy Rev.* **2016**, *54*, 1085–1091. [CrossRef]
50. Wang, Z.; Horseman, T.; Straub, A.P.; Yip, N.Y.; Li, D.; Elimelech, M.; Lin, S. Pathways and challenges for efficient solar-thermal desalination. *Sci. Adv.* **2019**, *5*, eaax0763. [CrossRef] [PubMed]
51. Zhao, F.; Zhou, X.; Shi, Y.; Qian, X.; Alexander, M.; Zhao, X.; Mendez, S.; Yang, R.; Qu, L.; Yu, G. Highly efficient solar vapour generation via hierarchically nanostructured gels. *Nat. Nanotechnol.* **2018**, *13*, 489–495. [CrossRef] [PubMed]
52. Alaidaros, A.M.; AlZahrani, A.A. Thermal performance of parabolic trough integrated with thermal energy storage using carbon dioxide, molten salt, and oil. *J. Energy Storage* **2024**, *78*, 110084. [CrossRef]
53. Boretti, A.; Castelletto, S.; Al-Zubaidy, S. Concentrating solar power tower technology: Present status and outlook. *Nonlinear Eng.* **2019**, *8*, 10–31. [CrossRef]
54. Dowling, A.W.; Zheng, T.; Zavala, V.M. Economic assessment of concentrated solar power technologies: A review. *Renew. Sustain. Energy Rev.* **2017**, *72*, 1019–1032. [CrossRef]
55. Kiwan, S.; Khammash, A.L. Optical Performance of a Novel Two-Receiver Solar Central Tower System. *J. Sol. Energy Eng.* **2020**, *142*, 011005. [CrossRef]
56. Göttsche, J.; Hoffschmidt, B.; Schmitz, S.; Sauerborn, M.; Buck, R.; Teufel, E.; Badstübner, K.; Ifland, D.; Rebholz, C. Solar concentrating systems using small mirror arrays. *J. Sol. Energy Eng.* **2010**, *132*, 011003. [CrossRef]
57. Barlev, D.; Vidu, R.; Stroeve, P. Innovation in concentrated solar power. *Sol. Energy Mater. Sol. Cells* **2011**, *95*, 2703–2725. [CrossRef]
58. Mpodi, E.K.; Tjiparuro, Z.; Matsebe, O.; Namoshe, M. *A Review of Solar Tracking Mechanisms for Photovoltaic Systems*; Botswana International University of Science and Technology: Palapye, Botswana, 2017.
59. He, C.Y.; Li, Y.; Zhou, Z.H.; Liu, B.H.; Gao, X.H. High-Entropy Photothermal Materials. *Adv. Mater.* **2024**, *36*, 2400920. [CrossRef]
60. Kongtragool, B.; Wongwises, S. A review of solar-powered Stirling engines and low temperature differential Stirling engines. *Renew. Sustain. Energy Rev.* **2003**, *7*, 131–154. [CrossRef]
61. Mancini, T.; Heller, P.; Butler, B.; Osborn, B.; Schiel, W.; Goldberg, V.; Buck, R.; Diver, R.; Andraka, C.; Moreno, J. Dish-Stirling systems: An overview of development and status. *J. Sol. Energy Eng.* **2003**, *125*, 135–151. [CrossRef]
62. Meijer, R. The STM4-120 Stirling Engine for Solar Applications. In *Advances in Solar Energy Technology*; Elsevier: Amsterdam, The Netherlands, 1988; pp. 1390–1395.
63. Singh, U.R.; Kumar, A. Review on solar Stirling engine: Development and performance. *Therm. Sci. Eng. Prog.* **2018**, *8*, 244–256. [CrossRef]
64. Mari, M.A.; Memon, Z.A.; Shaikh, P.H.; Mirjat, N.H.; Soomro, M.I. A review study on mathematical modeling of solar parabolic d ish-Stirling system used for electricity generation. *Int. J. Energy Res.* **2021**, *45*, 18355–18391. [CrossRef]
65. Pheng, L.G.; Affandi, R.; Ab Ghani, M.R.; Gan, C.K.; Jano, Z.; Sutikno, T. A review of Parabolic Dish-Stirling Engine System based on concentrating solar power. *TELKOMNIKA (Telecommun. Comput. Electron. Control)* **2014**, *12*, 1142–1152. [CrossRef]
66. Zayed, M.E.; Zhao, J.; Elsheikh, A.H.; Li, W.; Abd Elaziz, M. Optimal design parameters and performance optimization of thermodynamically balanced dish/Stirling concentrated solar power system using multi-objective particle swarm optimization. *Appl. Therm. Eng.* **2020**, *178*, 115539. [CrossRef]
67. Zayed, M.E.; Zhao, J.; Elsheikh, A.H.; Li, W.; Sadek, S.; Aboelmaaref, M.M. A comprehensive review on Dish/Stirling concentrated solar power systems: Design, optical and geometrical analyses, thermal performance assessment, and applications. *J. Clean. Prod.* **2021**, *283*, 124664. [CrossRef]
68. Senthil, R.; Cheralathan, M. Enhancement of the thermal energy storage capacity of a parabolic dish concentrated solar receiver using phase change materials. *J. Energy Storage* **2019**, *25*, 100841. [CrossRef]

69. Baum, V.; Aparasi, R.; Garf, B. High-power solar installations. *Sol. Energy* **1957**, *1*, 6–12. [CrossRef]
70. Francia, G. Pilot plants of solar steam generating stations. *Sol. Energy* **1968**, *12*, 51–64. [CrossRef]
71. Feuermann, D.; Gordon, J. Analysis of a two-stage linear Fresnel reflector solar concentrator. *J. Sol. Energy Eng.* **1991**, *113*, 272–279. [CrossRef]
72. Grena, R.; Tarquini, P. Solar linear Fresnel collector using molten nitrates as heat transfer fluid. *Energy* **2011**, *36*, 1048–1056. [CrossRef]
73. Taramona, S.; González-Gómez, P.Á.; Briongos, J.V.; Gómez-Hernández, J. Designing a flat beam-down linear Fresnel reflector. *Renew. Energy* **2022**, *187*, 484–499. [CrossRef]
74. Grena, R. Geometrical Aspects of the Optics of Linear Fresnel Concentrators: A Review. *Energies* **2024**, *17*, 3564. [CrossRef]
75. Zhai, H.; Dai, Y.; Wu, J.; Wang, R.; Zhang, L. Experimental investigation and analysis on a concentrating solar collector using linear Fresnel lens. *Energy Convers. Manag.* **2010**, *51*, 48–55. [CrossRef]
76. Khan, M.; Alsaduni, I.N.; Alluhaidan, M.; Xia, W.-F.; Ibrahim, M. Evaluating the energy efficiency of a parabolic trough solar collector filled with a hybrid nanofluid by utilizing double fluid system and a novel corrugated absorber tube. *J. Taiwan Inst. Chem. Eng.* **2021**, *124*, 150–161. [CrossRef]
77. Alamdari, P.; Khatamifar, M.; Lin, W. Heat loss analysis review: Parabolic trough and linear Fresnel collectors. *Renew. Sustain. Energy Rev.* **2024**, *199*, 114497. [CrossRef]
78. Xu, Y.; Pei, J.; Yuan, J.; Zhao, G. Concentrated solar power: Technology, economy analysis, and policy implications in China. *Environ. Sci. Pollut. Res.* **2021**, *29*, 1324–1337. [CrossRef]
79. Du, E.; Zhang, N.; Hodge, B.-M.; Kang, C.; Kroposki, B.; Xia, Q. Economic justification of concentrating solar power in high renewable energy penetrated power systems. *Appl. Energy* **2018**, *222*, 649–661. [CrossRef]
80. Ummadisingu, A.; Soni, M.S. Concentrating solar power—Technology, potential and policy in India. *Renew. Sustain. Energy Rev.* **2011**, *15*, 5169–5175. [CrossRef]
81. Elmohlawy, A.E.; Ochkov, V.F.; Kazandzhan, B.I. Thermal performance analysis of a concentrated solar power system (CSP) integrated with natural gas combined cycle (NGCC) power plant. *Case Stud. Therm. Eng.* **2019**, *14*, 100458. [CrossRef]
82. Bansal, N.; Pany, P.; Singh, G. Visual degradation and performance evaluation of utility scale solar photovoltaic power plant in hot and dry climate in western India. *Case Stud. Therm. Eng.* **2021**, *26*, 101010. [CrossRef]
83. Benrezkallah, A.; Marif, Y.; Soudani, M.E.; Belhadj, M.M.; Hamidatou, T.; Mekhloufi, N.; Aouachir, A. Thermal performance evaluation of the parabolic trough solar collector using nanofluids: A case study in the desert of Algeria. *Case Stud. Therm. Eng.* **2024**, *60*, 104797. [CrossRef]
84. Abdelhady, S. Performance and cost evaluation of solar dish power plant: Sensitivity analysis of levelized cost of electricity (LCOE) and net present value (NPV). *Renew. Energy* **2021**, *168*, 332–342. [CrossRef]
85. Bellos, E.; Tzivanidis, C.; Papadopoulos, A. Optical and thermal analysis of a linear Fresnel reflector operating with thermal oil, molten salt and liquid sodium. *Appl. Therm. Eng.* **2018**, *133*, 70–80. [CrossRef]
86. Zhang, C.; Xu, Q.; Zhang, Y.; Arauzo, I.; Zou, C. Performance analysis of different arrangements of a new layout dish-Stirling system. *Energy Rep.* **2021**, *7*, 1798–1807. [CrossRef]
87. Kumar, K.R.; Chaitanya, N.V.K.; Kumar, N.S. Solar thermal energy technologies and its applications for process heating and power generation–A review. *J. Clean. Prod.* **2021**, *282*, 125296. [CrossRef]
88. Bellos, E.; Bousi, E.; Tzivanidis, C.; Pavlovic, S. Optical and thermal analysis of different cavity receiver designs for solar dish concentrators. *Energy Convers. Manag. X* **2019**, *2*, 100013. [CrossRef]
89. Modi, A.; Bühler, F.; Andreasen, J.G.; Haglind, F. A review of solar energy based heat and power generation systems. *Renew. Sustain. Energy Rev.* **2017**, *67*, 1047–1064. [CrossRef]
90. Aboelwafa, O.; Fateen, S.-E.K.; Soliman, A.; Ismail, I.M. A review on solar Rankine cycles: Working fluids, applications, and cycle modifications. *Renew. Sustain. Energy Rev.* **2018**, *82*, 868–885. [CrossRef]
91. Schnitzer, H.; Brunner, C.; Gwehenberger, G. Minimizing greenhouse gas emissions through the application of solar thermal energy in industrial processes. *J. Clean. Prod.* **2007**, *15*, 1271–1286. [CrossRef]
92. Farjana, S.H.; Huda, N.; Mahmud, M.A.P.; Saidur, R. Solar process heat in industrial systems—A global review. *Renew. Sustain. Energy Rev.* **2018**, *82*, 2270–2286. [CrossRef]
93. Ramaiah, R.; Shekar, K.S. Solar thermal energy utilization for medium temperature industrial process heat applications—A review. In *IOP Conference Series: Materials Science and Engineering*; IOP Publishing: Bristol, UK, 2018; p. 012035.
94. Kalogirou, S. The potential of solar industrial process heat applications. *Appl. Energy* **2003**, *76*, 337–361. [CrossRef]
95. Kumar, L.; Hasanuzzaman, M.; Rahim, N. Global advancement of solar thermal energy technologies for industrial process heat and its future prospects: A review. *Energy Convers. Manag.* **2019**, *195*, 885–908. [CrossRef]
96. Kylili, A.; Fokaides, P.A.; Ioannides, A.; Kalogirou, S. Environmental assessment of solar thermal systems for the industrial sector. *J. Clean. Prod.* **2018**, *176*, 99–109. [CrossRef]

97. Koçak, B.; Fernandez, A.I.; Paksoy, H. Review on sensible thermal energy storage for industrial solar applications and sustainability aspects. *Sol. Energy* **2020**, *209*, 135–169. [CrossRef]
98. Atkins, M.J.; Walmsley, M.R.; Morrison, A.S. Integration of solar thermal for improved energy efficiency in low-temperature-pinch industrial processes. *Energy* **2010**, *35*, 1867–1873. [CrossRef]
99. Gajendiran, M.; Nallusamy, N. Application of Solar Thermal Energy Storage for Industrial Process Heating. *Adv. Mater. Res.* **2014**, *984–985*, 725–729. [CrossRef]
100. Malinowski, M.; Leon, J.I.; Abu-Rub, H. Solar Photovoltaic and Thermal Energy Systems: Current Technology and Future Trends. *Proc. IEEE* **2017**, *105*, 2132–2146. [CrossRef]
101. Mekhilef, S.; Saidur, R.; Safari, A. A review on solar energy use in industries. *Renew. Sustain. Energy Rev.* **2011**, *15*, 1777–1790. [CrossRef]
102. Phani Kumar, K.K.; Mallick, S.; Sakthivel, S. Nanoparticles based single and tandem stable solar selective absorber coatings with wide angular solar absorptance. *Sol. Energy Mater. Sol. Cells* **2022**, *242*, 111758. [CrossRef]

energies

MDPI

Article

Day-Ahead Scheduling of IES Containing Solar Thermal Power Generation Based on CNN-MI-BILSTM Considering Source-Load Uncertainty

Kun Ding [1], Yalu Sun [1], Boyang Chen [2], Jing Chen [3,*], Lixia Sun [3], Yingjun Wu [3] and Yusheng Xia [3]

1 Economic and Technological Research Institute of State Grid Gansu Electric Power Company, Lanzhou 730030, China
2 State Grid Gansu Electric Power Company, Lanzhou 730030, China
3 School of Electrical and Power Engineering, Hohai University, Nanjing 211100, China; leexiasun@hhu.edu.cn (L.S.); ywu_ee@vip.163.com (Y.W.); x13952772074@163.com (Y.X.)
* Correspondence: chenjingsoo@163.com

Abstract: The fluctuating uncertainty of load demand as an influencing factor for day-ahead scheduling of an integrated energy system with photovoltaic (PV) power generation may cause an imbalance between supply and demand, and to solve this problem, this paper proposes a day-ahead optimal scheduling model considering uncertain loads and electric heating appliance (EH)–PV energy storage. The model fuses the multi-interval uncertainty set with the CNN-MI-BILSTM neural network prediction technique, which significantly improves the accuracy and reliability of load prediction and overcomes the limitations of traditional methods in dealing with load volatility. By integrating the EH–photothermal storage module, the model achieves efficient coupled power generation and thermal storage operation, aiming to optimize economic targets while enhancing the grid's peak-shaving and valley-filling capabilities and utilization of renewable energy. The validity of the proposed model is verified by algorithm prediction simulation and day-ahead scheduling experiments under different configurations.

Keywords: load demand; uncertainty handling; photovoltaic power plant; EH

1. Introduction

Concentrated Solar Power (CSP), as an emerging renewable energy technology, utilizes solar energy resources and converts them into electricity through heat collection and storage devices, which has the advantages of environmental protection, high efficiency, and sustainability. However, the output of photovoltaic power generation is affected by a variety of factors, such as weather and the environment. It is characterized by a certain degree of uncertainty, so optimizing day-ahead scheduling has become the key to ensuring the economical and safe operation of new energy sources, such as photovoltaic power, in the grid [1].

At present, day-ahead scheduling considering new energy has become a research hotspot. In terms of scheduling objects, the research mainly focuses on the coordinated scheduling of traditional energy and new energy, or it focuses on new energy scheduling. Reference [2] proposed a two-phase robust optimal scheduling model for a pumped storage–wind–photovoltaic–thermal co-generation system, which realizes the optimal allocation of a power generation plan through the optimization of multiple energy sources and suppresses power fluctuations in the power grid. Reference [3] developed a model for a wind–photovoltaic–hydroelectric–thermal pumped storage system and established a

 https://doi.org/10.3390/en18092160

short-term optimal scheduling model with multiple optimization objectives to improve the economic stability of the system. Reference [4] explored the integration of Concentrated Solar Power (CSP) with other renewable sources to enhance system stability and economic performance.

In terms of scheduling objectives, studies usually cover economic, technical, and low-carbon dimensions, among others. Reference [5] integrated security and economic objectives and proposed a two-stage scheduling framework which realizes the double balance of economic benefits and security synergy. Reference [6] proposed a low-carbon dispatch model for the stochastic nature of renewable energy and a low-carbon modulation mechanism, which enhanced the economic efficiency of the power system under low-carbon-emission conditions. Reference [7] proposed that enhanced sustainability assessment and system flexibility are essential for achieving the efficient and economic operation of integrated energy systems and accurate matching of supply and demand.

In terms of uncertainty, scheduling models can be categorized into two main types: deterministic and uncertain. Reference [8] effectively reflected on the uncertainty of renewable energy by constructing a robust uncertainty set. Reference [9] proposed an optimal scheduling strategy based on a deep reinforcement learning algorithm, which solves the challenge of high-dimensional decision spaces. Reference [10] proposed to construct a two-stage robust optimization model of producers and consumers for the multiple uncertainties of wind power generation and load demand in order to promote the effective use and management of sustainable energy.

Load fluctuation is a key factor to be considered in the day-ahead scheduling of photovoltaic-containing power systems. Reference [11] utilized the WWO algorithm to improve the efficiency of dynamic optimal scheduling in microgrids. Reference [12] proposed a probabilistic power future prediction tool based on time-series clustering, which is used in short-term load forecasting to make forecasts more accurate and intelligent. However, existing studies often ignore the uncertainty of load fluctuations, which may lead to scheduling plan failures. Therefore, the uncertainty of load fluctuation needs to be fully considered in the scheduling of complex integrated energy systems containing photovoltaic power generation [13].

Forecasting technology is an effective means to solve the uncertainty of load fluctuation. Reference [14] investigated the proposed adaptive load forecasting model by customizing AI algorithms and cloud-side collaboration to analyze the accuracy and resource usage, adapting different hardware environments to meet the specific needs of microgrids. Reference [15] proposed a short-term load forecasting model based on EEMD-LN-GRU according to the uncertainty and nonlinear characteristics of electric load. Reference [16] improved the accuracy of capturing short-term changes in electric load through the improved Convolutional Neural Network–Bidirectional Long- and Short-Term Memory (CNN-BILSTM) model. In addition, reference [17] fused mutual information (MI) and BILSTM to dynamically evaluate the importance of features to fit the load fluctuation characteristics. Reference [18] proposed a robust optimization-based scheduling framework to deal with the dual uncertainties of loads and renewable energy sources through interval uncertainty sets. Given this, this paper synthesizes multiple forecasting algorithms in the CNN-MI-BILSTM model, adopts multi-interval uncertainty sets to portray load variability, and improves forecasting accuracy with the help of historical data.

To address the problem that existing studies neglect the impact of load uncertainty on the scheduling of solar thermal storage systems, reference [19] developed a day-ahead co-scheduling method for concentrating solar photovoltaic–wind power that takes into account the uncertainty of source loads, characterizes the uncertainty of intraday source loads with a trapezoidal fuzzy number equivalence model, and carries out the day-ahead optimal dispatch based on the set of day-ahead wind-power output prediction combination scenarios. Reference [20] took into consideration the uncertainty of source loads for electricity–heat conversion and constructed a stochastic optimal dispatch model for a wind–solar complementary fire system with the objective of minimizing the comprehensive operating cost of the combined system. Such models not only optimize the economic objective but also incorporate robustness and stochastic approaches. Reference [21] analyzed the effectiveness of the combined peaking of TPU and CSP power plants with EH and analyzed the principle of a low-carbon power supply for the proposed strategy during peak and off-peak periods. Therefore, it has become necessary to integrate uncertainty-handling methods in the day-ahead scheduling of integrated energy systems containing EH-CSP modules.

To address the challenges posed by uncertain loads in the context of a new energy grid-connected environment, this paper proposes a day-ahead optimal scheduling model that incorporates uncertain loads and EH-CSP (Energy Hub–Concentrated Solar Power) storage. The objective is to achieve supply–demand balance with optimal economic efficiency. This study specifically tackles the limitations of traditional scheduling models in dealing with the volatility of renewable energy sources and the complex coupling between power and heat systems. By accurately simulating load uncertainties, the model closely aligns with real-world operating conditions and provides an in-depth analysis of their impact on the day-ahead scheduling of renewable energy hybrid plants, as well as the operational mechanisms of the EH-CSP system. This improves both the accuracy of scheduling decisions and the overall stability of the system. The main contributions of this paper are as follows:

(1) An optimal day-ahead scheduling model is developed, considering uncertain loads and EH-CSP storage, with the goal of achieving economic efficiency, grid stability, and operational reliability. Through refined modeling and optimization techniques, the model enhances the overall economic performance of the system and the robustness of grid operations.

(2) A multi-interval uncertainty set and a CNN-MI-BILSTM model are introduced to improve the accuracy of load forecasting. By overcoming the limitations of traditional models in handling data correlations, the approach significantly enhances prediction reliability and enables rapid, precise forecasting of fluctuating loads.

(3) A modular modeling approach for EH–photovoltaic power plants is proposed, along with a coupled operation model of power generation and thermal storage modules. This captures the dynamic processes of energy conversion and storage, accounts for environmental impacts on system efficiency, and improves the model's generality and accuracy in reflecting the operational characteristics of various field stations in a renewable energy co-generation system.

The rest of the article is structured as follows: the second part elaborates the research problem, the third part describes the establishment of the variable-load model for the improved day-ahead scheduling model in the description of the prediction method that constantly improves the data relevance as well as the prediction accuracy, the fourth part proposes a model for the EH–photovoltaic and thermal co-generation system corresponding to the power generation–storage operation relationship, the fifth part presents the solution method of the optimization model, the sixth part provides an analysis through examples, and the seventh part concludes the entire paper with an overview.

2. Problem Formulation

2.1. Day-Ahead Scheduling Model for Integrated Energy Systems

The integrated energy system in this section covers wind, photovoltaic, and solar thermal power generation, and its day-ahead scheduling model aims to minimize the operating cost of the integrated daily plan, which is expressed as follows (Please refer to Appendix A for detailed module-by-module costs)

$$\min: C = C_1 + C_2 + C_3 + C_4 \tag{1}$$

In the formula, C is the daily average cost; C_1 represents the daily operating cost; C_2 represents the start–stop cost of wind, PV, and solar thermal power generation; C_3 represents the equipment maintenance cost; and C_4 is the penalty for wind and light abandonment.

The constraints include generation constraints for each unit, creep rate constraints, start- and stop-state constraints, line constraints, and power-balance constraints:

$$0 \leq P_w(t) \leq P_w^N \tag{2}$$

$$R_w^{down} \leq P_w(t) - P_w(t-1) \leq R_w^{up} \tag{3}$$

$$0 \leq P_{\text{p}}(t) \leq P_p^N \tag{4}$$

$$R_p^{down} \leq P_{\text{p}}(t) - P_{\text{p}}(t-1) \leq R_p^{up} \tag{5}$$

$$0 \leq P_{\text{pb}}(t) \leq P_{pb}^{\max} \mu_{PB,t}^{su} \tag{6}$$

$$R_{pb}^{down} \leq P_{\text{pb}}(t) - P_{\text{pb}}(t-1) \leq R_{pb}^{up} \tag{7}$$

$$0 \leq \mu_t^{su} + \mu_t^{sd} \leq 1 \tag{8}$$

$$0 \leq P_{g,t} \leq P_{g,\max} \tag{9}$$

$$P_w(t) + P_{\text{p}}(t) + P_{pb}(t) - P_{load} \geq 0 \tag{10}$$

In the formulas, P_w^N is the rated capacity of wind turbines; R_w^{down}, R_p^{down}, R_{pb}^{down}, R_w^{up}, R_p^{up}, and R_{pb}^{up} are the upper and lower limits of the creep rate of the wind, photovoltaic, and solar thermal power generation system; $P_{pb}^{\max}$ is the maximum output power of the solar thermal power generation system; μ_t^{su} and μ_t^{sd} are the start-up and release efficiencies of the solar thermal power generation, respectively; $P_{g,\max}$ is the maximum allowable power of the liaison line; and P_{load} is the power of the load demand. Remark: The current model overlooks load uncertainty, renewable energy consumption efficiency, and CSP storage economics. Ignoring these factors may result in power imbalances due to mismatched generation and demand. Integrating EH-CSP joint optimization into day-ahead scheduling enables coordinated energy utilization, enhances wind and solar absorption, and improves grid reliability through thermal storage's peak-shaving capability. Thus, incorporating load uncertainty and EH-CSP optimization is essential for secure and economic system operation.

2.2. Day-Ahead Optimal Scheduling Model Considering Uncertain Loads and EH-CSP

Thermal energy storage (TES) can store the excess heat of the concentrating heat collection system when the load demand is low and release energy through thermoelectric conversion when the load demand is high to improve the system's operation and energy utilization efficiency. Considering the CSP storage, Formula (1) can be improved as follows:

$$C' = C_1' + C_2' + C_3' + C_4 \tag{11}$$

$$C_1' = \sum_{t=1}^{T} \left[c_w P_w(t) + c_p P_p(t) + c_{pb} P_{pb}(t) + c_{tes} P_{tes}(t) \right] \tag{12}$$

$$C_2' = C_2 + \sum_{t=1}^{T} [SU_{TES}\mu_{TES,t}^{su} + SD_{TES}\mu_{TES,t}^{sd}] \quad (13)$$

$$C_3' = \sum_{t=1}^{T} [OM_{WT}P_w(t) + OM_{PV}P_p(t) + CM_{PB}P_{pb}(t) + OM_{TES}H_{tes}(t) + OM_{EH}E_{EH}(t)] \quad (14)$$

$$P_{tes}(t) = H_{tes}(t)\eta_2 \quad (15)$$

where c_{tes} represents the unit operating cost of the TES module; $P_{tes}(t)$ is the thermal power for charging/discharging the TES at time t; SU_{TES} is the start-up cost for thermal-to-electric conversion; SD_{TES} is the start-up cost for charging the TES; $\mu_{TES,t}^{su}$ = 1 and $\mu_{TES,t}^{sd}$ = 1, respectively, indicate the discharge operation and the start-up cost of discharge for the TES module at time t; OM_{TES} is the maintenance cost coefficient for the TES system; $H_{tes}(t)$ represents the charge/discharge thermal energy of the TES system at time t; and η_2 is the thermal-to-electric conversion efficiency of the solar thermal storage system.

The load in Formula (10) is a deterministic expression, while, in essence, the load is a random fluctuation state with substantial uncertainty, and considering the uncertain load is conducive to optimizing the day-ahead scheduling and reducing the day-ahead operating cost. Considering the TES module at the same time, Formula (10) is improved:

$$P_w(t) + P_p(t) + P_{pb}(t) + P_{tes}(t) - P_{load}^{un}(t) > 0 \quad (16)$$

where $P_{tes}(t)$ is the conversion power of the TES module at the corresponding time and $P_{load}^{un}(t)$ is the uncertain load power.

The addition of the energy storage constraint to the constraints, as well as the improvement of the supply–demand coordination demand constraints and the EH operation constraints by taking the load uncertainty into account, is written as follows:

$$H_{tes}(t) = \mu_{tes}^{ch}H_{tes}^{ch}(t) + \mu_{tes}^{dis}H_{tes}^{dis}(t) + E_{EH}(t) \quad (17)$$

$$0 \le H_{tes}^{ch}(t) \le H_{tes}^{ch,\max}\mu_{tes}^{ch} \quad (18)$$

$$0 \le H_{tes}^{dis}(t) \le H_{tes}^{dis,\max}\mu_{tes}^{dis} \quad (19)$$

$$0 \le u_{tes}^{ch} + \mu_{tes}^{dis} \le 1 \quad (20)$$

$$H_{csp} = H_{tes}^{ch}(t) + H_t^{pb} \quad (21)$$

$$E_{EH}(t) = P_w(t) + P_p(t) + P_{pb}(t) - P_{load}^{un}(t) \quad (22)$$

$$0 \le E_{EH}(t) \le E_{EH}^{\max} \quad (23)$$

where $H_{tes}(t)$ is the charging/discharging heat of the TES system at time t; $H_{tes}^{ch}(t)$ and $H_{tes}^{dis}(t)$ are the charging and discharging heats of the TES μ_{tes}^{ch} is a binary variable; $\mu_{tes}^{ch} = 1$ means the heat storage unit for heat storage; μ_{tes}^{dis} is a binary variable; $\mu_{tes}^{dis} = 1$ represents the heat storage unit for exothermic power; $P_{pb}(t)$ is the output power of the solar thermal power generation at time t; $H_{tes}^{ch,\max}$ and $H_{tes}^{dis,\max}$ are the thermal units of the maximum heat storage and exothermic power, respectively; H_{csp} is the total heat absorbed by the concentrating light collector system; and $E_{EH}^{\max}$ is the maximum value of the electric heating device at the time of the conversion.

According to the above-improved model, the optimal scheduling model considering load uncertainty with day-ahead CSP storage is expressed as follows:

$$\min: f(11) \quad (24)$$

The constraints are as follows:

$$f\ (2)\text{–}(10)$$

$$f\ (17)\text{–}(23)$$

Remark 1. *The uncertain load power referred to in Formula (16) of the day-ahead optimal dispatch model concerning the supply–demand balance will be modeled in Section 3.*

Remark 2. *The outputs of the power generation module, $P_{pb}(t)$, and the heat storage module, $P_{tes}(t)$, of the photovoltaic power plant are described in Formula (16) in terms of power balance. In this paper, EH is introduced to improve the capacity of wind and solar energy consumption. The energy of the heat storage module in the EH-CSP power plant is not only derived from the concentrating heat and power collection system but also from the energy converted from the abandoned wind and light. Therefore, Section 4 will take the concentrating solar collector system as well as the EH output as the pivot to establish the relationship model of $P_{pb}(t)$ and $P_{tes}(t)$.*

3. Load-Power Uncertainty Model Based on CNN-MI-BILSTM

Uncertainty in pure electricity loads affects the balance between supply and demand. If the planned output for each day far exceeds the load demand, there will be an economic loss. If it is not possible to cope with changes in load, the system becomes unstable.

3.1. Uncertainty-Considering Load-Power Model Based on Baseload Prediction

In this section, a multi-interval uncertainty modeling approach is used to construct the load-power uncertainty set.

Multi-interval uncertainty modeling describes the predicted load power by incorporating a probability distribution. Firstly, the predicted power range $\left[P_{load}^{pre}(t) - \omega_{\partial,t}^{-}, P_{load}^{pre}(t) + \omega_{\partial,t}^{+}\right]$ is divided into N_b intervals, and the sum of the corresponding times of the N_b intervals is Π_{∂}^{ω}, and the corresponding time of each interval, $\Pi_{\partial}^{b,\omega}$, depends on the deviation ratio, $\omega_{\partial,t}^{b+} + \omega_{\partial,t}^{b-} / \omega_{\partial,t}^{+} + \omega_{\partial,t}^{-}$, and the probability distribution, ρ_{∂}^{b}; the multi-interval probability distribution is plotted as Figure 1.

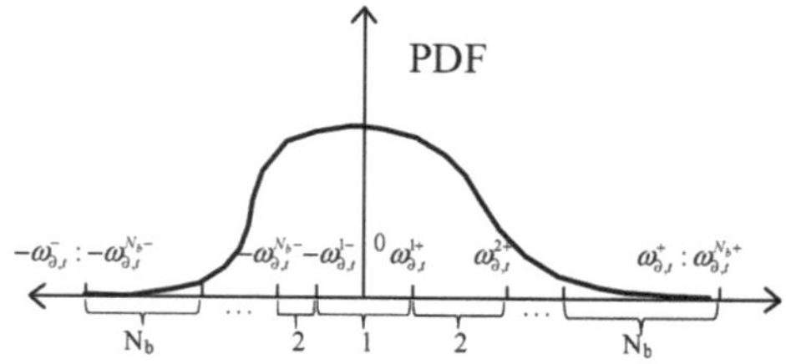

Figure 1. Multi-interval probability distributions for uncertain loads.

$$\sum_{b=1}^{N_b} (\Pi_{\partial}^{b,\omega} \cdot \frac{\omega_{\partial,t}^{b+} + \omega_{\partial,t}^{b-}}{\omega_{\partial,t}^{+} + \omega_{\partial,t}^{-}}) = \Pi_{\partial}^{\omega} \tag{25}$$

$$\frac{\Pi_{\partial}^{b,\omega}}{\Pi_{\partial}^{b+1,\omega}} = \frac{\rho_{\partial}^{b}}{\rho_{\partial}^{b+1}} \tag{26}$$

Formulas (25) and (26) enable the division of a single interval, $\left[P_{load}^{pre}(t) - \omega_{\partial,t}^{-}, P_{load}^{pre}(t) + \omega_{\partial,t}^{+}\right]$, into upper and lower error limits and probability distributions. Then, the uncertainty load model is expressed as follows:

$$P_{load}^{un}(t) = P_{load}^{pre}(t) + \sum_{b=1}^{N_b} (\omega_{\partial,t}^{b+}\varepsilon_{\partial,t}^{b+} - \omega_{\partial,t}^{b-}\varepsilon_{\partial,t}^{b-}) \tag{27}$$

$$0 \leq \sum_{b=1}^{N_b} (\varepsilon_{\partial,t}^{b+} + \varepsilon_{\partial,t}^{b-}) \leq 1 \tag{28}$$

$$\sum_{t=1}^{N_t} (\varepsilon_{\partial,t}^{b+} + \varepsilon_{\partial,t}^{b-}) \leq \Pi_{\partial}^{b,\omega} \tag{29}$$

$$-\Lambda_{\partial} \leq \frac{\sum_{t=1}^{N_t} \sum_{b=1}^{N_b} (\omega_{\partial,t}^{b+}\varepsilon_{\partial,t}^{b+} - \omega_{\partial,t}^{b-}\varepsilon_{\partial,t}^{b-})}{\sum_{t=1}^{N_t} P_{load}^{pre}(t)} \leq \Lambda_{\partial} \tag{30}$$

where $P_{load}^{pre}(t)$ is the load prediction value; $\omega_{\partial,t}^{b+}$ and $\omega_{\partial,t}^{b-}$ are the upper and lower error limits corresponding to the period b; the introduced variables $\varepsilon_{\partial,t}^{b+}$ and $\varepsilon_{\partial,t}^{b-}$ limit the uncertain load power within the range $\left[P_{load}^{pre}(t) - \omega_{\partial,t}^{-}, P_{load}^{pre}(t) + \omega_{\partial,t}^{+}\right]$; Formula (28) considers that only one deviation will occur at any period from a spatial point of view; Formula (29) ensures that the total uncertainty period for each interval, N_t, does not exceed, from a temporal point of view, $\Pi_{\partial}^{b,\omega}$; and the constraint (30) controls the range of deviation of the actual output from the predicted value, $[-\Lambda_{\partial}, \Lambda_{\partial}]$.

The load uncertainty model described above is based on load predictions, $P_{load}^{pre}(t)$. Therefore, the load prediction model is given in Sections 3.2–3.4

3.2. CNN-LSTM-Based Baseload Prediction Models

Considering the complexity and non-smoothness of the load data, a CNN-LSTM network short-term load prediction method is proposed to accomplish the prediction of the load, $P_{load}^{pre}(t)$.

Figure 2 illustrates the prediction flow of the CNN-LSTM network.

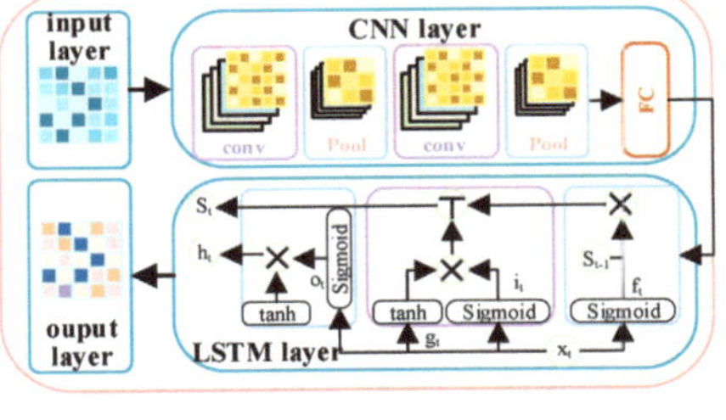

Figure 2. Prediction schematic for CNN-LSTM networks.

The modeling of each module of the CNN-LSTM network is explained below.

3.2.1. Input Layer

The input layer realizes the transfer of load data to the CNN layer.

3.2.2. CNN Layer

The CNN layer mainly includes the convolutional layer and the pooling layer, which uses the convolutional layer to realize the extraction of static features from the input data through the sliding-window operation of the convolutional kernel and then reduces the dimensionality of the extracted features by using the scale invariance of the key features through the pooling layer, which makes the key features further highlighted and reduces the complexity of the network through parameter sharing. Usually, the key features are extracted from the input data by convolutional and pooling layers, and the dimensionality of the features can be reduced.

The process of the CNN layer is represented as follows:

$$C_1 = \mathrm{ReL}U(X \otimes W_1 + b_1) \tag{31}$$

$$P_1 = \max(C_1) + b_2 \tag{32}$$

$$C_2 = \mathrm{ReL}U(X \otimes W_2 + b_3) \tag{33}$$

$$P_2 = \max(C_2) + b_4 \tag{34}$$

$$X_C = Sigmoid(P_2 \times W_3 + b_5) \tag{35}$$

where the outputs of convolutional layer 1 and convolutional layer 2 and pooling layer 1 and pooling layer 2 are C_1 and C_2 and P_1 and P_2, respectively; W_1, W_2, and W_3 are the weight matrices of the CNN layers, b_1, b_2, b_3, b_4, and b_5 are the deviation values; $\otimes$ is the convolution operation; and the activation function *Sigmoid* is chosen for the fully connected layer. The output feature vector of the CNN layer is denoted as X_C, and the length of the output of the CNN layer is denoted as $X_C = [x_1 \cdots x_{i-1}, x_i \cdots, x_t]$.

3.2.3. LSTM Layer

The basic unit model of the LSTM network is as follows:

$$f_t = \sigma(W_{fx}x_t + W_{fh}h_{t-1} + b_f) \tag{36}$$

$$i_t = \sigma(W_{ix}x_t + W_{ih}h_{t-1} + b_i) \tag{37}$$

$$g_t = \phi(W_{gx}x_t + W_{gh}h_{t-1} + b_g) \tag{38}$$

$$S_t = g_t \odot i_t + S_{t-1} \odot f_t \tag{39}$$

$$h_t = \phi(S_t) \odot o_t \tag{40}$$

$$o_t = Sigmoid(W_{ox}x_t + W_{oh}h_{t-1} + b_o) \tag{41}$$

where x_t is the forgetting gate input; h_{t-1} is the intermediate output; S_{t-1} is a status memory unit; f_t, i_t, g_t, h_t, S_t, and o_t are the states of the forgetting gate, the input gate, the input node, the intermediate output, and the state unit output gate; W is the matrix weight of the corresponding gate multiplied by the input, x_t, and the intermediate output, h_{t-1}; b_f, b_i, b_g, and b_o are the offset terms of the corresponding doors; $\odot$ represents the bitwise multiplication of vector elements; and ϕ indicates the change in the tanh function.

3.2.4. Output Layer

Take the output data of the LSTM-layer output gate as the result of the load prediction data:

$$P_{load}^{pre}(t) = o_t \tag{42}$$

Remark 3. *The hybrid model short-term load forecasting method based on the CNN-LSTM network can effectively extract the potential relationship between continuous data and discontinuous data in the feature map to form the feature vector and fully mine the internal correlation between time-series data. However, the output feature vector, X_C, of the CNN layer lacks consideration of data correlation, which will cause prediction deviation, so MI is introduced to improve the data correlation.*

3.3. MI Reconstruction Method of Input-Layer Data Considering Data Relevance

To solve the problem of prediction deviation, this section introduces eigenvalues to represent the importance value of the input characteristics and constructs an improved load data input correction model.

The characteristic matrix, X_C', of the new input LSTM network is obtained by introducing the eigenvalue vector, M, as follows:

$$X_C' = X_C \cdot M \tag{43}$$

$$X_C = \begin{bmatrix} x_{1,1} & \cdots & x_{i,1} & \cdots & x_{t,1} \\ x_{1,2} & \cdots & x_{i,1} & \cdots & x_{t,2} \\ \vdots & & \vdots & & \vdots \\ x_{1,z} & \cdots & x_{i,z} & \cdots & x_{t,z} \end{bmatrix} \tag{44}$$

$$M = \begin{bmatrix} M(x_{1,1}, o_1) & \cdots & M(x_{i,1}, o_i) & \cdots & M(x_{t,1}, o_t) \\ M(x_{1,2}, o_1) & \cdots & M(x_{i,2}, o_i) & \cdots & M(x_{t,2}, o_t) \\ \vdots & & \vdots & & \vdots \\ M(x_{1,z}, o_1) & \cdots & M(x_{i,z}, o_i) & \cdots & M(x_{t,z}, o_t) \end{bmatrix} \tag{45}$$

where X_C is the input characteristic matrix corresponding to the CNN output load data set obtained based on (35) and M is the time-varying importance value fluctuation matrix obtained by normalizing the output or input characteristic matrix, X_C, of the CNN layer based on MI. It contains important information on input features under different dimensions. The stronger the correlation between the two variables, the greater the M value. When the two variables are independent of each other, the M value is 0. o_t is the output load data obtained at time t based on the input characteristic matrix, X_C.

Remark 4. *The critical value extraction of the CNN-layer output data can accurately predict the data fluctuation, but the training speed of the MI-LSTM method is slow. The load sequence has the characteristics of correlation and strong randomness. The LSTM neural network can only extract and encode the sequence information in one direction. It cannot learn the forward and reverse information rules of the load data. The BI method neural network is proposed to consider the forward and backward sequences at the same time to solve the problem of data dependence.*

3.4. Bidirectional BILSTM Improvement Considering Prediction Accuracy

Section 3.2. The LSTM model is used to learn the output characteristic data of the CNN layer. However, the prediction accuracy of this method is not high, and the training speed is slow. Therefore, this section uses the BILSTM with two-way time information to bidirectionally mine the internal relationship between long-time data to improve the training speed and prediction accuracy. The specific structure is shown in Figure 3.

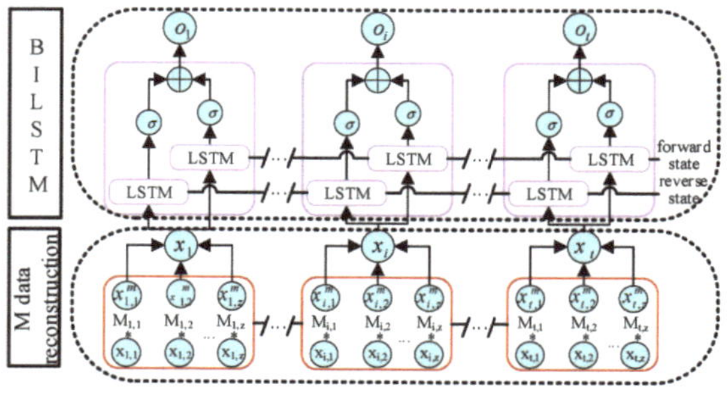

Figure 3. Structure of the prediction model of MI-BILSTM.

X'_C is the input data of the new LSTM network, and $X'_C = [x'_1 \cdots x'_{i-1}, x'_i \cdots, x'_t]$ defines a new network final output power prediction value, O_t, as follows:

$$f_t = \sigma(W_{fx}x'_t + W_{fh}h_{t-1} + b_f) \tag{46}$$

$$i_t = \sigma(W_{ix}x'_t + W_{ih}h_{t-1} + b_i) \tag{47}$$

$$\widetilde{g}_t = \phi(W_{gx}x'_t + W_{gh}h_{t-1} + b_g) \tag{48}$$

$$\widetilde{S}_t = \widetilde{g}_t \odot i_t + S_{t-1} \odot f_t \tag{49}$$

$$h_t = \phi(\widetilde{S}_t) \odot o_t \tag{50}$$

$$o_t = Sigmoid(W_{ox}x'_t + W_{oh}h_{t-1} + b_o) \tag{51}$$

$$\overrightarrow{o_t} = \overrightarrow{LSTM}(o_{t-1}, x'_t, c_{t-1}) \tag{52}$$

$$\overleftarrow{o_t} = \overleftarrow{LSTM}(o_{t+1}, x'_t, c_{t+1}) \tag{53}$$

In the formulas, $\overrightarrow{o_t}$ is the forward LSTM output predictive value calculated by inputting the input data, X'_C, of Formula (43) into the LSTM neural network (46)–(51); $\overleftarrow{o_t}$ is the forward LSTM input predictive value calculated according to the calculation of the X'_C reverse input LSTM neural network (45)–(51); a_t and b_t represent the forward and backward output weights, respectively; and c_t is the offset optimization parameter.

The model realizes LSTM training in the forward and reverse directions of X'_C and effectively improves the comprehensiveness and integrity of feature selection. The forward LSTM-layer output, $\overrightarrow{o_t}$, is connected to the backward LSTM-layer output, $\overleftarrow{o_t}$, and the final power prediction output value, o_t, is obtained through weighted fusion.

The structure of the prediction model of CNN-MI-BILSTM is shown above.

The algorithm of load prediction in Sections 3.2–3.4 was improved, and the results were input to Section 3.1, which accurately described the uncertain load according to the multi-interval uncertainty set.

4. Consider the Coupling Operation Model of the EH-CSP Combined CSP System

In Section 2.2, considering the mutual coupling between the power operation of the power generation, $P_{pb}(t)$, and the heat storage module, $P_{tes}(t)$, of the CSP power station in the EH-CSP combined system, the excess wind and solar capacity can be effectively transformed into heat energy by integrating EH in the CSP heat exchange platform. Secondly, different from the traditional heat storage–power generation process, this section divides the energy storage and power generation output into two independent modules. The following describes the coupling output of the power generation module, $P_{pb}(t)$, and the heat storage module, $P_{tes}(t)$, of the optical thermal power station assisted by the EH device from the perspective of the internal operation.

4.1. Photothermal Power Generation and TES Coupling Operation Model

This section mainly describes the operation relationship between the CSP power generation module and the energy storage module: the core of this section is to treat the load demand as a connecting bridge and further elaborate the interaction and coupling mechanism between the CSP power generation and the EH energy storage system.

When the load demand is low, the concentrated heat collection system transfers heat to the power generation module and converts it into electric energy for load demand. At the same time, the energy storage module stores heat energy:

$$P_{load}^{un}(t) \leq P_w(t) + P_p(t) + P_{pb}(t) \tag{54}$$

As an electric heat transfer element in a CSP power plant, the EH conversion efficiency can be close to 100%. The amount of abandoned air and light mainly determines the output of EH. The output model is expressed as follows:

$$P_{EH}(t) = P_w(t) + P_p(t) - P_{load}^{un}(t) \tag{55}$$

$$E_{EH}(t) = \eta_{EH} P_{EH}(t) \tag{56}$$

where $E_{EH}(t)$ is the heat transferred to the heat storage module, $P_{EH}(t)$ is the wind and solar residual power, and η_{EH} refers to the electrothermal conversion efficiency when EH works stably.

The energy stored in the energy storage system is expressed as follows:

$$P_{tes}(t) = P_{csp}(t) - P_{pb}(t) + \eta_d E_{EH}(t) \tag{57}$$

$$P_{total}(t) = \eta_d (Q_{pb}(t) - P_{tes}^{ch}(t)/\eta_c - E_{EH}(t)) \tag{58}$$

where $P_{csp}(t)$ is the total absorbed power of the CSP concentrator and collector system. When the load demand is low, one part of $P_{csp}(t)$ flows into the PC system to heat the steam to drive the turbine for power generation, $P_{pb}(t)$, and the other part flows into the TES to store heat energy, $P_{tes}(t)$. $P_{tes}^{ch}(t)$ is the charging power of TES; $P_{total}(t)$ represents the total power of the power generation module and the heat storage module; $Q_{pb}(t)$ is the heat transferred to the power generation system by concentrating and collecting heat at time t; η_d is the thermoelectric conversion efficiency; and η_c is the thermal storage efficiency of the TES (%).

When the load demand is high, the concentrated heat collection system transfers heat to the power generation module, while the energy storage module releases the stored heat energy: the heat released by the energy storage module is used by the power generation module alone.

$$P_{total}(t) = \eta_d(Q_{pb}(t) + P_{tes}^{dis}(t)\eta_f + E_{EH}(t)) \tag{59}$$

$$P_{load}^{un}(t) \leq P_w(t) + P_p(t) + P_{total}(t) \tag{60}$$

where $P_{tes}^{dis}(t)$ is the power released by the heat storage system. When the load demand is high, $P_{csp}(t)$ flows into the PC system, heating steam to drive the turbine for power generation, $P_{pb}(t)$, while releasing the heat energy stored in the TES, $P_{tes}(t)$. η_f is the heat release efficiency of the TES (%).

4.2. Operation Model of the Solar Thermal Power Generation Module Based on Direct Connection and High Efficiency

The thermal storage module and the power generation module of the CSP power generation system are entirely independent; that is, the thermal storage module is specially responsible for storing heat energy and generating power independently, while the output of the power generation module entirely depends on the real-time concentrating and collecting heat system and does not directly use the heat storage, so it is considered that the output of the power generation module is directly connected to the concentrating and collecting heat module, reducing the intermediate steps of energy form conversion.

Because the energy of the CSP power generation and energy storage charging comes from the concentrating heat collection system, the concentrating heat collection system is modeled first. The CSP power station uses the concentrating heat collection system to convert the reflected light energy of the mirror field into heat energy. The thermal power is as follows:

$$Q_{csp}(t) = \eta_2 S D_t \tag{61}$$

where $Q_{csp}(t)$ is the thermal power of the concentrating and collecting device, η_2 is the total optical efficiency, S is the mirror field area of the CSP power station, and D_t is the direct irradiation index of light at time t.

When the load demand is low, part of $Q_{csp}(t)$ is used for power generation to meet the load demand, and another part is stored through the heat storage system.

$$Q_{csp}(t) = Q_{pb}(t) + Q_{tes}^{ch}(t) \tag{62}$$

When the load demand is high, $Q_{csp}(t)$ is fully used for power generation to meet the load demand, and the heat storage system releases the stored heat:

$$Q_{csp}(t) = Q_{pb}(t) + Q_{tes}^{dis}(t) \tag{63}$$

where $Q_{pb}(t)$ is the thermal power transmitted from the concentrating and collecting device to the power generation system, $Q_{tes}^{ch}(t)$ refers to the thermal power transmitted to the heat storage system by the concentrator and collector, and $Q_{tes}^{dis}(t)$ is the thermal power released by the heat storage system.

CSP power generation is defined as the transfer of energy only by the concentrating and collecting system, and the modeling is as follows:

$$P_{pb}(t) = \eta_d Q_{pb}(t) \tag{64}$$

where $P_{SF}(t)$ is the thermal power directly used by the concentrating and collecting device for power generation and η_d is the thermoelectric conversion efficiency.

4.3. Operation Model of CSP Storage Module Considering EH-CSP Combination

The independent TES–power generation and direct generation thermoelectric modules enable the power output to be flexibly adjusted according to the actual demand and resource conditions. The heat storage module can store heat energy when the Sun is sufficient for use at night or on cloudy days, while the direct-generation thermal power module adjusts the power generation according to the real-time sunlight conditions, and the combination of the two forms is complementary.

When the load demand is low, from Formula (61), the concentrator and collector will store the heat through the CSP storage system, and the heat energy stored in the energy storage module not only comes from this path but also converts the excess wind and solar output through the electrical heat transfer capacity of the EH, and the heat energy is released when the system needs to operate to drive the steam turbine to generate electricity. This is modeled as follows:

$$P_{tes}^{ch}(t) = \eta_c(Q_{tes}^{ch}(t) + E_{EH}(t)) \tag{65}$$

When the load demand is high, the heat stored in the CSP storage system is released, which is expressed as follows:

$$P_{tes}^{dis}(t) = (Q_{tes}^{dis}(t) + \eta_d E_{EH}(t))/\eta_f \tag{66}$$

where $P_{tes}^{ch}(t)$ and $P_{tes}^{dis}(t)$ are the TES storage and release power and η_c and η_f are the heat storage and heat release efficiency of the TES (%).

The power output of the TES at time t is expressed as follows:

$$P_{tes}(t) = \eta_d(-P_{tes}^{ch}(t)/\eta_c + P_{tes}^{dis}(t)\eta_f + E_{EH}(t))^{vv} \tag{67}$$

where $P_{tes}(t)$ represents the power output of the TES.

5. Solution

The day-ahead optimal scheduling model of shared energy storage considering uncertain loads constructed above can be defined as a mixed-integer nonlinear planning problem. Because it has an NP-hard property, it can be approximated by converting it into a mixed-integer linear programming problem. In this section, Benders + MOPSO is used to solve the optimal scheduling problem.

The decision variables of the day-ahead optimal dispatch model considering uncertain loads and EH solar thermal energy storage include wind power, photovoltaic power, solar thermal output power, and solar thermal energy storage power.

First, rewrite the model (23) in compact mode:

$$\begin{aligned} &\min_{pint,ptes} C^{int}(P^{int}) + C^{tes}(P^{tes}) \\ &s.t. \mathbf{A}P^{int} \leq \mathbf{b} \\ &\quad \mathbf{D}P^{int} + \mathbf{E}P^{tes} \leq \mathbf{f} \end{aligned} \tag{68}$$

In Formula (63), the response variable output power of the wind, photovoltaic, and solar thermal power generation modules is expressed as P^{int}, and the power change of the solar thermal unit energy storage module is described as P^{tes} in Formulas (A1)–(A3) and (A5). The scheduling response cost, $C^{int}(P^{int})$, is a function of P^{int}. The cost of CSP storage, $C^{tes}(P^{tes})$, is a function of P^{tes}, which is composed of the power change of the CSP storage module contained in Formulas (11)–(13). Constraint $\mathbf{A}P^{int} \leq \mathbf{b}$ represents the constraint only related to P^{int}, i.e., Formulas (2)–(9), and constraint $\mathbf{D}P^{int} + \mathbf{E}P^{tes} \leq \mathbf{f}$ represents P^{int} and P^{tes}. The related coupling constraint is expressed in Formulas (16)–(22).

5.1. MOPSO Operational Flow

The Benders decomposition and multi-objective particle swarm optimization (MOPSO) algorithm are combined to optimize the combined output of day-ahead scheduling.

Algorithm 1 describes the process of solving the Pareto-optimal solution by MOPSO using the main decision variable group decomposed by benders as particles. Firstly, the algorithm initializes the particles. At this time, each group of decision variables includes the output power values of the wind, photovoltaic, and solar thermal power generation modules. $P[i]$ is the cost corresponding to each group of decision variables. In Section 2.1, the proposed traditional cost function, C, is the optimal solution, Pbest, of the decision variable; in Section 2.2, the cost function considering the load uncertainty and CSP storage proposed in Section 2.1 is used as the group extremum, Gbest. The inertia weight, velocity, and position of particles are continuously updated, and the Pbest and Gbest corresponding to particles in each scene are solved until the end of the iteration to obtain the global optimal solution in the Pareto solution set.

Algorithm 1: MOPSO algorithm framework

Input: Particle swarm size N (N = 50); dimension, M, of the objective function; maximum number of iterations, MaxIter (MaxIter = 100); acceleration constants, c1 and c2 (usually close to 2); inertia weight, W; initialize particle position, X[N][D]; initialize particle velocity, V[N][D]; initialize personal best position, Pbest[0], and global best position, Gbest[0].

Termination condition: Maximum number of iterations reached.

For t = 1 to MaxIter, do

$Pbest[0] = \min C_0$
$Gbest[0] = \min C'_0$
For i = 1 to N, do
$V[i] = \omega * V[i] + c_1 * rand() * (\min C_0 - P[i]) + c_2 * rand() * (\min C'_0 - P[i])$
$P[i] = P[i] + V[i]$
Calculate the current particle fitness, Fi
If Fi (X_i) > Fi ($\min C_0$), then
$\min C = X_i$
Add $\min C$ **to Pareto frontier set, p**
$\min C' = selectParetoFront(P)$
End for
$\min C'$
End

5.2. Benders+MOPSO Algorithm Solving

In view of the complex correlation between P^{int} and P^{tes}, it is not easy to directly solve the model (68). This paper proposes a multi-objective optimization scheme with a decomposition structure based on Benders decomposition and MOPSO, which decomposes the integrated model (68) into a response optimization problem (69) and energy storage optimization problem (70).

$$\begin{aligned} &\min_{p^{int}} C^{int}\left(p^{int}\right) \\ &s.t. \mathbf{A}P^{int} \leq \mathbf{b} \end{aligned} \tag{69}$$

$$\begin{aligned} &\min_{p^{int},p^{tes}} C^{tes}(P^{tes}) \\ &s.t.\mathbf{D}P^{int} + \mathbf{E}P^{tes} \le \mathbf{f} \\ &P^{int} = P^{int*} \end{aligned} \tag{70}$$

In each iteration, Formula (69) optimizes the response quantity and transfers the boundary variable, P^{int*}, to Formula (70).

Benders decomposition divides the complex problem into a main problem and subproblems and approximates the optimal solution through an iterative cutting plane; MOPSO is applied to the main problem for multi-objective optimization and searches for the Pareto-optimal solution set using group intelligence. The combination of the two effectively solves the NP-hard problem.

6. Results and Discussion

This section consists of two parts for the calculation example: one part details the construction of the load uncertainty of the prediction algorithm and the accuracy of the calculation analysis, and the other part addresses the consideration of the EH-CSP thermal storage module on the day before the day scheduling economy and the wind power consumption capacity in the calculation analysis.

6.1. Background

Firstly, the prediction algorithm was evaluated to ensure the feasibility and accuracy of the simulation experiment. This paper selected a specific region of China to choose any day with an interval of 15 min and obtained the maximum load characteristics, the daily load data, and the daily meteorological data.

In this paper, the mean absolute percentage error (MAPE) and root mean square error (RMSE) evaluation indexes were selected to evaluate the prediction algorithm. The calculation formula for each evaluation index is as follows:

$$e_{\mathrm{MAPE}} = \frac{1}{n}\sum\nolimits_{i}^{n}\left|\frac{y_i - y_r}{y_r}\right| * 100\% \tag{71}$$

$$e_{\mathrm{RMSE}} = \sqrt{\frac{1}{n}\sum\nolimits_{i}^{n}(y_i - y_r)^2} \tag{72}$$

where y_r denotes the actual value, y_i denotes the predicted value, and n is the sample size; MAPE is used to measure the percentage of the mean absolute error between the predicted value and the actual value, and RMSE is the mean of the squared prediction error squared and then squared, which is more sensitive to the effect of large errors.

Secondly, the results of the day-ahead scheduling example were analyzed. In this paper, 500 MW doubly fed wind farms, 600 MW photovoltaic power plants, and 200 MW tower-type photovoltaic power plants were used to form a wind–heat co-generation system. The power and direct sun index of wind power, PV power, and load predicted by the proposed algorithm are shown in the following figure, and the relevant parameters of the photovoltaic and thermal power plant are shown in Table 1. The cost coefficient of wind and photovoltaic power generation was $c_w = c_p = 90$ CNY/MW, the cost coefficient of maintenance was $OM_{WT} = OM_{PV} = 30$ CNY/MW, the cost of starting and stopping was $SU_W = SD_W = SU_P = SD_P = 15$ CNY/MW, the cost coefficient of the power generation of the photovoltaic power plant and the power generation cost coefficient of the storage module were $c_{pb} = c_{tes} = 60$ CNY/MW and $OM_{PB} = OM_{TES} = 20$ CNY/MW, the cost coefficient of O&M was $OM_{PB} = OM_{TES} = 20$ CNY/MW, the costs of starting and stopping were $SU_{PB} = SD_{PB} = SU_{TES} = SD_{TES} == 10$ CNY/MW and $OM_{PB} = OM_{TES} =$

20 CNY/MW, the cost coefficient of the O&M was $OM_{EH} = 10$ CNY/MW of the EH unit, and the maximum wind power (50 MW) and solar abandonment cost factor was $c_{wc} = c_{pc} = 108$ CNY/MW.

Table 1. Main parameters of 100 MWCSP power station.

Parameter	Value
Maximum output power of CSP plant/MW	100
Photothermal conversion efficiency/%	40
Thermal–electrical conversion efficiency/percent	40
Solar field area/m^2	1.3×10^6
Charging (discharging) efficiency of heat storage device/%	98.5
CSP power plant output upper (lower) limit/MW	100
Maximum (small) heat storage capacity of heat storage device/MWh	1500

The predicted power and direct solar insolation (DNI) indices for wind power and photovoltaic power generation are shown in the Figure 4, and the relevant parameters for the photovoltaic power plant are shown in Table 1.

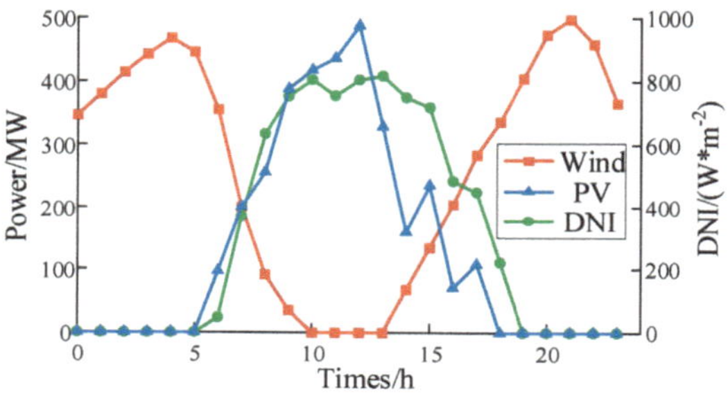

Figure 4. Predicted reference power and DNI index of wind power and PV.

6.2. Model Validation

6.2.1. Validation of the CNN-MI-BILSTM Algorithm in Predicting the Accuracy of Load Data

In this section, to validate the effectiveness and accuracy of the proposed CNN-MI-BILSTM algorithm, the mean absolute percentage error (MAPE), as well as the root mean square error (RMSE), were used as the evaluation metrics, and a certain number of samples were generated based on the city's past load data, where 70% of the samples were used for the training of the unknown coefficients and 30% were used to check the model's accuracy and detection rate. The prediction results obtained from the CNN-MI-BILSTM algorithm were compared with those of the BPNN, CNN-GRU, and conventional CNN-LSTM algorithms, as well as the CNN-MI-LSTM algorithm.

In order to compare the uncertain pure electric-load forecasting modeling methods, this experiment first used the control variable method for stepwise model adjustment. Based on the CNN-LSTM network, the importance value of input features was characterized by the MI value between the input features and the load. Completing the processing of correlated data and adding bidirectional moment information, BILSTM was used to mine the intrinsic connection between long-time data bidirectionally. It was compared with the BPNN and CNN-GRU algorithms; firstly, the load data were preprocessed, including operations such as normalization and processing of missing values and outliers, and then the data were fed into the deep learning model.

By applying the sliding-window technique, the original time-series data were sliced. Here, the window width was set to 24 h, which means that the model considered the data from the past full day as input to predict the load value at a point in the future each time. The step size was set to 1, which means that the window slid forward one unit at a time, so that the continuity of the time series could be fully utilized to capture the temporal dynamic features in the data while ensuring that the data utilization was maximized with no omissions.

Figure 5 compares the correlation between the prediction results produced by the adopted CNN-MI-BILSTM algorithm and four other mainstream prediction algorithms with the actual situation. It demonstrates the performance of different algorithms in predicting the daytime power-load variation in Gansu, clearly reflecting that the load fluctuates significantly within a day, especially peaking during the traditional peak period of power consumption, while there is a large drop during the trough period. All the prediction algorithms can fit the trend of the actual load profile well. To accurately quantify the performance, Table 2 provides the corresponding MAPE and RMSE metrics for each prediction algorithm.

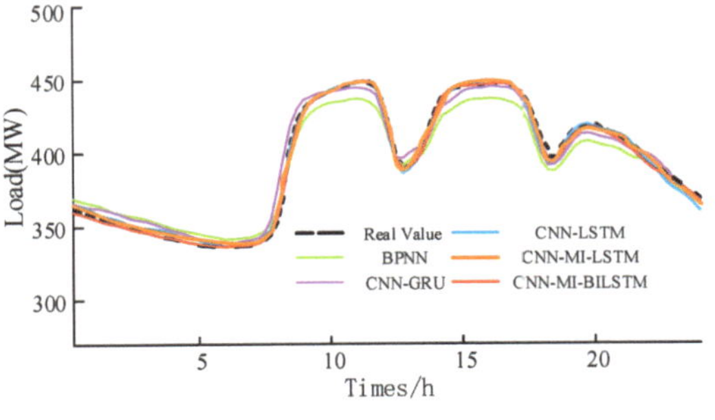

Figure 5. Load prediction curves for different algorithms.

Table 2. Predictors for different algorithms.

Model \ Norm	e_{MAPE}	e_{RMSE}
BPNN	3.22	140.04
CNN-GRU	2.85	127.83
CNN-LSTM	2.67	118.04
CNN-MI-LSTM	2.48	113.27
CNN-MI-BILSTM	2.40	109.83

By inputting 96 data points into different neural networks during the feature day, the final prediction results were compared with the actual data on the second day to obtain e_{MAPE} and e_{RMSE} metrics. These are shown in the following table:

Given that the lower of these two metrics represents a higher prediction accuracy, the data in Table 2 directly prove that the CNN-MI-BILSTM algorithm outperforms the compared algorithms in terms of prediction accuracy. First, regarding prediction accuracy, the CNN-MI-BILSTM algorithm can provide more accurate results in predicting the actual data in a particular location in Gansu. A smaller MAPE value means that the model has a low percentage error between the predicted and actual values, reflecting the strong ability of the predicted values to follow the changes in the actual data closely. Second, regarding error magnitude control, a lower RMSE value means that the overall magnitude of the prediction error is small, reducing the risk of uncertainty due to prediction bias. Finally, regarding algorithm effectiveness validation, by comparing it with other mainstream prediction algorithms, the CNN-MI-BILSTM algorithm is confirmed to be more accurate and reliable in dealing with this kind of time-series prediction task.

6.2.2. Validation of the Impact of the Proposed Scheduling Model in Promoting Wind Energy Consumption and Improving Economics

In order to study the impact of the EH-CSP combined solar thermal system integrated into the integrated energy system on the day-ahead optimal scheduling results, this paper set up three models to analyze the scheduling results by comparing the results of the other two scheduling models to verify the effectiveness of the proposed scheduling model in promoting wind energy consumption and reducing the integrated cost of the system. The three comparison models were as follows:

1. In the traditional day-ahead dispatching model, only the role of wind power was considered, and the wind–heat co-generation system without CSP was used, in which five 100 MW wind turbines and six 100 MW PV plants were used.
2. In conventional power generation, three 100 MW PV plants were replaced by two 100 MW PV plants to form a wind–heat co-generation system.
3. Using the Benders + MOPSO solution, in conventional power generation, three 100 MW PV plants were replaced by one 150 MW PV plant, and the EH device was introduced to form the EH–wind–thermal co-generation system.

Figure 6 shows the comparison of the three scheduling models in terms of day-ahead output, and Table 3 reveals the differences in the cost-effectiveness for each scenario: Scenario 1 only considers the role of wind and solar, without considering the impact of the solar thermal factor and EH on day-ahead scheduling, and the corresponding cost of wind and solar abandonment is significantly higher than that of Scenarios 2 and 3. Scenario 2 considers the impact of the solar thermal factor on top of Scenario 1, but does not take into account the effect of EH, and the corresponding cost of dispatching is lower than that of Scenario 1, and the cost of wind and solar abandonment is higher than that in Scenario 3. Scenario 3 takes the impact of combining the EH-CSP with Scenario 2, effectively reducing wind and solar abandonment while further cutting the dispatching cost. Scenario 2 takes into account wind and heat, but not EH, based on Scenario 1, and the corresponding dispatch cost is lower than that of Scenario 1, and the cost of the wind and light abandonment rate is higher than that of Scenario 3. Scenario 3 takes into account the effect of the EH-CSP combination based on Scenario 2, which effectively reduces wind and light abandonment and, at the same time, further cuts down dispatch costs and improves economic efficiency.

The dispatch results of the above three comparative models for a few days previously were as follows:

The total costs of day-ahead scheduling for the three comparison models and the rates of wind and light abandonment were compared.

Table 3. Costs and abandonment rates for each model.

Model	Cost (CNY)	Energy Abandonment Rate
1	1,552,928	11.6%
2	1,357,023	6.15%
3	1,306,027	2.366%

Aiming to achieve the goal of the lowest average daily cost, the combined scheduling of wind, photovoltaic (PV), and photothermal power generation was optimized for an integrated energy system. While ensuring that each generation unit operated within its respective constraints, an ideal pre-daily schedule was explored to achieve the best balance between economic efficiency and resource utilization.

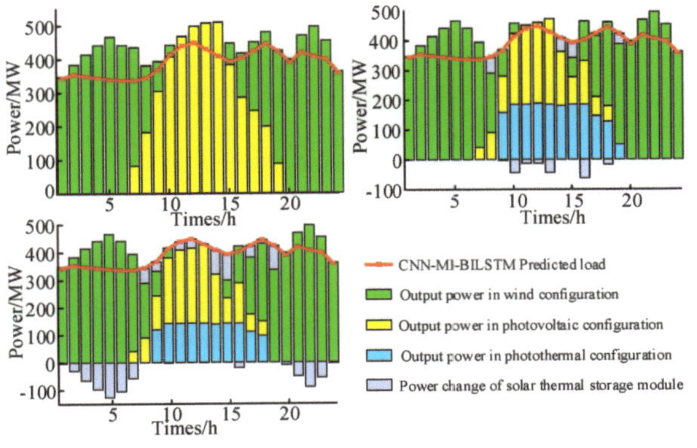

Figure 6. Scheduling results for each model day.

6.3. Analysis of Relevant Factors

6.3.1. Analyzing the Effectiveness of the CNN-MI-BILSTM Algorithm with Different Sample Sizes

A certain number of samples were generated based on the load data of the city, where 70% of the samples were used to train the unknown coefficients and 30% were used to test the accuracy and detection rate of the model. The BPNN, CNN-GRU, and regular CNN-LSTM algorithms, the CNN-MI-LSTM algorithm, and the CNN-MI-BILSTM algorithm can make predictions. However, the accuracies of different prediction algorithms corresponding to different sample sizes deviate.

Therefore, Figure 7 shows in detail the accuracy of the CNN-MI-BILSTM algorithm compared with the other four prediction methods under different sample sizes (100, 400, 700, 1000, and 1300). The results clearly show that the CNN-MI-BILSTM algorithm exhibits the highest level of accuracy regardless of the sample size, and the advantage of its prediction accuracy is especially prominent in challenging scenarios with small sample sizes (e.g., 100 samples). As the number of samples grows to 400, 700, and even 1000 and 1300, although the prediction performance of all algorithms generally improves, the CNN-MI-BILSTM algorithm has optimal stability and robustness under different data sizes. Taken together, the proposed CNN-MI-BILSTM algorithm has optimal prediction accuracy under various sample sizes.

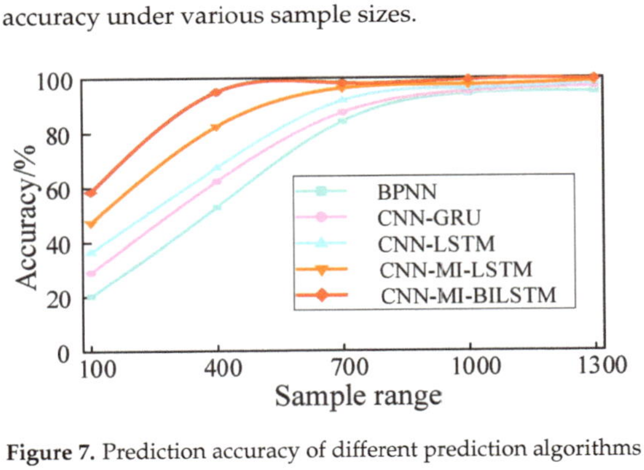

Figure 7. Prediction accuracy of different prediction algorithms with varying sample sizes.

6.3.2. Analysis of the Reliability and Economy of the Optimized Configuration of the Combined EH–Wind–Heat System

To study the impact of different capacity configurations on the reliability and economy of the system, this paper verified the reliability of the optimized configuration by analyzing the proportion of the load demand that cannot be met due to insufficient generating resources to the total demand, i.e., the proportion of the critical load loss, under different stochastic configurations, and verified the impact of the various configurations on the economy of the whole system utilizing the cost curves.

Firstly, the reliability of the random configuration was analyzed. Figures 8 and 9 show the analysis of the critical load loss ratio based on the existence of the EH device, a fixed CSP capacity, and the step of wind power according to 20 MW and the PV capacity according to 15 MW, respectively. As shown in the Figure 8, the smaller the generating capacity is than the optimized configuration generating capacity, the larger the corresponding critical load loss ratio, and the critical load loss ratio is zero when the generating capacity is larger than the optimized configuration generating capacity; with a reduction in the wind power and PV unit capacity, the larger the critical load loss ratio; and the generating system with a higher than optimized capacity configuration can achieve zero load loss.

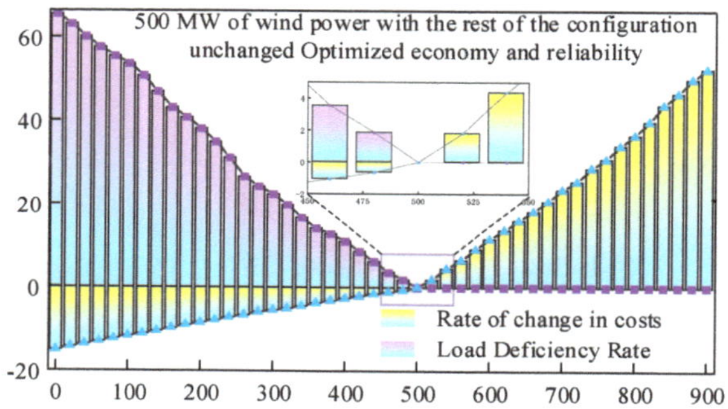

Figure 8. Corresponding load loss rates and cost change rates for stochastic wind power configurations.

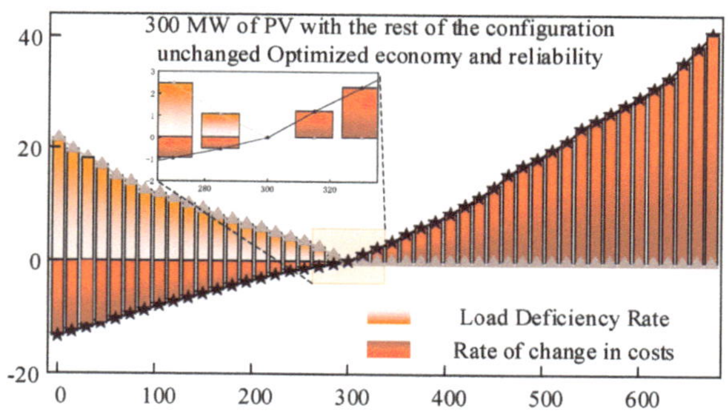

Figure 9. Corresponding load loss rates and cost change rates under stochastic PV configurations.

Secondly, to analyze the economics under the random configuration, the cost curve of power generation under the random configuration is given. As shown in the Figure 9, when the power generation capacity is larger than the optimized configuration capacity, the cost gradually increases with the increase in the configuration capacity, and the rate of wind and light abandonment increases. When the generation capacity is smaller than the optimized capacity, the cost gradually decreases as the configuration capacity decreases, and the rate of wind and light abandonment decreases.

The above verifies that the optimized configuration of the EH–wind thermal system can maintain high power supply reliability and economy in the face of the intermittency and uncertainty of renewable energy generation.

7. Conclusions

The paper innovatively proposes a day-ahead optimal scheduling model considering uncertain loads and EH-CSP storage. The model first adopts the multi-interval uncertainty set to portray the uncertainty of loads, and the prediction algorithm is continuously improved to accurately predict the load data using the CNN-MI-BILSTM algorithm. By combining the feature extraction capability of the convolutional neural network, the feature selection advantage of MI, and the powerful capture of time series by BILSTM, the method effectively handles the nonlinear relationships and long- and short-term dependencies in the data, which significantly improves the accuracy and stability of prediction. The accuracy of the proposed algorithm is verified by comparing the prediction results of different prediction algorithms.

In this paper, the CNN-MI-BILSTM algorithm obtains smaller MAPE and RMSE values of 2.40 and 109.83, respectively, compared to other prediction algorithms. The smaller the MAPE value, the smaller the percentage of error between the predicted value and the actual value. The smaller the RMSE value, the smaller the overall magnitude of prediction error, and thus the smaller the risk of uncertainty. The cost of the model, considering the inclusion of EH-CSP and adopting the Benders + MOPSO solution, is CNY 1306027, and the energy abandonment rate is 2.366%, which is significantly lower than that of the model without considering EH-CSP.

The proposed day-ahead optimal dispatch model integrates the output characteristics of new energy power plants, aiming to optimize the balance between the day-ahead dispatch cost and adaptability in the face of uncertain loads, taking into account the flexibility and complementarity of EH and solar thermal storage technologies. Integrating them into the model not only smoothes intraday load fluctuations but also effectively stores and dispatches intermittent wind and solar energy, reducing the scheduling challenges caused by renewable energy uncertainties. A combination of Benders decomposition and MOPSO determines the optimal capacity configuration. Scheduling simulations of the power system in Gansu Province using different unit configurations are conducted, and it is verified that the proposed day-ahead optimal scheduling model considering uncertain loads and EH-CSP storage significantly improves the economic efficiency and environmental sustainability of the scheduling process.

Author Contributions: Conceptualization, K.D.; data curation, K.D., Y.S. and B.C.; formal analysis, J.C.; methodology, K.D., L.S. and Y.W.; validation, L.S. and Y.X.; writing—original draft, J.C.; writing—review and editing, K.D., Y.S., B.C. and J.C. All authors have read and agreed to the published version of the manuscript.

Funding: This research is partially funded by the Science and Technology Project of the State Grid Corporation of China (5108-202218280A-2-300-XG).

Data Availability Statement: The original contributions presented in the study are included in the article; further inquiries can be directed to the corresponding author.

Acknowledgments: This work was supported by the Science and Technology Project of the State Grid Corporation of China (5108-202218280A-2-300-XG).

Conflicts of Interest: Authors Kun Ding and Yalu Sun were employed by the Economic and Technological Research Institute of the State Grid Gansu Electric Power Company. Author Boyang Chen was employed by the State Grid Gansu Electric Power Company. The remaining authors declare that the research was conducted in the absence of any commercial or financial relationships that could be construed as a potential conflict of interest.

Appendix A

$$C_1 = \sum_{t=1}^{T} [c_w P_w(t) + c_p P_p(t) + c_{pb} P_{pb}(t)] \tag{A1}$$

$$C_2 = \sum_{t=1}^{T} [SU_W \mu_{W,t}^{su} + SD_W \mu_{W,t}^{sd} + SU_P \mu_{P,t}^{su} + SD_P \mu_{P,t}^{sd} + SU_{PB} \mu_{PB,t}^{su} + SD_{PB} \mu_{PB,t}^{sd}] \tag{A2}$$

$$C_3 = \sum_{t=1}^{T} [OM_{WT} P_w(t) + OM_{PV} P_p(t) + OM_{PB} P_{pb}(t)] \tag{A3}$$

$$P_{pb}(t) = H_t^{pb} \eta_1 \tag{A4}$$

$$C_4 = \sum_{t=1}^{T} \left[c_{wc} P_w^c(t) + c_{pc} P_p^c(t) \right] \tag{A5}$$

In the formulas, c_w, c_p, and c_{pb} are the unit operating costs of wind, PV, and solar thermal power generation, respectively; $P_w(t)$, $P_p(t)$, and $P_{pb}(t)$ are the output power values for wind, photovoltaic, and solar thermal power generation at time t; SU_W, SU_P, and SU_{PB} are the start-up costs of wind, photovoltaic and solar thermal power generation, respectively; SD_W, SD_P, and SD_{PB} are the stopping costs for wind, photovoltaic, and solar thermal power generation, respectively; $\mu_{W,t}^{su}$ ($\mu_{P,t}^{su}$, $\mu_{PB,t}^{su}$) = 1 when there is an energy block at the time of start-up; $\mu_{W,t}^{sd}$ ($\mu_{P,t}^{sd}$, $\mu_{PB,t}^{sd}$) = 1 when there is an energy block at the time of stopping; OM_{WT}, OM_{PV}, and OM_{PB} are the maintenance cost coefficients for wind, photovoltaic, and solar thermal power generation; H_t^{pb} is the heat transferred to the power generation module by the concentrating solar collector system; η_1 is the thermoelectric conversion efficiency, where each MW penalty is taken to be 1.2 times the operating cost of the wind power unit; c_{wc} and c_{pc} are the penalty cost coefficients for wind and light abandonment; and $P_w^c(t)$ and $P_p^c(t)$ are the amounts of wind and light abandonment at time t.

References

1. Yang, Z.; Xiong, W.; Wang, P.; Shen, N.; Liao, S. A Day-Ahead Economic Dispatch Method for Renewable Energy Systems Considering Flexibility Supply and Demand Balancing Capabilities. *Energies* **2024**, *17*, 5427. [CrossRef]
2. Luo, Y.; Wang, Y.; Liu, C.; Fan, L. Two-stage robust optimal scheduling of wind power-photovoltaic-thermal power-pumped storage combined system. *IET Renew. Power Gener.* **2022**, *16*, 2881–2891. [CrossRef]
3. Wang, K.; Zhu, H.; Dang, J.; Ming, B.; Wu, X. Short-term optimal scheduling of wind-photovoltaic-hydropower-thermal-pumped hydro storage coupled system based on a novel multi-objective priority stratification method. *Energy* **2024**, *309*, 133190. [CrossRef]
4. Ma, Z.; Wang, D.; Zhang, H.; Wang, Y.; Dong, H. Two-stage optimal dispatching based on wind-photovoltaic—Pumped storage-thermal power combined power generation system. In Proceedings of the 2019 IEEE 3rd Conference on Energy Internet and Energy System Integration (EI2), Changsha, China, 8–10 November 2019; IEEE: Piscataway, NJ, USA, 2019; pp. 1714–1719.
5. Zhang, Z.; Xian, W.; Tan, W.; Li, J.; Liu, X. Research on Safe-Economic Dispatch Strategy for Renewable Energy Power Stations Based on Game-Fairness Empowerment. *Energies* **2024**, *17*, 6146. [CrossRef]
6. Shi, Y.; Li, W.; Fan, G.; Zhang, L.; Yang, F. Low-Carbon Dispatch of an Integrated Energy System Considering Confidence Intervals for Renewable Energy Generation. *Energy Eng.* **2024**, *121*, 461–482. [CrossRef]

7. Han, J.; Han, K.; Wang, Y.; Han, Y.; Ye, Z.; Lin, J. Multi-objective planning and sustainability assessment for integrated energy systems combining ORC and multi-energy storage: 4E (economic, environmental, exergy and emergy) analysis. *Case Stud. Therm. Eng.* **2025**, *65*, 105674. [CrossRef]
8. Wang, P.; Zheng, L.; Diao, T.; Huang, S.; Bai, X. Robust Bilevel Optimal Dispatch of Park Integrated Energy System Considering Renewable Energy Uncertainty. *Energies* **2023**, *16*, 7302. [CrossRef]
9. Zhou, X.; Wang, J.; Wang, X.; Chen, S. Optimal dispatch of integrated energy system based on deep reinforcement learning. *Energy Rep.* **2023**, *9*, 373–378. [CrossRef]
10. Zhou, Q.; Zhang, J.; Gao, P.; Zhang, R.; Liu, L.; Wang, S.; Cheng, L.; Wang, W.; Yang, S. Two-Stage Robust Optimization for Prosumers Considering Uncertainties from Sustainable Energy of Wind Power Generation and Load Demand Based on Nested C&CG Algorithm. *Sustainability* **2023**, *15*, 9769. [CrossRef]
11. Huynh, D.C.; Ho, L.D.; Pham, H.M.; Dunnigan, M.W.; Barbalata, C. Water Wave Optimization Algorithm-Based Dynamic Optimal Dispatch Considering a Day-Ahead Load Forecasting in a Microgrid. *IEEE Access* **2024**, *12*, 48027–48043. [CrossRef]
12. Masood, Z.; Gantassi, R.; Choi, Y. Enhancing Short-Term Electric Load Forecasting for Households Using Quantile LSTM and Clustering-Based Probabilistic Approach. *IEEE Access* **2024**, *12*, 77257–77268. [CrossRef]
13. Rubasinghe, O.; Zhang, T.; Zhang, X.; Choi, S.S.; Chau, T.K.; Chow, Y.; Fernando, T.; Iu, H.H.-C. Highly accurate peak and valley prediction short-term net load forecasting approach based on decomposition for power systems with high PV penetration. *Appl. Energy* **2023**, *333*, 120641. [CrossRef]
14. Zhao, Y.; Shi, J.; Wang, D.; Jiang, H.; Zhang, X. An adaptive load forecasting model in microgrids: A cloud-edge orchestrated approach tailored for accuracy, real-time response, and privacy needs. *Int. J. Electr. Power Energy Syst.* **2025**, *165*, 110490. [CrossRef]
15. Lian, H.; Wang, S.; Gao, N.; Qu, F.; Wang, H.; Xie, C.; Yang, B. A Short-Term LOAD forecasting Method Based on EEMD-LN-GRU. In Proceedings of the 2021 3rd Asia Energy and Electrical Engineering Symposium (AEEES), Chengdu, China, 26–29 March 2021; IEEE: Piscataway, NJ, USA, 2021; pp. 666–673.
16. Tang, C.; Zhang, Y.; Wu, F.; Tang, Z. An Improved CNN-BILSTM Model for Power Load Prediction in Uncertain Power Systems. *Energies* **2024**, *17*, 2312. [CrossRef]
17. Hu, S.; Zhang, T.; Yang, F.; Gao, Z.; Ge, Y.; Zhang, Q.; Sun, H.; Xu, K. Short-Term Load Forecasting Based on Mutual Information and BI-LSTM Considering Fluctuation in Importance Values of Features. *IEEE Access* **2024**, *12*, 23653–23665. [CrossRef]
18. Zhang, X.; Liu, Y. Two-stage stochastic robust optimal scheduling of virtual power plant considering source load uncertainty. *Eng. Rep.* **2024**, *6*, e13005. [CrossRef]
19. Xu, M.; Li, W.; Feng, Z.; Bai, W.; Jia, L.; Wei, Z. Economic Dispatch Model of High Proportional New Energy Grid-Connected Consumption Considering Source Load Uncertainty. *Energies* **2023**, *16*, 1696. [CrossRef]
20. Sun, J.; Song, X.; Hua, D.; Liao, M.; Li, Z. Stochastic optimal scheduling of a combined wind-photovoltaic-CSP-fire system accounting for electrical heat conversion. *Energy Inform.* **2024**, *7*, 89. [CrossRef]
21. Yun, Y.; Zhang, D.; Li, L.; Li, Y.; Shi, Y.; He, J. Optimal operation strategy of peak regulation combined thermal power units and concentrating solar power plant with energy conversion and spinning reserve. *J. Clean. Prod.* **2023**, *424*, 138864. [CrossRef]

energies

MDPI

Article

Development of a Visualization Platform for Power Generation Analysis in Urban Building-Integrated Photovoltaic Systems

Xi Chen [1,2,3], Hai Long [4] and Ye Xia [5,*]

1 School of Design, Huazhong University of Science and Technology, Wuhan 430074, China; xichen2019@hust.edu.cn
2 Hubei Engineering Research Center for Tech of Digital Lighting, Wuhan 430074, China
3 The Key Laboratory of Lighting Interactive Service and Technology Ministry for Ministry of Culture and Tourism, Wuhan 430074, China
4 China-EU Institute for Clean and Renewable Energy, Huazhong University of Science and Technology, Wuhan 430074, China; longhai@hust.edu.cn
5 Evaluation Statistics and Information Technology Department, Hubei Academy of Scientific and Technical Information, Wuhan 430071, China
* Correspondence: xiay2020@outlook.com

Abstract: Urban high-density planning and the rise of super-high-rise buildings have significantly limited the development of distributed photovoltaic (PV) systems, creating an urgent need for optimized three-dimensional (3D) layout strategies within urban building spaces. Given that PV power generation is influenced by environmental factors and building spatial configurations, a 3D panoramic visualization tool is essential to intuitively display relevant data and support decision-making for government planners and PV operators. To address this, we developed a visualization platform to assess the integrated PV power generation potential of buildings at both city and single-building levels. The platform enables a 3D spatial panoramic display, where building surfaces are color-coded to clearly represent key performance metrics, such as power generation capacity, installation costs, and potential electricity savings. This intuitive visualization allows stakeholders to identify optimal PV installation areas and evaluate economic benefits effectively. This article details the implementation of the visualization platform across four key aspects: data generation and input, power generation and economic calculation, building model creation and data mapping, and visual interface design, aiming to facilitate the efficient planning and deployment of distributed photovoltaic systems in complex urban environments.

Keywords: urban building-integrated photovoltaic; power generation analysis; Unreal Engine; 3D visualization; building model

1. Introduction

Limited land availability and the complexity of urban environments present significant challenges to the development of urban photovoltaic systems. Compared to centralized photovoltaics, distributed photovoltaic systems offer a more efficient solution by maximizing the use of available urban spaces, delivering higher carbon emission reduction potential, and achieving lower per-kilowatt-hour costs [1–3]. With the continuous decline in photovoltaic module costs and the growing availability of materials suitable for building facades, facade photovoltaics are becoming increasingly competitive with rooftop installations [4–6]. As a result, integrating photovoltaic systems into building facades has emerged as a key direction for advancing urban renewable energy utilization. However, photovoltaic power generation is highly sensitive to environmental variables such as climate conditions, building designs, lighting availability, and urban spatial layouts, which vary considerably across different cities [7–10]. These factors introduce significant uncertainties regarding power generation performance, economic viability, and the investment return cycle of photovoltaic systems [11,12]. Traditional visualizations relying solely on charts or two-dimensional maps are insufficient to comprehensively and intuitively represent these dynamic and spatial changes [13–15]. To address this challenge, the development of a city-level three-dimensional visualization platform for integrated-building photovoltaics has become particularly critical.

Currently, a variety of building photovoltaic visualization tools are available, such as PVsyst, Easy PV, PVcase, HelioScope, SolarGIS, and SolarMapper. PVsyst provides city-scale data primarily through charts while integrating individual-building photovoltaic planning with 3D modeling software [16]. SolarGIS focuses on large-scale photovoltaic planning using 2D GIS-based visualization techniques [17]. Easy PV, PVcase, SolarMapper, and HelioScope are designed for smaller regions or single buildings, with PVcase and HelioScope relying heavily on 2D GIS visualization methods [18–20]. With advancements in drone technology and remote sensing data, software for urban photovoltaic potential assessment increasingly incorporates high-resolution imagery and 3D modeling techniques. For example, SolarMapper [21] leverages remote sensing data combined with LiDAR technology to deliver city-wide visualization of solar potential, while Easy PV [22] uses satellite remote sensing and radiation modeling for predicting building-level photovoltaic generation capacity.

The growing influence of the gaming industry has also contributed to the evolution of building photovoltaic visualization software. Game engines such as Unity and Unreal Engine provide advanced visualization capabilities at a significantly reduced cost [23,24]. These engines enable features such as highly realistic 3D scene construction, real-time rendering, interactive simulations, terrain modeling, and physical environment simulations [25,26].

Compared with GIS, BIM, or energy simulation software that offer similar functionalities, the key contributions of this platform are as follows:

An integrated building–energy–economy coupling model: Unlike tools such as URBANopt or OpenStudio, which focus primarily on regional-building energy system simulation, this platform further incorporates economic evaluation indicators. By setting thresholds for economic parameters, the platform enables the generation of visualized investment planning schemes.

User-friendly interface design: Compared with traditional software such as DesignBuilder or EnergyPLAN, this platform offers interactive functionalities for users. Without the need to manually construct building or energy system models, users can quickly access and understand the BIPV potential—including economic power generation and installed capacity—at various spatial scales, such as individual buildings, neighborhoods, and cities. This makes the platform easier to use, with a relatively smooth learning curve.

This platform innovatively combines the high-precision solar radiation calculation of Ladybug with the real-time 3D rendering capabilities of Unreal Engine, supporting dynamic sunlight simulation and interactive perspective adjustment. It realizes "roaming-style" visualization simulation for city-level building photovoltaics. This technological breakthrough allows users to intuitively observe the changes in photovoltaic performance across different time periods and seasons, greatly enhancing the realism of the simulation and the interactive experience. It achieves integrated decision support for 'technology-economics-esthetics'.

The proposed platform enables a city-scale panoramic visualization of photovoltaic power generation potential at both the building level and city level. It can intuitively display key parameters such as power generation, installation costs, electricity savings, and economic returns using color mapping on building models. The platform provides data-driven insights, actionable solutions, and visual representations to government agencies and investors, thereby supporting informed decision-making for urban renewable energy development. This paper describes the four main components of the proposed visualization platform: data resources, calculation methods, building model generation and parameter mapping, and visual interface design. The platform is then validated using Wuhan, China, as the case study to demonstrate its effectiveness and applicability.

User Demand Survey

The primary user groups of the platform include related commissioners in government departments, experts in the power industry, and photovoltaic investors. To gather insights, we selected 15 representative users, comprising 5 individuals from each category, to explore the platform's purpose of use, data requirements, display preferences, and interaction mechanisms. To gain a comprehensive understanding of user needs, we conducted semi-structured interviews using a semi-open format. The consolidated findings are presented in Table 1. Across all user categories, the overarching goal is to support regional photovoltaic investment planning. The data content requirements are generally consistent across groups, although specific needs vary. Government and power-sector decision-makers prioritize reports and presentation capabilities, photovoltaic investors emphasize the need for precise installation planning, and power industry experts value efficient access to detailed data.

Table 1. User demand survey for the platform.

		Government Department Commissioner	**Power Industry Expert**	**Photovoltaic Investor**	**Total**
Purpose of Use	Regional Photovoltaic Investment Planning	5	5	5	15
	Large-Screen Report Presentation	5	3	2	10
Data Content	Estimated Power Generation	5	5	5	15
	Estimated Installation Cost	5	5	5	15
	Estimated Installed Capacity	5	5	5	15
	Estimated Electricity Cost Savings	5	5	5	15
	Building Data Information	5	5	5	15
Display Preferences	Panoramic 3D Visualization	5	4	4	13
	High-Precision Models for Significant Buildings	3	3	5	11
	Color-Coded System with Clear Data Correlations	5	5	5	15
	Interface Design for Large Screens	5	3	2	10
	Interface Design for Laptops	2	4	3	9
	Customized Photovoltaic Layout Plan	5	4	5	14

Table 1. *Cont.*

		Government Department Commissioner	Power Industry Expert	Photovoltaic Investor	Total
Interaction Methods	Investment Plan Generation	5	4	5	14
	Scene Roaming and Query	5	4	4	13
	Scene Display Switching	5	5	4	14
	Individual Buildings' Detailed Information	5	5	5	15
	Adjustment of Installation Plans for Key Buildings	2	3	5	10
	Dynamic Effects for Interface Interaction	5	3	3	9

2. Platform Architecture

Based on the findings from platform research and software selection, we established a clear workflow for platform development. As illustrated in Figure 1, the development process consists of three main components: basic data calculation, building surface data mapping, and interface display and interaction. The workflow begins with generating and acquiring data, which includes obtaining basic parameters, calculating solar radiation intensity, analyzing photovoltaic power generation potential, and converting these data for use in Unreal Engine (UE). Next, building models are generated using UE and Cesium for Unreal, followed by building surface segmentation and data mapping to achieve panoramic 3D visualization. This step also incorporates mapping data onto building surfaces through color representation. Finally, the platform's information architecture, user interaction flow, and interactive interface display are developed to complete the visualization platform.

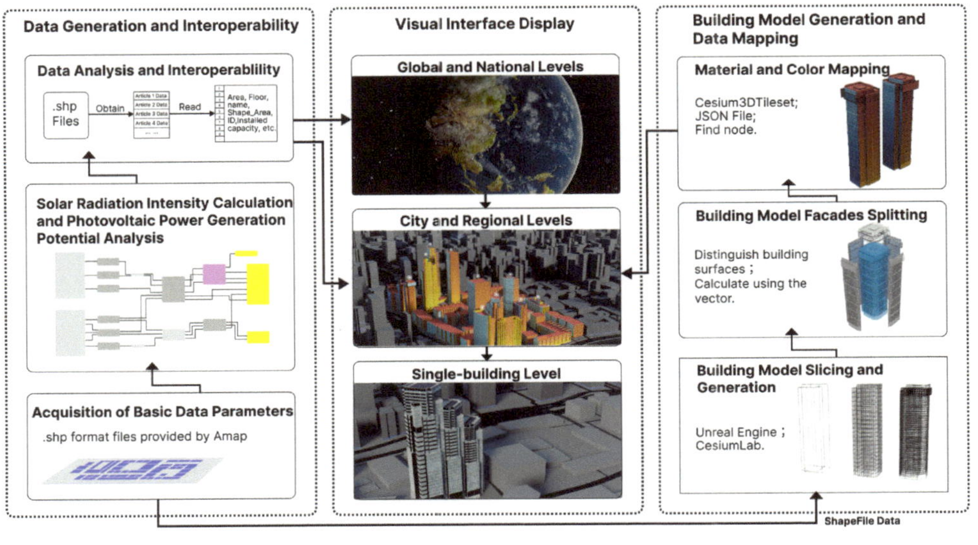

Figure 1. Workflow of urban building-integrated photovoltaic system visualization platform.

2.1. Data Sources

In large-scale urban environments, the construction of 3D building models and solar radiation simulation calculations become significantly more challenging. Therefore, this study divided the research area based on the fourth-level administrative boundaries (i.e., neighborhood blocks). Within each block, meteorological data collection and radiation simulation were conducted separately, which significantly improved the computational efficiency of the model. To simulate urban building-integrated photovoltaic (BIPV) power generation, this study utilized high-resolution meteorological data from NREL's NSRDB 2020 dataset and detailed building data from Amap.

The NSRDB dataset provides hourly meteorological data for China in 2020 with a spatial resolution of 2 km. In this study, based on the latitude and longitude coordinates of the centroid of each street, Python 3.9 programming was used to call the API provided by the official NSRDB website to download hourly meteorological data—including latitude, longitude, elevation, annual solar radiation, temperature, and humidity—in CSV format. Then, the CSV files were further converted into EnergyPlus Weather (EPW) files according to the EPW format specifications. Finally, the "import EPW" component in Grasshopper was used to import the EPW files.

Building information was extracted from official Shapefiles (.shp) at 1 m spatial resolution, containing attributes such as coordinates, height, area, number of floors, building names, Shape_Area, and Shape_Length. After preprocessing, these data enabled the construction of detailed 3D urban models for accurate PV performance analysis.

Overall, this study achieves detailed segmentation at the building level and parallel partitioned computation at the urban level, making the visualization of high-spatiotemporal-resolution BIPV simulation results at the city scale more feasible.

2.2. Software and Plugin Selection

Based on data analysis and visualization requirements, specialized toolchains were selected for different tasks (Figure 2). For building environment simulations, Rhino's Grasshopper and the Ladybug plugin were used to calculate photovoltaic generation and assess thermal performance, offering precise solar radiation mapping and adjustable window-to-wall ratios for diverse building types. Unreal Engine (UE) served as the core platform for high-fidelity 3D interactive visualization, selected for its real-time rendering capabilities, cross-platform support, and blueprint visual programming features. To address geospatial data processing challenges, the Cesium Lab toolchain was introduced for efficient conversion of .shp files into UE-compatible formats, ensuring accurate large-scale city model integration. Blender was used for detailed modeling of key buildings, while City Engine facilitated rapid generation of roads, water bodies, and urban base layers. High-fidelity UI/UX prototypes were developed in Figma to ensure consistent user experience across the platform.

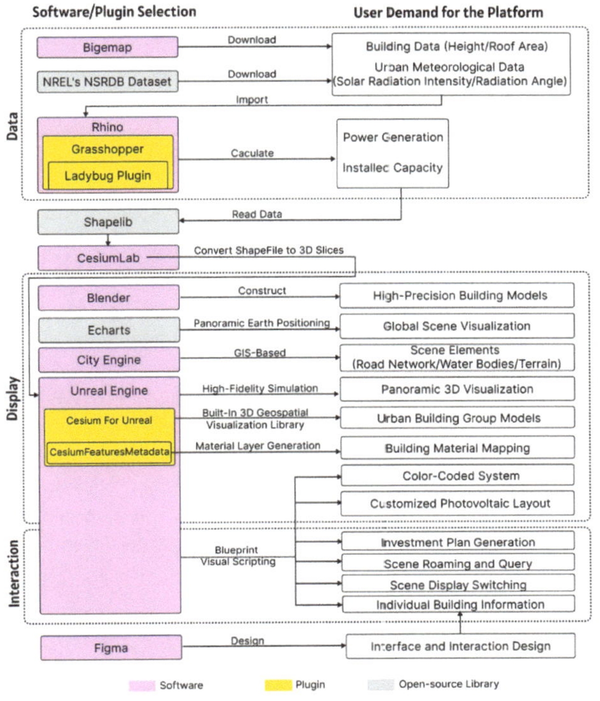

Figure 2. Software and plugin selection for the platform.

2.3. Interface Interaction

The photovoltaic (PV) visualization platform adopts a multi-scale interactive design to address the decision-making needs of users at different levels.

At the macro scale, government agencies can use the 3D globe navigation interface to intuitively grasp the global distribution of solar energy potential. By clicking on a specific city, users can access detailed solar radiation data and theoretical power generation estimates, supporting early-stage policy planning and renewable energy target setting.

At the city and regional scale, the platform currently covers data for 77 cities across China. Government authorities and investors can conduct in-depth analyses within highly detailed 3D urban environments. The platform supports free roaming, customized area selection, and administrative boundary overlays, allowing users to quickly assess the PV generation potential of specific districts. For example, to support a city's 2035 renewable energy targets, municipal planners can select the relevant administrative area, input parameters such as investment budget, expected electricity savings, and target PV generation, and submit them through the platform. The system then recommends suitable buildings for deployment based on estimated rooftop potential and cost-effectiveness. Users receive visual feedback with highlighted buildings and can view detailed attributes for each structure and export the building list to assist in investment prioritization.

At the building scale, individual property owners and designers can perform detailed PV planning. Users input target power outputs, budget constraints, and solar panel specifications, and the system generates customized simulation reports, including installation layouts and projected performance.

To ensure a seamless user experience, the platform incorporates several technological innovations. CesiumLab's streaming terrain technology ensures rapid loading of large-scale urban models, while Level of Detail (LOD) techniques dynamically adjust model resolution to maintain smooth performance across different hardware configurations. Cross-platform data sharing allows users to view reports generated on PCs directly on mobile devices. These design choices make complex PV data analysis intuitive and accessible, truly enabling data-driven decision-making across multiple user groups.

3. Methods

3.1. Solar Radiation Intensity Calculation and Photovoltaic Power Generation Potential Analysis

This study constructs 3D models of urban buildings on the Rhino-Grasshopper platform based on building footprint data and planar segmentation methods. By integrating meteorological data and ray tracing techniques, a BIPV solar radiation and photovoltaic power generation model is developed, which accounts for mutual shading between buildings. The model calculates both the power output and the installed capacity of the BIPV system. Finally, based on the cost of photovoltaic systems and economic parameters such as electricity prices, an economic evaluation indicator for the BIPV system is established to quantify its economic potential. It is worth noting that Xie et al. [27] have already demonstrated the accuracy of using Ladybug for calculating solar radiation potential. Therefore, this study does not validate the simulation results against real façade PV power generation data or other standard simulation models. The detailed modeling process is described as follows (Figure 3):

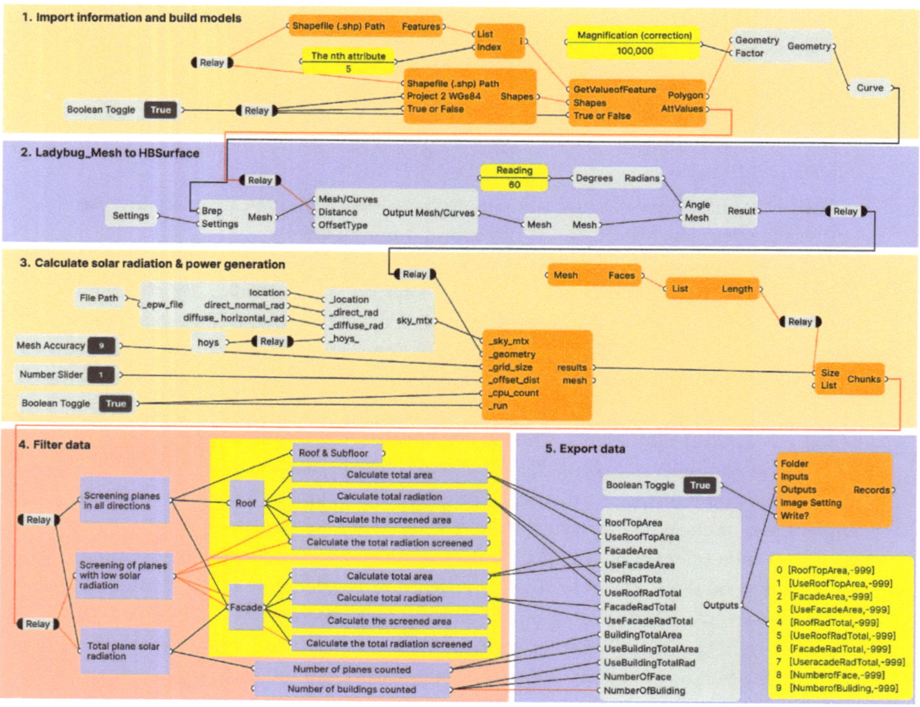

Figure 3. Grosshopper visual script for analyzing solar radiation intensity and photovoltaic power generation potential.

3.1.1. Construction of City-Scale 3D Building Models

Building footprint data in Shapefile format, containing height attributes, is first imported into the Rhino platform. Using the "Weaverbird's Mesh Thicken" component in Grasshopper, the footprint data are extruded according to their height to generate 3D building models represented by mesh surfaces (Figure 4a). Subsequently, the "Mesh Brep" component is applied to subdivide the building models into finer planar segments by setting constraints on the maximum and minimum edge lengths (Figure 4b).

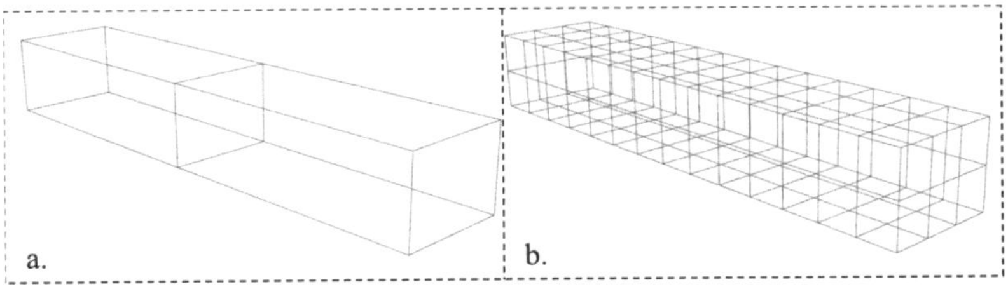

Figure 4. Construction of city-scale 3D building models. (**a**) 3D Building Extrusion; (**b**) Planar Surface Subdivision.

For each building surface, the tilt angle and azimuth angle are the two most critical parameters. In our model, the tilt angle of vertical façades was set to 90°, while the tilt angle of rooftops was set equal to the local latitude. The azimuth angle of each surface was computed based on its normal vector.

3.1.2. Solar Radiation Model with Shading Consideration

First, a sky matrix was generated based on meteorological data to determine the solar position for each of the 8760 h in a year. Then, direct solar radiation rays between the sun and the centroid of each building surface were constructed as vectors, as illustrated in Figure 5.

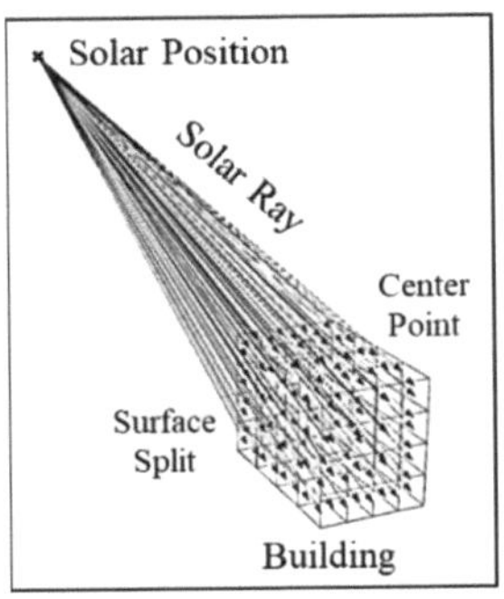

Figure 5. Schematic diagram of direct solar radiation ray tracing.

Without considering shading, the total solar radiation received by each surface at time t can be calculated as follows:

$$Total\_Rad_t = Dir\_Rad_t * cosq_t + Dif\_Rad_t + Ground\_Rad_t \tag{1}$$

where Dir_Rad, Dif_Rad_t, and $Ground\_Rad_t$ represent the direct, diffuse, and ground-reflected solar radiation at time t, respectively, and $cos\theta_t$ is the correction factor for the angle between the direct solar incidence direction and the surface normal at time t.

To account for shading effects between buildings, all surrounding buildings near the target building are added into the 3D model as shading elements. Ray tracing is used to determine whether the direct solar radiation ray to each building surface intersects with any of the shading elements. If an intersection is detected, it indicates that the surface is shaded and does not receive direct solar radiation at time t. If the ray is blocked, its contribution to the surface's received radiation is set to 0; otherwise, it is set to 1. This can be expressed as follows:

$$State\_Ray_t = \begin{cases} 1, & noshade \\ 0, & shade \end{cases} \quad (2)$$

After accounting for shading, the total solar radiation received by each surface at time t is modified as follows:

$$Total\_Rad_t = Dir\_Rad_t * State\_Ray_t * cosq_t + Dif\_Rad_t + Ground\_Rad_t \quad (3)$$

3.1.3. Photovoltaic Power Generation Potential Analysis

For the photovoltaic power generation potential analysis, the conversion efficiency of photovoltaic panels is set, and the expected annual photovoltaic power generation on both building roofs and facades is simulated. This simulation incorporates solar radiation intensity, panel power, and installation parameters. The installed capacity is estimated based on the panel area and power, and the corresponding installation cost is calculated by multiplying the installed capacity by the unit capacity price.

The power output of a single PV module is calculated as follows [28]:

$$P = I \cdot V \cdot \beta_{inv} \cdot \beta_{sys} \quad (4)$$

where I and V are the operating current and voltage of the PV module, respectively, determined by ambient temperature, wind speed, atmospheric pressure, and the actual solar radiation received by the module [29]. The meteorological data used are sourced from the NSRDB dataset developed by NREL [30]. Here, β_{inv} represents the inverter efficiency, and β_{sys} denotes the photovoltaic system efficiency, assumed to be 80.96% [31].

The installed capacity and actual power generation of the building-integrated PV system are expressed as follows:

$$Cap_{PV}^{In} = \sum\left(Cap_{single}^{In} \cdot \beta_i^{Use} \cdot A_i / A_{single}\right) \quad (5)$$

$$Power_t^{PV} = \sum\left(Cap_{single}^{In} \cdot \beta_i^{Use} \cdot A_i / A_{single}\right) \cdot Power_{i,t}^{PV} \quad (6)$$

where Cap_{single}^{In} is the rated capacity of a single PV module, A_i is the area of building surface i, A_{single} is the area of a single PV module (using the SolarWorld Sunmodule Pro-Series 250 W polycrystalline solar panel [32]), and β_i^{Use} is the utilization factor of building surface i, assumed to be 80% for all surfaces. $Power_{i,t}^{PV}$ represents the photovoltaic power generation of surface i at time t, which can be calculated by Equation (4).

3.1.4. Economic Assessment

The payback period and the self-sufficiency rate are used to evaluate the economic performance of the photovoltaic system. First, the system's annual revenue and initial investment cost can be calculated as follows:

$$Cost_{invest} = Cost_{cap} \cdot Cap_{PV}^{In} \quad (7)$$

$$Income_{year} = Cost_{power} \cdot Power^{PV} - Cost_{OM} \cdot Cap_{PV}^{In} \quad (8)$$

where $Cost_{cap}$, $Cost_{power}$, and $Cost_{OM}$ represent the unit capital cost of installed capacity, electricity purchase price, and unit operation and maintenance (O&M) cost, respectively. The economic parameters of the photovoltaic system are derived from China's 2050 Photovoltaic Development Outlook (2019) [33], while the electricity price data are obtained from the 2020 report published by the Provincial Development and Reform Commission [34,35].

Accordingly, the payback period and the photovoltaic self-sufficiency rate can be calculated as follows:

$$PBP = Cost_{invest} / Income_{year} \quad (9)$$

$$SelfUsage = Power^{PV} / Power^{Load} \quad (10)$$

Finally, the cost per unit of electricity is estimated based on the approximate investment cost. All resulting data, including solar radiation, power generation potential, and economic metrics, are exported in bulk to .shp format files for further use.

3.1.5. Simulation Parameter Settings

A series of assumptions were made in this study. For instance, it was assumed that the photovoltaic system's power generation efficiency and the inverter conversion efficiency remain constant, and that the available photovoltaic installation area coefficient is consistent across building surfaces of different orientations. The detailed system simulation parameter settings are shown in Table 2.

Table 2. Simulation parameter settings.

Parameters	Value
β_{inv}	90%
β_{sys}	80.96%
β_i^{Use}	80%
$Cost_{cap}$	3.16 RMB/kW
$Cost_{OM}$	0.047 RMB/kW
$Cost_{power}$ for Residential	0.5580 RMB/kW
$Cost_{power}$ for Commercial	0.6907 RMB/kW

3.2. Data Analysis and Data Interoperability

Since the platform was developed using UE, the batch-exported data must be accurately imported and processed within UE. This is achieved by utilizing the SHPOpen function from the GIS open-source library Shapelib to open .shp files. The SHPGetInfo and SHPReadObject functions are used to retrieve data entries within these files, while the DBFGetFieldInfo function iteratively reads field data for each entry, including attributes such as area, floor count, building name, Shape_Area, ID, object ID, Shape_Length, installed cost, electricity cost savings, effective installed capacity, and kWh cost.

The retrieved data are categorized into three groups: geographical information for reproducing urban architectural models, basic data for calculating power generation, and computed results for display mapping. UE reproduces urban building models based on geometric data from the .shp file, leveraging latitude and longitude coordinates to generate models in the correct geographic location.

Basic data for calculating power generation and installed capacity, as detailed in Table 2, include values such as CenterXYZ for plane centering, VectorXYZ for plane orientation, and power generation calculations using "Rad × Area × efficiency coefficient" for each plane. Installed capacity is computed by dividing Area by the area of a single photovoltaic panel and multiplying it by the panel's capacity.

Data for display mapping include ten calculated values from RoofCap to SouthPower (shown in Table 3), which are directly mapped in UE. Electricity cost savings and installation costs are determined based on whether the building type is residential or commercial, with calculations performed directly within UE.

Table 3. Analysis of batch-exported .shp data entries.

Classification	Data Header	Specific Value	Representative Meaning
Geographic Data for Reproducing Urban Architectural Models	Geometry Information Stored in .shp file	The longitude and latitude of each vertex of the building.	Different 3D modeling software uses this longitude and latitude information to generate models of the same geographical location.
	ID	1	Identity Number of Each Building
	Elevation (m)	3	Building Height (determine the height of each building during white mold model slice generation in CesiumLab v0.25, Cesium GS, Inc., Salt Lake City, UT, USA.)
Basic Data for Power Generation Calculations	FaceNum	8	Number of Building Planes
	CenterX/Y/Z	11,355,522.75, 11,355,506.25, 11,355,522.75, 11,355,506.25, 11,355,523.5, 11,355,531.0, 11,355,522.0, 11,355,498.0, 2,215,085.1875, 2,215,092.125, 2,215,085.1875, 2,215,092.125, 2,215,086.75, 2,215,081.75, 2,215,083.625, 2,215,095.625, 2,215,093.75, 2,215,090.5, 0.0, 0.0, 3.0, 3.0, 1.5, 1.5, 1.5, 1.5, 1.5, 1.5	Plane Center Point Coordinates
	VectorX/Y/Z	0.0, 0.0, 0.0, 0.0, 0.3807, 0.8321, −0.3887, −0.8517,0.0, 0.0, 0.0, 0.0, 0.9247, −0.5547, −0.9214, 0.5241, 0.9162, −0.9247, −1.0, −1.0, 1.0, 1.0, 0.0, 0.0, 0.0, 0.0, 0.0, 0.0	Plane Normal Vector
	Rad (kWh/m^2)	283.7222, 283.7222, 1598.8638, 1598.8384, 425.9791, 922.2407, 960.1941, 665.7827	Solar Radiation Results for Each Plane (Considering occlusions)
	Area (m^2)	61.875, 64.125, 61.875, 64.125, 55.1543, 10.8167, 52.0967, 11.4483	Total Area of Each Plane
	Direction	Ground, Ground, Roof, Roof, North, East, South, West, North, South	The Orientation of Each Plane (Identified using VectorXYZ in Python 3.9)

Table 3. *Cont.*

Classification	Data Header	Specific Value	Representative Meaning
Computed Data for Display Mapping	RoofCap (kW)	16.30588235	Roof Facade Installed Capacity
	NorthCap (kW)	13.9178470588235	North Facade Installed Capacity
	EastCap (kW)	1.39980823529412	East Facade Installed Capacity
	WestCap (kW)	1.48154470588235	West Facade Installed Capacity
	SouthCap (kW)	13.8795411764706	South Facade Installed Capacity
	RoofPower (kWh)	37384.48032	Roof Power Generation
	NorthPower (kWh)	8008.86552001925	North Facade Power Generation
	EastPower (kWh)	1852.73434408509	East Facade Power Generation
	WestPower (kWh)	1404.61523382367	West Facade Power Generation
	SouthPower (kWh)	19,217.9880733783	South Facade Power Generation
	BuildType	1	1 for Residential Buildings, 2 for Public Buildings.

3.3. Building Model Generation and Data Mapping

The building information generated in UE was obtained from .shp files, which were processed in Rhino to create the computational results for 3D slices. The Cesium for Unreal plugin was then used to generate the urban buildings. As shown in Table 2, each building has a unique ID corresponding to a data entry, including electricity generation and installed capacity for the east, west, south, north, and roof surfaces. Therefore, each building must be divided into five surfaces in UE to apply the appropriate color materials.

3.3.1. Building Model Slicing and Generation

To create urban building models in UE, the ShapeFile data are converted into basic 3D model grid slices using the CesiumLab toolkit and its geographic data processing platform. First, the basic white model grid slice is selected, ensuring the index field attributes are verified and vertex data are not compressed. The Prj file is then loaded, and the model height is bound to the Floor or Height field attributes. Once the field attributes are successfully bound, the model slices can be accessed using the Cesium for Unreal plugin in Unreal Engine. The relevant field data are subsequently obtained through UE's Blueprint Visual Scripting material system for further processing and visualization.

3.3.2. Building Model Facade Splitting

Since the basic white model generated by CesiumLab treats each building as a whole, the model needs to be split into five surfaces. In the material script, the direction of each surface is determined by calculating the dot product between the surface normal and the unit vector representing the basic directions (east, west, north, south, and vertical up). The dot product operation compares the surface normal with the reference direction vector, setting a threshold of 0.5 to determine the surface orientation (for example, for the east-facing surface: $\text{normal}\cdot\{1, 0, 0\} > 0.5$) and using the world coordinate system to ensure accuracy in the direction judgment.

3.3.3. Material and Color Mapping

This study achieves the dynamic visualization of photovoltaic data through Unreal Engine's Cesium plugin. The key steps include adding the Cesium3DTileset component to import the city model, using CesiumFeaturesMetadata to automatically associate building attribute data, and developing a dedicated material system to map parameters such as electricity generation into color gradients via linear interpolation (Lerp nodes). This method supports querying of photovoltaic data when clicking on buildings, enabling the efficient rendering of hundreds of thousands of building instances.

4. Case Study

4.1. Overview

Wuhan, China, was selected as the research location due to its representative climate and urban layout. The researchers are based in Wuhan, a city located in the central plains of China. Wuhan's climate is characterized by cold winters and hot, humid summers, making it ideal for studying solar energy potential. Meteorological data were sourced from typical-year datasets available through EPW maps, with the Wuhan weather station located at (30.60° N, 114.05° E) at an altitude of 34.4 m. Wuhan, situated in central China's plains, experiences cold winters and hot summers, with annual solar radiation ranging between 1163 and 1393 kWh/m^2/y. Typical-year meteorological records show temperatures varying from −8 °C to 39 °C and daily radiation levels ranging from 0 to 8 kWh/m^2 [36,37].

In this study, Wuhan City is divided into research areas according to the fourth-level administrative boundaries (neighborhood blocks), resulting in a total of 182 building blocks and over 200,000 buildings, with a commercial-to-residential building ratio of approximately 5.5:4.5. Since Wuhan lies in the northern hemisphere, solar radiation is generally higher on south-facing surfaces and lower on north-facing surfaces. The platform displays critical metrics such as estimated annual power generation, total installed cost, estimated electricity savings, effective installed capacity, and average electricity cost per kilowatt-hour for all buildings. Users can also generate optimized power generation plans by setting electricity cost savings targets, budget constraints, and photovoltaic panel models.

4.2. Results

The platform calculations show that the total installed capacity of rooftop and facade photovoltaic systems for all buildings in Wuhan is 70.59 GW, with an annual electricity generation of 46.17 TWh, covering 61.1% of the city's electricity demand. Among this, rooftop photovoltaics contribute the most (22.94 TWh), followed by the east facade (10.54 TWh). The total project investment is 223 billion CNY, with an annual revenue of 42.2 billion CNY. The investment payback period is 5.3 years, showing good economic feasibility, with a self-consumption rate of 61.2%, indicating a high-energy-self-sufficiency potential (Table 4).

Table 4. Total values for Wuhan.

Generation of Each Face (TWh/year)					**Capacity of Each Face (GW)**				
Roof	**South**	**East**	**West**	**North**	**Roof**	**South**	**East**	**West**	**North**
22.94	4.52	10.54	4.20	3.97	17.98	17.75	17.71	8.61	8.54
Building PV Costs (Billion RMB)		Building PV Incomes (Billion RMB/year)		PBP (year)		Power Demand (TWh/year)		Self-Use Rate (%)	
223		42.2		5.3		75.5		61.2	

The annual daily photovoltaic generation of a single solar panel in Wuhan shows a trend of lower generation in summer and higher generation in winter, as shown in Figure 6. The highest daily generation is 3.85 kWh/m^2, occurring on March 2. Despite the strong solar radiation in Wuhan during the summer, the high working temperature reduces its efficiency, leading to lower power generation.

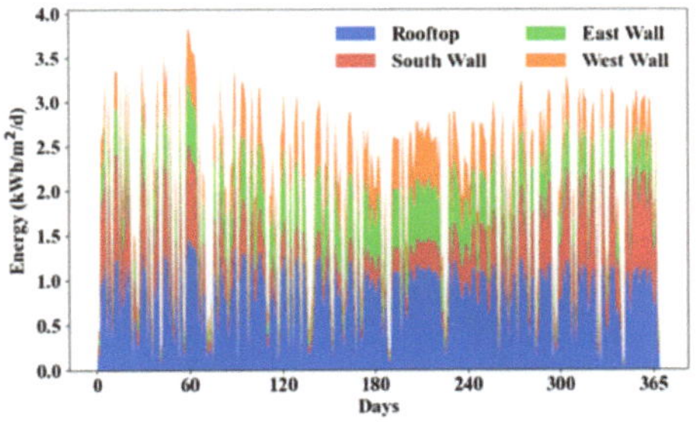

Figure 6. Daily PV power production of single PV panel in Wuhan.

The seasonal electricity generation of a single photovoltaic panel on each wall of Wuhan is shown in Figure 7. Due to temperature and radiation effects, the rooftop generation is relatively balanced throughout the seasons (70–80 kWh/m^2). The electricity generation characteristics of the east and west walls are similar in each season, with higher generation in spring and summer, and the lowest generation in winter. The south wall exhibits more significant seasonal fluctuations in power generation, with the lowest generation in summer (only 25 kWh/m^2). This is due to the severe deviation of the solar incidence angle from the optimal tilt angle during summer in low-latitude regions. In winter, the generation is highest (60 kWh/m^2), approximately 2.4 times that of summer generation, and the difference from the rooftop winter generation is smaller. This is because the winter temperature is lower, and the solar incidence angle on the south wall in winter is closer to the optimal tilt angle, leading to higher direct solar radiation reception.

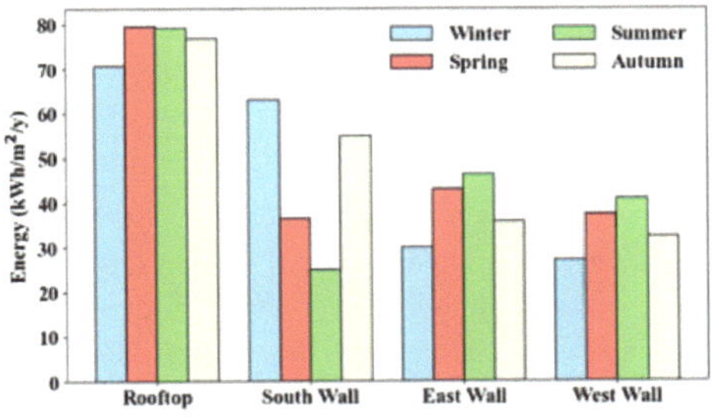

Figure 7. Seasonal Variation in PV Power Generation (Wuhan, Single Panel).

4.3. Interface Display Content and Effects

When users click on Wuhan in the two-dimensional map of China, the interface transitions to a three-dimensional visualization of Wuhan's spatial layout and buildings. As illustrated in Figure 8, users can freely navigate the virtual environment to explore the estimated annual power generation for the rooftops and four facades of each building. The annual power generation is represented using a color gradient ranging from blue (low values) to red (high values). Hovering the mouse over any building reveals detailed information, including building name, coordinates, estimated annual power generation, total installed cost, estimated electricity savings, effective installed capacity, and average electricity cost. Additionally, the platform identifies and highlights the top 100 buildings with the highest total power generation, and the relevant detailed information of each building can be exported (Table 5).

(a) (b)

Figure 8. Wuhan City interface. (**a**) Panorama view. (**b**) Detailed building information.

Table 5. Information on the top 10 buildings in Wuhan.

Index	Height (m)	Area (m^2)	PBP (Year)	Power (GWh/Year)					Capacity (MW)				
				Roof	South	East	West	North	Roof	South	East	West	North
1	54	478,122	4.87	79.16	4.73	2.95	2.91	2.20	56.92	5.33	4.04	4.03	5.43
2	54	478,122	4.88	79.16	4.48	2.95	2.89	2.20	56.92	5.33	4.04	4.03	5.43
3	18	148,161	4.79	24.52	1.64	0.28	0.29	0.65	17.64	1.87	0.43	0.42	1.87
4	15	106,328	4.82	17.31	0.39	0.33	0.23	0.32	12.66	0.71	0.44	0.44	0.91
5	12	94,659	4.71	15.64	0.56	0.31	0.22	0.19	11.27	0.70	0.42	0.42	0.70
6	12	91,042	4.68	14.99	0.50	0.33	0.33	0.19	10.84	0.59	0.47	0.48	0.59
7	21	76,910	5.67	12.66	1.51	0.27	0.62	0.54	9.16	1.92	0.61	0.88	2.25
8	21	81,569	3.73	12.81	0.54	0.44	0.45	0.81	9.71	0.95	1.06	0.58	3.14
9	3	86,288	4.35	14.28	0.30	0.00	0.00	0.17	10.27	0.33	0.00	0.00	0.33
10	12	80,151	4.70	13.24	0.53	0.20	0.19	0.26	9.54	0.68	0.31	0.29	0.68

Users can also access a selection box to switch between different data visualizations. For instance, as shown in Figure 9, when the "Estimated Installed Cost" option is selected, the buildings are color-coded based on their total installation cost, using a gradient from yellow (lower cost) to red (higher cost). Alternatively, clicking the "Estimated Electricity Savings" option changes the building colors accordingly, distinguishing between commercial buildings (blue) and residential buildings (pink).

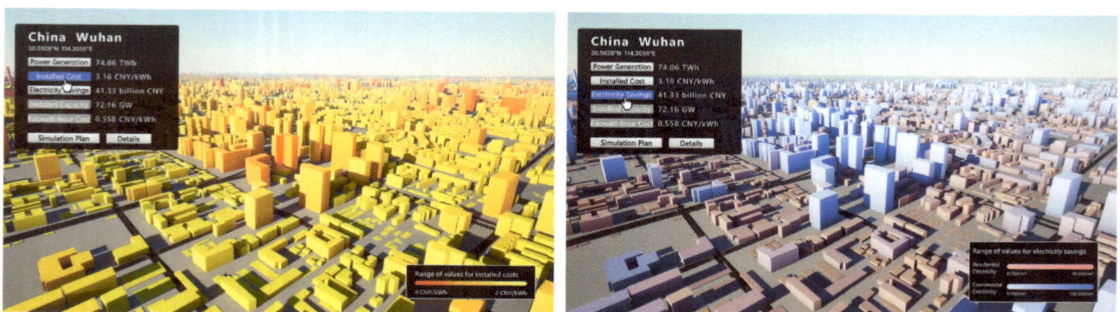

(a) (b)

Figure 9. Information selection interface. (**a**) Estimated installation cost. (**b**) Electricity savings visualization.

To optimize power generation within a specific budget, users can access the "Simulation Plan" feature (Figure 10). After entering parameters such as electricity savings target, budget, and preferred photovoltaic panel model, the platform calculates the optimal solution. The recommendations consider building power consumption types, annual power generation potential, and installation costs. The interface highlights the relevant buildings and displays specific details of the proposed plan.

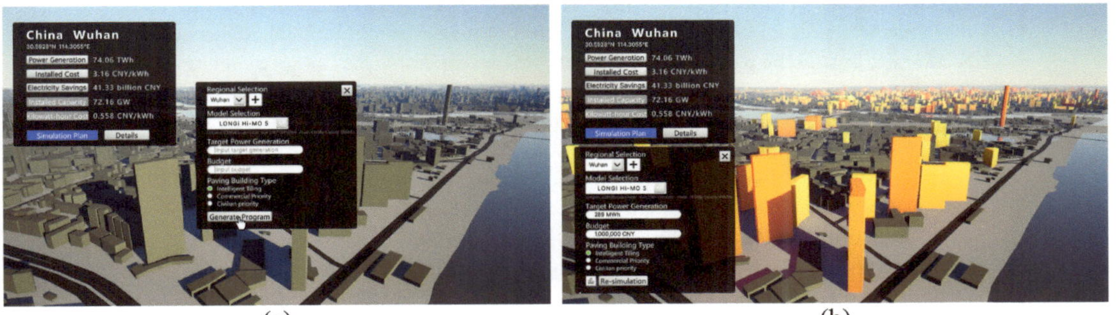

(a) (b)

Figure 10. Simulation plan interface. (**a**) User inputs filter conditions. (**b**) Platform displays optimal recommendations.

For critical or landmark buildings, the platform supports importing high-precision three-dimensional models. As shown in Figure 11, users can interactively select and adjust photovoltaic panel paving areas on the building's surfaces to achieve more accurate performance estimates.

(a) (b)

Figure 11. Single-building interface. (**a**) 3D fine model. (**b**) User-adjusted paving plan.

5. Discussion and Conclusions

The platform integrates solar radiation, building models, and economic data through 3D visualization technology, providing electricity generation, installed capacity, and investment return analysis for urban-level photovoltaic deployment. It supports macro-level decision-making and investment optimization. Its core advantage lies in efficient urban-scale evaluation, but due to the research focus, it does not consider factors such as the type of photovoltaic components (thin-film modules, photovoltaic bricks, photovoltaic tiles, photovoltaic curtain walls, etc.), degradation of photovoltaic systems, performance of photovoltaic materials (monocrystalline silicon, polycrystalline silicon, etc.), rooftop layout details, and the impact of facade photovoltaics on building energy consumption. These simplifications aim to balance computational efficiency and generality, focusing on rapid evaluations needed for policy-making rather than detailed analysis of individual buildings.

The platform's limitations primarily stem from its positioning for city-level planning, where it prioritizes overall data over micro-level variables. For instance, ignoring component differences and degradation effects reduces model complexity, while omitting the impacts of rooftop layouts and building energy consumption avoids coupling additional simulation tools. Future upgrades could gradually incorporate more details through modular enhancements, but a balance must be struck between computational efficiency and accuracy to ensure the platform remains practical and operable for energy planning.

The platform pre-calculates the relevant data for each building in all cities that can be viewed, significantly improving the platform's computational speed. For example, Wuhan spent 1 h pre-calculating data in Rhino, which were then transferred to UE to accelerate display speed in UE. Tests showed that on a computer configured with Intel i5-13600KF/AMD RX 7900 XT/32 GB DDR5, running a city photovoltaic project took about 2 s to start, and loading an offline map level for a city with a road network took 3–4 s. Loading a level without a road network using an online map took less than 1 s, with most scenes running at a frame rate exceeding 90 frames per second. Switching between white model datasets took no more than 1 s.

Future upgrades to the platform will focus on three areas: first, integrating building electricity loads, photovoltaic component performance, and degradation models to optimize generation-demand matching; second, adding energy consumption impact assessments and rooftop layout optimization tools to address current limitations; third, enabling batch display of photovoltaic arrays and dynamic regional analysis to improve decision-making accuracy. In the future, the platform will incorporate more diverse analytical scenarios. For instance, at the single-building level, users will be able to set different PV panel types or installation coverage ratios to explore how different photovoltaic system parameters affect investment decisions. At the regional level, the platform will provide BIPV potential analyses across different building types—including residential, industrial, and commercial buildings—offering more realistic and policy-relevant insights for decision-makers. Through these multidimensional upgrades, the platform will balance macro-level planning with micro-level optimization needs, providing a more comprehensive solution for urban photovoltaic deployment.

Author Contributions: Conceptualization, X.C. and Y.X.; methodology, X.C. and H.L.; software, X.C. and H.L.; validation, Y.X.; formal analysis, X.C.; investigation, H.L.; resources, X.C.; data curation, H.L.; writing—original draft preparation, X.C.; writing—review and editing, Y.X.; visualization, X.C.; supervision, X.C.; project administration, X.C. All authors have read and agreed to the published version of the manuscript.

Funding: The research was supported by "the Fundamental Research Funds for the Central Universities" 2024WKYXQN074.

Data Availability Statement: The raw data supporting the conclusions of this article will be made available by the authors on request.

Conflicts of Interest: The authors declare that the research was conducted in the absence of any commercial or financial relationships that could be construed as a potential conflict of interest.

References

1. Sun, H.; Heng, C.K.; Tay, S.E.R.; Chen, T.; Reindl, T. Comprehensive feasibility assessment of building integrated photovoltaics (BIPV) on building surfaces in high-density urban environments. *Sol. Energy* **2021**, *225*, 734–746. [CrossRef]
2. Nørgaard, P.; Poder, S.H. Perspectives on solar power in dense urban areas–with Copenhagen International School as a case study. *J. Eng.* **2019**, *18*, 5134–5137. [CrossRef]
3. Vyas, M.; Chowdhury, S.; Verms, A.; Jain, V.K. Solar Photovoltaic Tree: Urban PV power plants to increase power-to-land occupancy ratio. *Renew. Energy* **2022**, *190*, 283–293. [CrossRef]

4. Azizkhani, M. *Assessment of Solar PV Potential on Facades: Toward Zero-Energy Buildings*; Springer: Cham, Switzerland, 2021.
5. Freitas, S.; Brito, M.C. Solar façades for future cities. *Renew. Energy Focus* **2019**, *31*, 73–79. [CrossRef]
6. Gholami, H.; Røstvik, H.N. Economic analysis of BIPV systems as a building envelope material for building skins in Europe. *Energy* **2020**, *204*, 117931. [CrossRef]
7. Liu, R.; Liu, Z.; Xiong, W.; Zhang, L.; Zhao, C.; Yin, Y. Performance simulation and optimization of building façade photovoltaic systems under different urban building layouts. *Energy* **2024**, *288*, 129708. [CrossRef]
8. Li, J.; Wang, Y.; Xia, Y. A novel geometric parameter to evaluate the effects of block form on solar radiation towards sustainable urban design. *Sustain. Cities Soc.* **2022**, *84*, 104001. [CrossRef]
9. Liu, J.; Liu, Z.; Wu, Y.; Chen, X.; Xiao, H.; Zhang, L. Impact of climate on photovoltaic battery energy storage system optimization. *Renew. Energy* **2022**, *191*, 625–638.
10. Kazem, H.A.; Chaichan, M. Effect of environmental variables on photovoltaic performance based on experimental studies. *Int. J. Civ. Mech. Energy Sci.* **2016**, *2*, 1–8.
11. Moraitis, P.; Kausika, B.B.; Nortier, N.; Sark, W.V. Urban environment and solar PV performance: The case of the Netherlands. *Energies* **2018**, *11*, 1333. [CrossRef]
12. Zhu, R.; Wong, M.S.; Kwan, M.-P.; Chen, M.; Santi, P.; Ratti, C. An economically feasible optimization of photovoltaic provision using real electricity demand: A case study in New York City. *Sustain. Cities Soc.* **2022**, *78*, 103614. [CrossRef]
13. Chen, X.; Chen, X. Data visualization in smart grid and low-carbon energy systems: A review. *Int. Trans. Electr. Energy Syst.* **2021**, *31*, e12889. [CrossRef]
14. Healy, K. *Data Visualization: A Practical Introduction*; Princeton University Press: Princeton, NJ, USA, 2024.
15. Telea, A.C. *Data Visualization: Principles and Practice*; CRC Press: Boca Raton, FL, USA, 2014.
16. Baqir, M.; Channi, H.K. Analysis and design of solar PV system using Pvsyst software. *Mater. Today Proc.* **2022**, *48*, 1332–1338. [CrossRef]
17. Romero-Ramos, J.A.; Gil, J.D.; Cardemil, J.M.; Escobar, R.A.; Ariasb, I.; Pérez-García, M. A GIS-AHP approach for determining the potential of solar energy to meet the thermal demand in southeastern Spain productive enclaves. *Renew. Sustain. Energy Rev.* **2023**, *176*, 113205. [CrossRef]
18. Berrian, D.; Linder, J. Impact of large format solar panels on the balance of system costs of photovoltaic power plants. In Proceedings of the 8th World Conference on Photovoltaic Energy Conversion, Milan, Italy, 26–30 September 2022. Available online: https://userarea.eupvsec.org/proceedings/WCPEC-8/4BV.5.13/ (accessed on 23 September 2024).
19. Ali, M.S.; Rima, N.N.; Sakib, M.I.H.; Khan, M.F. Helioscope-based design of a MWp solar PV plant on a marshy land of Bangladesh and prediction of plant performance with the variation of tilt angle. *GUB J. Sci. Eng.* **2018**, *5*, 1–5. [CrossRef]
20. Sharma, D. Review and analysis of solar photovoltaic software. *Int. J. Eng. Technol.* **2014**, *4*, 725–731.
21. Schupp, T. Workflow for solar irradiance concentration analysis. In Proceedings of the Illumination Optics VII, Proceedings of Spie Optical Systems Design, Strasbourg, France, 7–12 April 2024; Volume 13022, pp. 86–92.
22. Geh, N.; Emuze, F. Solar photovoltaic deployment acceleration model to advance the sustainability of buildings in public universities in South Africa. *Energy Build.* **2023**, *284*, 112855. [CrossRef]
23. Buyuksalih, I.; Bayburt, S.; Buyuksalih, G.; Baskaraca, A.P. 3D modeling and visualization based on the Unity game engine–advantages and challenges. *ISPRS Ann. Photogramm. Remote Sens. Spat. Inf. Sci.* **2017**, *4*, 161–166.
24. Laksono, D.; Aditya, T.; Riyadi, G. Interactive 3D city visualization from structure motion data using a game engine. *ISPRS Arch. Photogramm. Remote Sens. Spat. Inf. Sci.* **2019**, *XLII-4/W16*, 737–740. [CrossRef]
25. Xu, H.; He, B.; Li, Z.; Lin, H. Efficient visualization of 3D city scenes by integrating GIS and Unreal Engine. In Proceedings of the Fourth International Conference on Geoscience and Remote Sensing Mapping (GRSM 2022), Changchun, China, 4–6 November 2022; SPIE: Bellingham, WA, USA, 2023; Volume 12551.
26. Edler, D.; Keil, J.; Dickmann, F. From Na Pali to Earth—An 'Unreal' engine for modern geodata? In *Modern Approaches to the Visualization of Landscapes*; Springer VS: Wiesbaden, Germany, 2020.
27. Xie, M.; Wang, M.; Zhong, H.; Li, X. The impact of urban morphology on the building energy consumption and solar energy generation potential of university dormitory blocks. *Sustain. Cities Soc.* **2023**, *96*, 104644. [CrossRef]
28. Ghiasi, M.; Wang, Z.; Mehrandezh, M.; Paranjape, R. Enhancing efficiency through integration of geothermal and photovoltaic in heating systems of a greenhouse for sustainable agriculture. *Sustain. Cities Soc.* **2025**, *118*, 106040. [CrossRef]
29. Kratochvil, J.; Boyson, W.; King, D. *Photovoltaic Array Performance Model*; SAND2004-3535; Sandia National Laboratories: Albuquerque, NM, Mexico, 2003.
30. NSRDB. Available online: https://nsrdb.nrel.gov/ (accessed on 30 August 2023).
31. Shi, M.; Lu, X.; Jiang, H.; Mu, Q.; Chen, S.; Fleming, R.M.; Zhang, N.; Wu, Y.; Foley, A.M. Opportunity of rooftop solar photovoltaic as a cost-effective and environment-friendly power source in megacities. *iScience* **2022**, *25*, 104890. [CrossRef]
32. SolarWorld SunModule Pro 250 Watt, 20V Polycrystalline Solar Panel (SW250P). Available online: https://www.mrsolar.com/solarworld-sw250p-250w-24v-solar-panel/ (accessed on 10 October 2023).

33. National Development and Reform Commission Energy Institute. *Outlook for China's Photovoltaic Development in 2050*; National Development and Reform Commission Energy Institute: Beijing, China, 2019. Available online: http://sogrb.com/nd.jsp?id=32 (accessed on 17 December 2021).
34. China Photovoltaic Industry Association. *Roadmap for the Development of China's Photovoltaic Industry (2023–2024)*; China Photovoltaic Industry Association: Beijing, China, 2024. Available online: https://www.chinapv.org.cn/ (accessed on 18 April 2024).
35. Available online: https://fgw.hubei.gov.cn/fbjd/zc/zcwj/tz/202011/t20201126_3763550.shtml (accessed on 21 November 2022).
36. Chen, Z.; Yang, B.; Zhu, R.; Dong, Z. City-scale solar PV potential estimation on 3D buildings using multi-source RS data: A case study in Wuhan, China. *Appl. Energy* **2024**, *359*, 122720. [CrossRef]
37. Jun, Q. Multiple time scales analysis of climate variation in Wuhan during the last 100 years. *J. Cent. China Norm. Univ.* **2000**, *5*, 184–198.

MDPI AG
Grosspeteranlage 5
4052 Basel
Switzerland
Tel.: +41 61 683 77 34

Energies Editorial Office
E-mail: energies@mdpi.com
www.mdpi.com/journal/energies

www.ingramcontent.com/pod-product-compliance
Lightning Source LLC
Chambersburg PA
CBHW041412210926
51850CB00013B/3253
9783725844272